THE SECRET HISTORY

of the

WAR ON CANCER

ALSO BY DEVRA DAVIS

When Smoke Ran Like Water:
Tales of Environmental Deception and
the Battle Against Pollution

THE
SECRET
HISTORY
of the
WAR ON
CANCER

Devra Davis

BASIC
BOOKS

A Member of the Perseus Books Group

NEW YORK

DESIGN BY JANE RAESE
Text set in 13-point Perpetua

A CIP catalog record for this book is available from the Library of Congress.
ISBN-13: 978-0-465-01566-5
ISBN-10: 0-465-01566-2

4 6 8 10 9 7 5 3

For Richard

Cowardice asks the question, "Is it safe?"

Expediency asks the question, "Is it politic?"

And Vanity comes along and asks the question, "Is it popular?"

But Conscience asks the question, "Is it right?"

And there comes a time when one must take a position

that is neither safe, nor politic, nor popular,

but he must do it because Conscience tells him it is right.

—MARTIN LUTHER KING, JR.

Contents

Preface

Writing has to do with darkness, and a desire, perhaps a
compulsion, to enter it, and, with luck, to illuminate it, and
bring something back out to the light.[1]
——MARGARET ATWOOD

MY MOTHER ALWAYS SAID that G-d watches over little children because parents can't be everywhere all the time. I come from a long line
of well-watched children. When she was five, my great-grandmother
Molly once spent an entire day hiding under a stack of hay on a horse-
drawn cart until her mother could whisk her away from pogroms in
Transcarpathia. Molly grew up to be a very patient woman.

Sometimes survival traits that work in desperate circumstances can
lead to problems in other environments. As a boy of nine during the
First World War, Molly's son, my great-uncle Paul, roamed through
the woods of Hungary, eating as much as he could whenever he
could. About the ability to binge, he once said, "By the time the fat
ones were thin, the thin ones were dead."[2] This could explain why
Central European peasants who survive famine tend to be stout but
able to run like hell.

It was a good thing for me that Molly's daughter, my grandmother,
Bubbe Fanne, came from that stock. In the winter of 1924, a fiery explosion rocked the basement dry-cleaning factory below my grandparents' small wooden home in Monongahela. Bubbe Fanne raced
through the flames to grab her two toddlers, one of whom, Harry,
would become my dad less than twenty years later. Only years after
she died did I learn how Bubbe Fanne got the thick red scars that ran
down her arms and across her shoulders and chest. Long after his
mother had pulled him from a blazing building, my father eluded death
another time. As a drill sergeant, he snatched a live grenade from the

shaking hands of a green army recruit and tossed it away just before it blew to smithereens.

If my grandmothers or father had lacked good timing, they wouldn't have survived. My brothers and sister and I wouldn't be here. You could say that perseverance is a family trait.

It's taken me twenty years to write this book. My first try ended in 1986 after I explained to Frank Press, my boss at the National Academy of Sciences, that I had been offered a hefty advance to write about the fundamental misdirections of the war on cancer. With support from Press, the National Institute of Environmental Health Sciences, and universities and research institutions in Europe and the United States, my colleagues and I had published a series of papers showing that cancer had actually increased and it couldn't all be explained by smoking, improved diagnoses or aging. Judged by this standard, the then two-decade-long war on cancer wasn't going well.[3] Our work, released at conferences of Danish, Dutch, Swedish, German, British and American cancer researchers, made headlines. It felt like we had some serious scientific mojo.[4] The book contract seemed to confirm that judgment.

Press, an MIT professor and former science adviser to President Jimmy Carter, was a seasoned diplomat and not a person I would ever play poker with, even if I knew how. He nodded as I told him of my plans and then said gravely, "It had better be a good book."

I replied, "I guess they think it will be. They're offering me more than half my annual salary. That's quite a lot for a first-time author."

"It had better be a really, really good book," he said.

I didn't understand. "Of course they expect it to be good," I said. "So do I."

"Well," he explained, "it had better be, because you won't be able to work here after you write it."

He quickly added, "Of course, I'm not telling you what to do. That is completely your decision. You are free to do whatever you want. I'm just telling you that you can't write a book critical of the cancer enterprise and hold a senior position at this institution."

Frank Press had achieved positions of eminent authority by dint of remarkable diplomatic skills and impeccable timing. We were then living through a period that would later be termed the Reagan Revolution.

The nation's leaders bragged of lessening the power of government across the board. Under the charismatic but underestimated President Ronald Reagan, the White House set up an ambitious program aimed at easing the burdens of regulations across the board. Proposals to expand government's control of anything, even cancer-causing agents in the environment, had little chance of survival.

At about the same time that Press offered his reflections on my proposed book project, I got some friendly advice from a man who was temporarily running the National Institutes of Health. He called me into his spacious office overlooking what was then the green campus of the NIH.

"This work you've been publishing on cancer patterns is pretty interesting. You know, I started out my career interested in the environment and cancer. I'm pretty sure that some of the lung cancer we see in women in southwestern Pennsylvania, where you come from, has something to do with the environment. I actually tried to do a study on that when I started out doing research, but I decided against it."

"What made you change your mind?" I asked.

He leaned back in his chair and put his hands atop the back of his head, rocking in thought. "You ever hear of Wilhelm Hueper?"

I shook my head.

"Hueper started out like you. Lots of good ideas about the environment. He thought the exclusive focus on smoking would lead us away from other causes of cancer that were far more deadly. He was railroaded out of here. He wasn't the easiest fellow to work with and rubbed lots of people the wrong way, but not necessarily for the wrong reasons. I decided after seeing what happened to him that I was better off sticking to basic research. Somebody like you should think about that."

I did. I stayed with NAS for a decade, working with some of the most talented experts on some of the most fascinating and challenging problems in science at the time. We put out more than two dozen thoroughly referenced NAS reports, every one of which struggled to gauge evidence of the ways the world in which we live and work affects our health and environment. Whether about smoking in public spaces or the chlorination of drinking water, each volume navigated treacherous and uncertain waters, and each ended with the familiar message: we need more research before we can be sure. I watched the maturing of the science of doubt promotion—the concerted and well-funded effort to identify, magnify and exaggerate doubts about what we could say that we know as a way of delaying actions to change the way the world operates.

How did we get to this point? Since its formal launch more than thirty-five years ago, the war on cancer has been fighting many of the wrong battles with the wrong weapons and the wrong leaders. Officially declared by President Nixon in 1971, the American effort aggressively targeted the illness but left its myriad causes untouched. Less than a decade after the famed U.S. Surgeon General's report of 1964 indicted tobacco as a cause of lung cancer, the president announced a national attack on cancer. Left off the table completely were tobacco, radiation, asbestos and benzene—materials that for decades had been well understood to be hazardous.

Years before any modern industrial nation started an official war on the disease, in the 1930s, researchers in Germany, Japan, Italy, Scotland, Austria, England, Argentina, the United States and France had shown that where people lived and worked affected their chances of getting cancer.[5] Hueper published a sweeping synthesis of industrial, pharmaceutical and natural sources of cancer—at an especially inauspicious time, right after the Japanese attack on Pearl Harbor in 1941.[6] The war against those things that cause cancer has always been hampered whenever nations have traded metaphorical wars for real ones.

If some scientists had figured out nearly a century ago that the world around us affects the chance that we will develop cancer, why have we made so little headway in controlling these causes? My goal in this book is to explain when, how, why and by whom the spotlight has been kept away from many of the things that produce cancer. I will show how two radically different sets of standards have been applied to learning how to treat the disease on the one hand, and figuring out what produces it on the other. Where animal studies on the causes of cancer exist, they are often faulted as not relevant to humans. Yet when studies of almost identical design are employed to craft novel treatments and therapies, the physiological differences between animals and humans suddenly become insignificant.

Many people think that the reason large numbers of us no longer die from infectious diseases is the miraculous breakthroughs of scientific discovery. Not so. In fact, the decline of epidemics in the nineteenth century had nothing to do with breathtaking scientific advances; all of these came much later. Deaths from germ-fed contagious diseases began to ebb long before microscopes or drugs could find or kill them. This decline happened because dirty water, crowded housing, rotten food and dangerous jobs became much less common in developed nations. As a result, diphtheria, typhoid and tuberculosis claim far fewer lives in in dustrialized nations today than at any time in human history.

While some may question whether filling the world with iPods and text-messaging has made us better human beings, none can question that other achievements of modern life have allowed us to live longer and better than our grandparents. If medicine didn't vanquish lethal epidemics of the past, surely today the story is more nuanced. New medications and fast-paced information technology undoubtedly afford us the capacity to confront new ailments, like looming pandemics of bird flu, providing that governments don't lie or cover up early reports.

But what about cancer? Can modern medicine, with its reliance on finding and treating diseases one at a time, alter the ways that the disease presents itself? We know how to cure relatively rare cancers, like those of children. We have made spectacular advances against many forms of the disease. That's why in the U.S. alone, there are more than 10 million cancer survivors. Why, then, are the rates of many forms of cancer increasing, especially when fewer people are smoking?

The complexities of the real world make unequivocal evidence on the causes of cancer in humans quite hard to come by. In truth, there is much bona fide scientific uncertainty about such a complicated illness. The existence of this doubt is easily exploited. Since World War II, whenever and however information on the cancer hazards of the workplace and the environment has been generated, it has typically been discredited, dismissed, or disparaged.[7]

The tobacco companies' long struggle to obscure and muddy findings on the dangers of cigarettes, successful for many decades, serves as the model. Other, larger industries, following Big Tobacco's lead, continue to use a combination of deceptive advertising, sophisticated scientific spin and strongarm politics, and have been even more successful: they remain mostly unscathed to this day.[8] Scientists who tackled industrial causes of cancer often found themselves facing subtle and sometimes not so subtle warnings. Those who resisted pressure to back off often found their funding cut.[9] In some cases, scientific research was stopped in its tracks, and many careers, like Hueper's, were derailed.[10]

In retrospect it seems clear that Frank Press was correct about many things. It's not enough to write the right book. The world has to be ready to listen. I'm certainly not the first person to try to shine light on the lopsided nature of the effort against cancer, nor am I unique in commenting on the arrogance of environmental policies. But there are signs that the world may be more ready to listen.

The modern critique of our failure to ferret out and act on preventable causes of cancer goes back more than four decades, to Murray Bookchin and Rachel Carson.[11] Valiant but little heeded efforts were mounted right before or during the Reagan Revolution by Larry Agran, Sam Epstein and Janette Sherman.[12] In 1996, Robert Proctor published a book called *Cancer Wars,* adopting the title from my own waylaid effort at the time.[13] He chronicled the successes of the producers of tobacco and other cancer-causing materials in crafting scientific doubt about their hazards and the politically problematic efforts of the Carter administration to rein in tobacco and industrial chemicals.[14] Sandra Steingraber drew well-deserved attention with her haunting, sometimes humorous books *Living Downstream* and *Having Faith*—the latter about becoming a mother as a cancer survivor in a world full of chemical risks.[15] Mitchell Gaynor, one of America's top oncologists, lambastes environmental and industrial sources of cancer in his recent works.[16] More recently, David Michaels, Gerald Markowitz and David Rosner have used original industry records to detail the duplicity of researchers and companies in keeping the dangers of a number of industrial materials hidden.[17] While some of these works got critical accolades and even made it onto public television and radio, their impact on public policy has been limited.

One of the reasons I allow myself to think the time is right for this book is the response from the business community. As word got out about my intentions, I began to hear from people I'd never met and others I'd never imagined were sympathetic. They offered me stories I'd never heard before and documents I could never find in libraries or government dockets, some of which form the bones of this book. I thought I had a pretty fair notion of what went on behind the scenes, but I was stunned by what I found out.

- Some of the early leaders of the American Cancer Society and National Cancer Institute left their posts to work directly for the tobacco industry, where they funded major academic research programs throughout the world to foment uncertainty about the dangers of their product right up to the 1990s. While people may think of the ACS as a foremost supporter of research, in 2005 it reported spending less than 10 percent of its nearly billion dollar budget on independent scientific studies.[18]
- The life-saving test for cervical cancer, called the Pap smear, · was not put into widespread use until more than a decade after it had been proven to prevent this disease, because of fears that it would undermine the private practice of medicine. These delays led to unnecessary surgery or death for millions of women.
- Some of the first modern studies on workplace causes of cancer, the dangers of medical and environmental hormones, and the cancer-causing properties of tobacco were carried out and published by scientists around 1936, including many who worked in Nazi Germany. In June 1945, Robert R. Kehoe, an Army captain who was a member of the Office of Strategic Services, traveled throughout Germany gathering information on chemical and hormonal hazards for the U.S. Army Field Investigations Unit and the British Secret Service. Sixty years later, these files remain unpublished.[19]
- The punishments meted out to war criminals after World War II did not extend to senior officials of some U.S.-German shell firms, such as EthylGemeinschaft, which operated with some slave labor. (Ethyl Corporation was owned by Standard Oil of New Jersey and General Motors.) In the late 1930s,

Ethyl and other companies gave their German partners the know-how to produce leaded gasoline and synthetic rubber in direct contravention of U.S. War Department orders. Nazi scientists devised innovative and cruel methods for studying the cancer-causing properties of these and other compounds in their workers, many of whom died in concentration camps.

- From 1929 until the late 1960s, the founder of modern industrial hygiene in America, former Army captain and OSS operative Kehoe, worked directly for Ethyl, General Motors, American Cyanamid and many major chemical companies under a special agreement at the Kettering laboratories of the University of Cincinnati; the laboratories carried out secret studies on the hazards of workplace chemicals, including: lead in gasoline, materials used to coat cooking surfaces of pots and pans, residues of cancerous materials in paraffin wax used in milk cartons, the manufacture of rubber and coke, and many other major industrial chemicals. Like most contract research on worker health and safety—then as now—the results of this work were not released to workers or the public unless those funding it agreed. Worker health remains a matter that can be deemed a trade secret in many industrial nations.

- Over a span of two decades, up to the 1990s, millions of taxpayer dollars in the United States and Britain were spent trying to develop a "safe" cigarette—despite a broad consensus among scientists that such a thing was impossible.

- Some distinguished academic leaders in the efforts against cancer in the United States, England, Greece, Sweden and France, including Sir Richard Doll of Oxford University, Hans-Olav Adami of the Karolinska Institute, and Dmitri Trichlopoulos of the Harvard School of Public Health, secretly worked for the chemical industry for years, and didn't disclose these ties even when publishing or providing government advice on subjects of direct interest to their sponsors.

- Major chemical companies bought up and moved the contaminated Louisiana delta towns of Mossville and Reveilletown. These companies did this without admitting any responsibility for the pollution that had rendered these places uninhabitable. They then pointed to the absence of information on health harms in these areas as proof that no such damage occurred. The same firms have also mounted sophisticated public relation campaigns, masquerading as cutting-edge science, to undermine reports on the dangers of vinyl chloride, benzene, asbestos and other chemical residues for workers, their families and communities—a technique borrowed from the tobacco industry that remains vitally alive.

- In the first six years of the twenty-first century, America has tripled the amount of some asbestos products it imports from China, Brazil, Columbia and Mexico.[20] Along with Canada, America is one of the few industrial countries not to have banned asbestos. Today in France, only one in four cases of mesothelioma—a rare tumor believed to be uniquely tied with asbestos exposure—is compensated for workplace exposures.[21] In many industrial countries today, one in three men and half of all women with this disease have no known history of exposure to asbestos.[22]

- Until he was fired in November 1989, Meyron Mehlman served at Mobil Oil as director of toxicology and manager of its environmental health and science laboratory, responsible for the international firm's testing of chemicals. (He was later awarded $7 million under New Jersey's Conscientious Employee Protection Act for his wrongful termination, an act that the trial judge branded "outrageous.") Mehlman's records reveal that Mobil and other oil companies hid what they knew about the dangers of benzene. As this book goes to press, these companies have allotted $27 million to an effort under way in China intended to "prove" the safety of their product.

For decades, critics of the cancer establishment have protested—some thoughtfully, others stridently—the limited nature of the cancer war and the revolving door of cancer researchers in and out of cancer-causing industries.[23] If many of these critiques have been animated and angry, they were not necessarily, for that reason, wrong.

I know what cancer looks like, feels like and smells like. Like many of my generation, I am a cancer orphan. The disease cut short the lives of both my parents. I know that before they died they found respite in a way that cannot really be imagined by most people who have never been there. I also know what it means to live with unanswered questions. I understand the terror of waiting. I know the grace that cancer patients, their physicians and families can reach, when they get the best care available and make peace with their struggle. I have been blessed to experience the power of prayer, song, psalms, humor, meditation, yoga, acupuncture and other mysteries.

I have come to admire the bold and compassionate work that is being carried out by cancer caregivers and researchers today. There are remarkable efforts under way involving natural products and breakthrough approaches in clinical trials, pushed by cancer patients who often have nothing to lose and doctors who may be wrestling with the disease themselves.

There is no one who deals with the disease now who doubts that we need to open a new front. To reduce the burden of cancer today, we must prevent it from arising in the first place, and we have to find new ways to keep the millions of cancer survivors from relapsing. No matter how efficient we become at treating cancer, we have to tackle those things that cause the disease to occur or recur. I believe that if we had acted on what has long been known about the industrial and environmental causes of cancer when this war first began, at least a million and a half lives could have been spared, a huge casualty rate that those who have managed the war on cancer must answer for. This book explains how I have come to that reckoning.

THE
SECRET
HISTORY
of the
WAR ON
CANCER

LIFE

FRESH HOPE ON CANCER
12 PAGES ON THE NEWEST METHODS
TO SAVE YOU FROM MALIGNANCIES

2,000,000 VOLT
RADIATION FOR
CANCER PATIENT

MAY 5, 1958 **25** CENTS

1

The Secret History

"Those who want the future to be
different from the past, must study the past"
—SPINOZA

MY VERY GOOD FRIEND Andrea Martin lived through three bouts of breast cancer. She used to say, "The only way I will know I have really survived breast cancer is when I die of something else." She did: when she was fifty-six, a new and unrelated malignancy of the brain turned her into a breast cancer survivor.

Three years before a tangled web of glioblastoma multiforme invaded her brain, Andrea was in excellent health. As part of a pilot research study, she was tested for chemical contaminants.[1] She had never worked in a factory. She had no chemically intensive hobbies like boatbuilding or oil painting. Yet it turned out that Andrea was a walking toxic waste site. Her body contained nearly one hundred different chemical residues, half of which caused cancer when tested in experimental animals. Many of these toxins didn't exist when she was born in the middle of the past century. Had they played any role in causing either her breast or brain tumors? Did her frequent use of those clunky first-generation cell phones have anything to do with it? It is sad that we don't know. It is appalling that we can't find out.

In 1973, out of every 100,000 men aged fifty-five to fifty-nine, only five developed multiple myeloma, a cancer of the bone marrow. By 1983 that number had doubled and included my dad, a healthy middle-aged man who never smoked or drank much. He was one of about seven thousand people with the disease that year. I have to say *about* because there is still no national system for counting cases of cancer. Only in those parts of the country with statewide or regional registries can we say with any accuracy how many cases are occurring. These registries form the backbone of any effort to understand the causes of the disease. They allow us to ask whether rates vary in different locales or for people of different backgrounds and experiences. For more than half a century, countries like Denmark and Sweden have maintained nationwide cancer registries that record every single case of the disease, but the United States has not. We still don't have a system that covers the entire country. What we don't measure, we can't count.

We can't be sure why my dad's bone marrow stopped working. He lived through some remarkable encounters with agents that affect the bones' ability to produce blood. Men who work with metal fumes as machinists or welders, or with cutting oils and solvents in steel mills, or who have had regular radiation exposures as part of frequent diagnostic checks in the military—my dad did all of these things—suffer multiple myeloma cancer at higher rates than the rest of us. That he also survived a massive explosion of benzene from the family dry cleaning shop underneath his home when he was a toddler surely added to the lifelong burden on his ability to make healthy blood. Maybe that accident in 1924 contributed to his death many decades later. We can't say.

Look around and it seems that cancer has become the price of modern life. In America and England, one out of every two men and one out of every three women will develop cancer in their lifetime. In America alone, there are currently more than 10 million cancer survivors. Cancer is the leading killer of middle-aged persons, and, after accidents, is the second-leading killer of children.

How did this happen? How did a disease that was once so atypical become so ordinary? Are we simply talking more about an illness that has always been around? Some twenty-five hundred years ago, the Greek physician Hippocrates depicted a tumor as a muddled irritable cavity with spindly legs flaring out of control in all directions. Fascinated with its evil animal-like appearance, he termed cancer *karkinoma*, the Greek word for *crab*. Like Hippocrates, we are drawn to objects of menacing beauty.

Now we have tools like electron micrographs that allow us to find things much smaller than the tiniest crab—things we could not have seen or imagined even a decade ago. Couldn't cancer have been there for eons, disguised as ordinary life? Some of the growth in cancer comes about just because more of us are reaching ages of what used to be called some distinction. As we get older the body loses the ability to defend itself against the damaging effects of being alive. Sunlight, oxygen, and other naturally occurring hazards continually assault that complex acid that sits in the center of all living cells and makes us alive—our DNA. Whether mammals or fish, we inherit genes from our parents. Slivers of complex helical chains of DNA can fix or worsen damage that comes about just from being alive. Genes tell cells when to die, when to get fixed, and how to mesh with other signals in the body. Ordinarily, attacks on our genetic base are fixed day in and day out by a series of well honed repair processes that occur within the blink of an eye with a constancy that approaches the amazing. Without these ongoing repairs, none of us would long survive.

All of us contain a remarkable array of genes, proteins and enzymes that work to keep bad cells in check and tell dangerous ones to die. As we age, the vibrancy of these repair systems and the rate at which they keep springing into action declines, just like a rubber band that eventually gives out from overuse. But an aging population does not explain why five times more men and women get brain cancer in America than in Japan. Nor do we understand why rates of testicular cancer in men under age forty have risen 50 percent in one decade in most industrial nations, why women of generation X are getting twice as much breast cancer as their grandmothers did, or why young black women get and die from breast cancer in greater numbers than their white counterparts do.

Aging does not explain why so many more children have developed cancer. Can aging be involved in growing cases of childhood leukemias, kidney and brain cancer? In a strange way, yes. All of these cancers come about because some part of the body begins to act as though it were much older, losing its ability to repair itself, as deformed cells take over. Normal cells of healthy children have a spring-like ability to recover from shocks, stay in order, stay in line, as though following some intelligent design. Children with cancer have lost their springs. They are overrun by galloping growth in organs that have just started a developmental spurt—their bone marrow, brains and kidneys. These organs, which roughly double in size in the first years of life, tend to get hit by cancer when they begin to grow out of step during their normal time of expansion. During periods of fast growth, cells can double in size several times a day. The tiniest mistake acquired when they were just being formed in early pregnancy or just after birth gets magnified over and over.

To an epidemiologist like myself, such explanations of run-amok cancer processes address the *how* but not the *why*. They talk about how cells and organs behave when they spin out of control but say nothing about why these things happen to specific groups of people located in a certain area at a certain time. Why have so many types of cancer not known to be tied with smoking increased from decade to decade in industrial countries and in those areas of the developing world that are becoming industrialized? Why do one out of every three cases of colo-rectal cancer in Egypt occur in persons under age forty, a rate that is nearly ten times higher than in the U.S.?[2] Why are so many people in their thirties and forties in many industrial countries coming down with often fatal cancers of the bone marrow and pancreas—diseases that used to occur only in those in their sixties or older? What can we do to reverse the trend? How can we get better at keeping cancers from happening in the first place? Despite impressive progress in finding and treating some forms of the disease, more than half of all those diagnosed with cancer will not last a decade with their illness.

We have all been told what we are supposed to do to reduce the risk of cancer on our own. We are supposed to eat right and exercise. Even prayer and meditation are touted as good things to do. Smoking, of course, is forbidden. And we are certainly not to drink much alcohol or engage in dangerous sex.

But we all know people who lead perfectly clean, even exemplary lives and still get cancer. They take good care of themselves, and appear to be the very nicest of people, yet somehow cancer hits. The first thing most cancer patients, and their sometimes unthinking friends, ask is—what did I do to make this happen? The answer often is—not a bloody thing. Sometimes cancer is due to a genetic susceptibility that we get from our fathers or mothers, but mostly it isn't. We know that no matter how careful anyone is about their good and bad habits, where and when we are born and what we work and play with has a lot more to do with whether we get cancer than who our parents happen to be. For instance, inherited defects do not account for most breast cancers. Nine out of ten women who develop breast cancer are born with perfectly healthy genes. When I was a girl, one in twenty women got breast cancer in her lifetime; by the time my friend Andrea reached middle age, one in seven did. Nobody can explain why. We do know that we live in a sea of synthetic estrogens and other hormones and routinely are exposed to materials that never previously existed. The producers of these agents take comfort in the fact that any one of them, tested by itself, looks fairly benign when gauged by various scientific measures of carcinogenic potency.

Still, it defies common sense and basic biology to assume that just because a single agent looks all right when tested on its own, we can safely encounter hundreds of such materials all at once. You would never take all the different pills in your medicine chest in one swallow, even though ingesting one or a few is fine. Why, then, should we accept that there is no danger in being subjected to combinations of agents without precedent in human history? Biologist Tyrone Hayes of the University of California at Berkeley thinks the tadpoles of the seed-corn fields of York County, Nebraska, are trying to tell us something: one in every three exposed to mixtures of ordinary chemicals in those fields die.[3]

Everybody knows that cancer can run in families. Take the Steingrabers of Illinois, the family of the lyrical environmental writer Sandra Steingraber, author of *Living Downstream* and *Having Faith*. Hers was a cancer-prone family for sure. The writer-daughter, her mother, three uncles, and a first cousin all got the disease. Could their shared disease have something to do with the pesticide-sodden wheat and cornfields of Illinois, and the huge grain silos surrounding their small town? We can't say. But we know it had nothing to do with shared genes: Sandra and all those with cancer in her family are not related by blood.

Identical twins come from the same egg and share more chromosomes than any other two humans. They are as close to cloning as exists in nature. Despite their similar roots, identical twins do not have identical cancers. Cancer does run in families, but for those who are adopted, like Steingraber, the risk of the disease mirrors that of the families in which they grow up, not those into which they were born.

What causes cancer is a complicated matter of intense debate. Some of that debate has been fostered and led by folks with a strong interest in fanning uncertainties as a way to promote inaction. They have argued that there's only one proof that a given agent causes cancer in humans: enough people with definite exposures to a specific compound have to have become sick and died of the disease. They dismiss experimental studies of cancer-causing chemicals conducted in rats and mice. After all, we know we have to be different from those test rodents in laboratories. This type of reasoning is morally flawed and ignores one simple fact: the same basic structure of DNA is found in all mammals. If we cannot act to protect or prevent exposures to suspected cancer causes based on solid experimental reasoning, and if we insist on proof that humans have already been harmed, then we are treating people like experimental animals in a vast and largely uncontrolled study. If the same people who oppose the use of animal studies then go on to prevent or suppress research on how environmental chemicals are affecting humans, the reasoning becomes morally indefensible.

THE START OF my own scientific career coincided with a short-lived period during the presidency of Jimmy Carter in the late 1970s, when the federal government looked serious about uncovering the causes of cancer. The National Cancer Institute and other federal agencies began a series of programs to assess the true effects of tobacco and certain widely used industrial chemicals. In 1978 these programs became more than rhetoric.

Until that time the government pretty much took industry reports on the safety of chemicals at face value, without requiring any documentation. This changed when it was learned that the company doing much of the testing for industry, Industrial Bio-Test, could not even find or account for all the animals it had supposedly studied. Industrial

Bio-Test had tested one out of every three chemicals on which the government had any data. But without adequate records to show that the testing had been done properly (or at all), the reports of safety based on this work were worthless. In 1979 the government set up its own experimental laboratory to test the cancer-causing capacity of chemicals in specially bred homogenous rodents, under the U.S. National Toxicology Program. Animals were reared with well established body sizes, types and inclinations, so that their responses to potential cancer-causing agents could be studied carefully in order to predict and prevent cancer and other chronic ailments in humans.

Even before the Industrial Bio-Test scandal, Congress, in response to growing public pressure, had begun passing rules that appeared to require the government to act to keep cancer-causing hazards out of the market. With more than 80,000 chemicals in widespread use and complete toxicity test results available on fewer than 1,000, these laws, like the Toxic Substances Control Act of 1976, forced the government to come up with some rational way to review chemicals and separate the good from the bad, the ugly, and the ones we don't even know what to do about. This was supposed to lead to efforts to come up with standard methods for evaluating risks, for making sense of experimental information, and for estimating ways to protect public health. Instead, the law has generated so much talk and so little action that insiders refer to it as the "Toxic Substances Conversation Act."

After the 1980 presidential election, even these meager efforts began to unravel. The early Reagan administration followed the lead of the Carter administration in its tobacco-friendly positions and also jettisoned programs that sought to rein in cancer-causing industrial sources. The new administration also curtailed funding for testing chemicals under the National Toxicology Program, while federal support for scientific research aimed at designing "safe cigarettes" grew. The problem with "safe cigarettes" is that, well before the 1980s, it was well known that there is no such thing. Inhaling thick clouds of smoke into your lungs, raising levels of carbon monoxide in your blood and that of your children, spouses and office mates, whether from burning tobacco, wood or coal, is simply an unhealthy thing to do.

What of the war on cancer itself? Cancer is the only disease that merits its own war. In fact, many of the first leaders of that battle came from a number of large firms that produced cancer causing materials. Throughout the 1980s, for instance, the National Cancer Institute's advisory board was chaired by Armand Hammer, the Chief Executive Officer of Occidental Petroleum. While he served as a senior adviser to the NCI, Hammer's firm produced more than 100 billion tons of toxic chemicals, including those that created the superfund toxic waste site at Love Canal and led to the contamination of lush Mississippi River delta towns in Calcacieu Parish. Similar conflicts continue today.

Other global firms have been leaders in the war on cancer. Industrial Chemicals Inc. is a vast corporation. Divisions of the company make a number of pesticides and other cancer causing chemicals. Other parts of the same firm, such as AstraZeneca, are renowned for the development of drugs like tamoxifen, one of the most widely prescribed cancer drugs in the world today.

The best wars, to take an ironic line from President McKinley's secretary of state, are short, splendid little affairs, all pageantry and little fighting. The protracted war on cancer has been none of the above. How did we get to this point?

Nearly forty years and more than $69 billion of taxpayer monies since the war on cancer was officially launched in the United States, many forms of the disease remain devastating. Cancer deaths have dropped chiefly because fewer people are smoking and more are getting screened for survivable cancers, like those of the colon, cervix and breast. Yet some forms of the disease not tied with tobacco use continue to increase, and deaths are unconscionably higher in blacks than in whites with few exceptions. Accounts of breathtaking advances in research provide a steady rumble today, just as they have done for decades. Leaders come and go. Battles change. Rhetoric shifts. But the conflict is not ending. The early talk of imminent victory has grown muted. It is still easier for people to become cancer statistics than to understand them.

From the start, this national campaign was blocked from dealing with some cancer causes that were known at the time—tobacco, the workplace and the general environment. Proof that the world in which we live and work has a lot to do with whether or not we get the disease was either overlooked or kept out of sight altogether, often by folks who had major economic interests in seeing this happen. Instead, the entire project focused on devising ways to find, treat and cure the disease, rather than coming up with actions to keep the two hundred different types of illnesses that comprise cancer from occurring. We are spending more money than ever to find and treat cancer—some $100 billion in direct treatment costs in one year. But when it comes to ferreting out the root causes of the disease, we have limped along ineffectively. Why? Could the fact that many of the leading figures in the war on cancer profited both from producing cancer-causing chemicals and from producing anti-cancer drugs have anything to do with the fact that both the incidence of cancer and its treatment options keep steadily increasing?

Of course not. Remember that we live in a highly technological, interconnected world. It is safer, and better for your reputation in polite society, to keep reminding yourself that the disease is just so damned complex. . . .

A FEW YEARS AGO, I learned that some of the reasons we have done so little to prevent cancer are less mysterious than I had imagined. I was talking with Devra Breslow, an accomplished public health researcher, about the small town of Donora, Pennsylvania, where we had each spent some time as children. Donora was a proud little steel-making town of coal-darkened skies and streets as steep as rooftops. "It's hard for anyone to appreciate how tough life was," she was saying. "The mills ran 24/7, every day of the year."

We were sitting in the lovely sunlit garden of her house in the Westwood section of Los Angeles, enjoying tall glasses of iced tea. "The workers would come off their graveyard shifts at seven in the morning and head straight to a local bar for a shot and a beer. They said it would clear their throats. I always imagined that the work was so brutal, they needed something to give them a quick lift."

We talked about how tough Donora was even for those who did not work in the mills. The dangers of millwork did not stay within the factory gates. In October 1948, twenty folks in the town dropped dead after a smoggy haze settled over the horseshoe bend in the river valley and did not lift for five days. The event caused national headlines and made it clear that a lot of pollution in a short time can kill people. But what about those who survived to breathe in more years' worth of fumes, dust and grime?

Devra Breslow had wondered about her own family. "My Grandpa Miller ran the haberdashery on Main Street in Donora. He didn't die during the smog, but he keeled over a year later. Of course, this didn't count as one of those smog deaths, but surely living under all that pollution took a toll."

I replied, "It definitely did. You know my grandmother, Bubbe Pearl, made it through the smog, though she did have her second heart attack then. She didn't die until her twenty-fifth attack, some seven years later."

Devra's grandpa and my bubbe both died of heart disease. But what about cancer, we wondered—had anyone ever looked into whether people from Donora had higher rates of cancer? Possibly not, we decided: four out of every five cases of cancer occur in people age sixty-five or older. Neither Bubbe Pearl nor Grandpa Miller had lived that long.

Like heart disease, cancer has lots of different causes and can take years to develop. Could some of the same things be involved? Anything that inflames the blood vessels that feed the heart and lungs can also damage cells or cause disordered growth. The process by which vessels form is called *angiogenesis*, comprised of two Greek words that mean the generation of blood. How could producing blood cause disease? After all, blood carries nutrients, energy, oxygen, proteins, antibodies and iron throughout the body and also takes away wastes.

It turns out that Mae West was wrong: you can have too much of a good thing. Blood is one of those things that we need the right amount of at the right times and places. Agents that stimulate channels of blood to develop in places where they are not needed can provide unwelcome routes for cell growth that can clog the arteries that fuel the heart or spread cancer. Thus the same toxins in the environment that can put the body into overdrive to snuff out inflammation could raise the incidence of both kinds of disease. This, we decided, was one of the many possible connections between the environment and cancer that nobody was studying.

As Devra and I talked about all this, we moved into the dining room, where we were joined by her husband, Lester, for a splendid California Tuscan lunch, complete with handmade pottery, perfect tomatoes and exquisite whole-grain breads. Lester Breslow is a former dean of the UCLA School of Public Health as well as a former president of the American Public Health Association—positions one does not achieve by neglecting the details of good nutrition. Internationally revered and renowned, he has attended the birth of most major public health efforts for more than half a century. At the age of eighty-five, he walks a few miles a day, oversees the growth of those world-class tomatoes, and has the wry sense of humor of someone who does not suffer fools lightly. He jumped into our conversation.

"You want to know why we've done so little to control cancer? Let me tell you something. We failed in ways that few people even imagine."

He went on: "There's a report we wrote, oh, close to three decades ago. It lays it all out. See if you can find a copy. Even you will be shocked."

"What are you talking about?" I was intrigued. This was not a fellow given to hyperbole.

"Early in the 1970s," Lester explained, "I went to Diane Fink, who headed up cancer control at NCI, and said, 'Let's interview all the major figures in cancer control. Some of them are getting pretty old. It could be our last chance to get it all down, and it'll give you a good grasp of where we've been and where we need to go.'

"NCI gave us a contract to do this, so we put together a small team. The whole project was done in less than two years. Larry Agran was then a young public interest attorney who had worked in Sacramento; he became coordinator of our project. [Agran later became mayor of Irvine, California.] Devra was the major interviewer; she taped about eighty people. We basically talked with everybody who had any role in setting up NCI, the whole war on cancer, the industry folks, all the surgeon generals, you name it.

"By the time the report was finished, Carter was president, Joseph Califano was secretary of Health, Education and Welfare, and Julie Richmond was the assistant secretary for health. At first, nobody really saw the report beyond the NCI division we submitted it to. We had documented how paltry the efforts had been to control the disease, how much had been missed, and how smoking had been swept under the table. Then the reaction started. All the people Devra interviewed who were still alive got copies and started talking about what everybody else had said.

"The next year in Congress, Ben Byrd, the president of the American Cancer Society, denounced our report because it made the ACS look bad. They deserved to look bad. The ACS directly delayed many things that could have helped a lot of people back then. They kept reports on the hazards of smoking locked up a lot longer than they should have. They delayed getting the Pap smear in use. They had a lot to be embarrassed about, and they weren't the only ones."

"Any idea where I can get a copy?" I asked.

"Well, it never got published, that's for sure," Devra laughed. "But I think it should be on some shelves somewhere."

After many months I located a xeroxed copy of the Breslows' five-volume report in the private library of Daniel Teitelbaum, a professor of toxicology in Colorado.[4] More than two thousand pages long, with complete transcripts of eighty interviews, it is a treasure trove of forgotten information on the early days of the American war on cancer. Reading this old document explains a number of anomalies. Why did it take so long for the government to act against tobacco? Tobacco and many chemical industries actively supported research on finding and healing cancer, while hiding or stifling evidence that their own products caused the disease. Why have we made so little effort to control modern workplace causes of cancer or to evaluate possible new cancer risks?

The Breslows' report showed that the revolving door of industrial and government cancer experts had operated since the earliest efforts to deal with cancer nationwide. It also revealed that scientists in the United States, Europe, Latin America and Japan understood a great deal more about some of the major causes of cancer in the workplace decades before the United States, Sweden, England, France, Canada, or any other nation opted for an official war on the illness. For nearly a hundred years, we have known that smoking, sunlight, industrial chemicals, hormones, bad nutrition, alcohol, and bum luck, all affect the chance we will get cancer. The interviewees—all of whom had devoted their professional lives to studying the disease—recounted how experimental studies and case reports of sick workers had filled medical textbooks in the 1930s and 1940s with instances of cancers caused by work and life.

At the official launch of the war on cancer in the United States in 1971, proof that how and where we live and work affects the chances we may get cancer was basically ignored. Astonishing alliances between naive or far too clever academics and folks with major economic interests in selling potentially cancerous materials have kept us from figuring out whether or not many modern products affect our chances of developing cancer.

Lester knew whereof he spoke. I was shocked.

A sixteenth-century woodcut by geologist and physician Agricola depicting the hazards of underground mining, reproduced in an English translation of his book De Re Metallica (On Metals) *by Lou and Herbert Hoover in 1912.*

2

Natural and
<u>Other Experiments</u>

Tragic sins become moral failures only if we should
have known better from the outset.

—JARED DIAMOND

LATE IN 1936, *Time* magazine reported on the remarkable four-week
journey of Maud Slye, a pathologist at the University of Chicago. It
was her first vacation in twenty-six years. Though largely unknown today, Slye is sometimes referred to within her specialty as America's
Marie Curie. She devised an innovative program showing that mice
could be bred at will to have cancer or not. Her work remains a cor-
nerstone of cancer research to this day.

Slye spent half of the month traveling. To leave her mostly academic (and mostly white) enclave of Hyde Park on Chicago's South
Side, she probably scheduled a Checker cab. The uniformed driver
would have taken her along windswept Lake Shore Drive, right next
to Lake Michigan, to Union Station downtown. At the station she
must have boarded the Empire State Express and spent two nights in
one of the sleepers. White-gloved black waiters served her meals in a
well-appointed, walnut-paneled dining car. Some forty hours later,
she would have arrived in Manhattan's central train depot, the cav-
ernous, marble-floored Penn Station. Another cab would have taken
her to a passenger ship docked behind the pink granite facades of

Chelsea Pier in New York Harbor on the west side of Manhattan. The champion black athlete Jesse Owens and the American Olympic team had set sail from the same place just a month before, en route to the Olympics in Berlin, Germany. At those games, the racist Nazis were appalled at Owens' success. He set records in all but one of his events while becoming the first person in history to win four gold metals in a single Olympics. Twenty years before that, my immigrant grandparents were ferried from this same pier to Ellis Island, where they stood in lines to be screened to see if they were fit to enter the country.

If Slye had sailed on the 75,000-ton Queen Mary's first voyage from America, she would have landed a week later at Southhampton, in the south of England, and taken another ship to the European Continent. More likely, she boarded one of the older, slower ships that took ten days and landed in the north of Belgium at Antwerp, on the right bank of the River Scheldt by the Westerschelde. From there, Slye would have boarded yet another train for her destination—the country's capital city, Brussels.

The *Time* story made it clear that Slye did not make this journey alone. More than two hundred of the world's top cancer scientists convened in Brussels that summer to attend the Second International Congress of Scientific and Social Campaign Against Cancer. The meeting had the makings of a veritable Manhattan Project on cancer: the best minds available, poised to create something astonishing and new. The great experimentalist Isaac Berenblum later remembered it as "the most momentous cancer congress ever held."[1] The scientists sailed from Latin America, America or Japan, a journey that could have taken close to two weeks, or took sleeper-car trains from Russia and Europe. With the world clearly on the brink of war, such a trip required considerable courage, as well as a strong stomach. At least one of the participants (Wilhelm Hueper, whom we will meet in Chapter 4) had survived poison gas attacks in the Great War; no doubt several others had had similar experiences. They kept no secrets—government or industrial—but ironically this historic gathering has itself remained nearly secret for more than seventy years. Many of your late relatives and mine might still be with us if the things these eminent women and men of science knew about the causes of cancer in 1936 had entered mainstream medical practice.

But they didn't. Something mysterious happened over the course of the twentieth century. At that meeting in Brussels the accomplishments of several centuries of cancer research flashed onto the scene, ready to coalesce into a substantial and coherent body of scientific understanding about the environmental causes of cancer. Instead, many of these accomplishments were forgotten, their message ignored. Much knowledge that really mattered ended up in that dusty section of the library reserved for books that are never read and papers that are never cited. Today, we're locked in ferocious debates about matters that scientists thought they had solved more than three generations ago. What kinds of evidence tell us the causes of cancer that we can do something about? What passes for scientific proof, while ultimately founded in methods and measures, is not immune to changing political and economic forces.

I FIRST LEARNED of the Second International Congress of Scientific and Social Campaign Against Cancer from a brief reference to it in a memoir by Berenblum. It turned out that the proceedings—the collected papers presented there—were not to be found in any medical library in either Pittsburgh or Washington, D.C. My friend Carol Conners, the indefatigable reference librarian at Teton County Public Library in Wyoming, assured me that if a single copy existed anywhere in the world, she would find it. She did. I wasn't quite sure what to expect when I formally requested the three-volume set from the library in Belgium where it was stored. I thought I would learn how naive the world of cancer research had once been. But after the books arrived, I spent a sleepless night fascinated by the sophisticated drawings and advanced research techniques that were employed to unravel the causes of cancer before I was born. The next morning, I scanned some of the most critical reports and put them on a web site for my colleagues that you now can find on this book's web site. I knew they would be just as stunned as I was to see how much was known about the social and environmental causes of cancer before World War II, seventy years ago.

Table 2-1 Proportion of Gastric Cancer Out of All Cancer Mortality in Different Occupations of Bavarian Men, 1924–1928

Social Class	Description of Occupation	Stomach
I	Merchants, high officials, lawyers, doctors	38.8
II	Office workers	40.1
III	Publicans, brewers, small officials	48.5
IV	Skilled workmen	56.8
V	Unskilled workmen	63.4
VI	Agriculture	68.5

Many of the texts were written in several languages, English, Spanish, French, and German, all presumed to be understood by the multilingual scientific crowd. One speaker, Clarence C. Little (whom we will also meet later), then famous for creating ways to study the inheritance of cancer in mice, argued that animal studies proved that most cancer arose from inherited defects. But at this conference, the view that cancer was dictated by our genes was in the clear minority.

William Cramer of London's Imperial Cancer Research Fund carefully examined patterns of cancer in people over about a century. He was able to do this because at the time, the British had been keeping records of deaths and illness for more than three hundred years. Cramer noted that much of the recorded increase in cancer was nothing other than better record keeping and people living longer. He went on to present techniques for evaluating these patterns that took these facts into account. The numbers of cancer cases had almost doubled since the turn of the century. Taking into account the fact that more people were alive and older, cancer had become about one-third more common than at the beginning of the twentieth century.[2]

Cramer also pointed to other proof of the modern growth in cancer, noting a profoundly simple and important observation that has since been repeatedly confirmed. He thought it important to look at what happens in what he called "uniovular" twins—more commonly known as "identical" twins, those that arise when a single fertilized egg splits into two developing embryos. In 1936, he already had determined that in most of these genetically identical pairs, if one develops cancer, the other does not. Cramer concluded that "cancer as a disease is not inherited." He urged that patterns of cancer—especially those of the workplace—should be tracked in order to learn how to control and reduce the disease.[3]

Cramer understood that human cancers were the result of past exposures, some 20 or more years ago. If one wanted to make progress against cancer it would be important to rely on experimental research with animals. Animal tests provide an important way to learn whether chemical and physical agencies which produce cancer in animals also produce cancer in man. Cramer noted that cancer often develops in both rodents and humans in the same tissues. The time between exposure to a chemical and the time when a tumor shows up varies greatly, occurring within a year in rodents and after decades in humans. Yet this period of latency is remarkably similar if expressed in fractions of the usual span of life in each case. Cramer argued that cancer is one of the few diseases in which the experimental production in animals closely simulates the disease in man. He allowed that cancer in humans may, in fact, be considered an experiment carried out by people on themselves.[4]

The three volumes from this congress included surprisingly comprehensive laboratory and clinical reports showing that many widely used agents at that time were known to be cancerous for humans, including ionizing and solar radiation, arsenic, benzene, asbestos, synthetic dyes and hormones. Angel Honorio Roffo, the founding director of the Institute of Experimental Medicine in Buenos Aires, Argentina, described experiments showing that both invisible forms of radiation—ultraviolet and x-ray—could produce cancers in animals. He was one of several experts at the time to show that these tumors could be surgically cut out from one animal and made to grow in another, a method of tumor transplantation still in use today. Roffo's work referenced even earlier experiments by Andre Clunet, who had produced sarcomas in rats in 1910, and clinical reports by Bruno Bloch from 1923 finding that radiation induced cancer in animals and in workers.

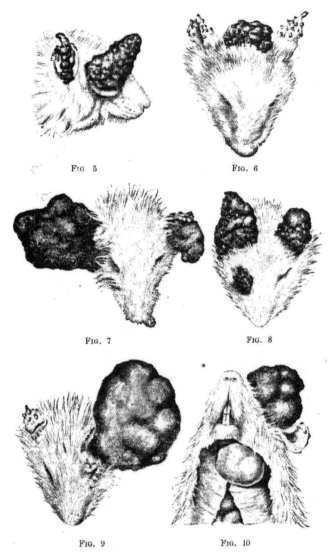

Figure 2-1 *Angel Honorio Roffo's detailed drawings of sprawling tumors growing from the heads, eyes, and ears of rats after 7–12 months of solar treatment.* Rat No. 1/Fig. 5: *A large tumor of the eye and papilomas in the ear after 7 months of solar radiation.* Rat No. 2/Fig. 6: *large tumor developed in the rat after 8 months of solar radiation.* Rat No. 3/Fig. 7: *Two large tumors developed in both ears and one in the nose after 7 months of solar radiation.* Rat No. 4/Fig. 8: *Two large tumors developed in both ears after 10 months of solar radiation, one fusocelluar sarcoma and the other carcinoma. A third tumor developed in the eye.* Rat No. 5/Fig. 9: *Large tumorlike mass (fusocellular sarcoma) developed in the ear after 12 months of solar radiation.* Rat No. 6/Fig. 10: *Large fusocellular sarcoma developed in the ear of the rat after 11 months of solar radiation, with gangliar metastasis of the neck.*

Roffo's own studies of workers showed that those with the greatest amount of time spent outdoors had the greatest vulnerability to skin cancer. His paper was accompanied by exquisitely detailed drawings of sprawling tumors growing from the heads, eyes, ears, and necks of rats following months of solar or x-ray treatment.[5] He also reported that combining some hydrocarbons with either sunlight or radiation produced much worse cancer damage than any one of these exposures alone. He advised avoiding radiation and sunlight, and reducing exposures to hydrocarbons. These are observations that the modern world didn't begin to take seriously until the 1980s.

Roffo was one of many experts to issue a strong statement against the fashionable view that a tanned skin signaled good health. At a time when movie stars and suntanned cowboys were seen as glamor figures, he concluded by "protesting strongly against excessive sunbathing which exposes the skin to intensive irradiations from the sun, placing individuals, victims of a ridiculous fashion, into a particularly dangerous state of receptivity to the development of skin cancer."[6]

I had expected to find amusing errors and preposterous assumptions in the conference volumes, but I didn't. The papers did not depict the dark ages of cancer research but rather an exhilarating time of lively and important work that seems to have come and gone like a comet. A review of carcinogenic chemical compounds by the noted researchers J. W. Cook and Edmund L. Kennaway and others with London's Royal Cancer Hospital reported that more than thirty different studies had found that regular exposure to the hormone estrogen produced mammary (breast) tumors in male rodents. The National Toxicology Program of the U.S. government did not formally list both estrogen and ultraviolet (sun) light as definite causes of human cancer until 2002.[7]

How did these scientists decide what was a cause of cancer in 1936? They combined autopsies with medical, personal and workplace histories of people who had come down with cancer. They reasoned that if they found tars and soots in the lungs of those who had worked in mining and showed that these same things caused tumors when placed on the skin or into the lungs of animals, that was sufficient to deem these gooey residues a cause of cancer that should be controlled. Their animal work was quite sophisticated in today's terms, extending from complex laboratory studies of rats, mice, rabbits, monkeys, dogs and cats to various physical and chemical agents that left clear marks of cancer.

In many ways the 1936 congress was the culmination of centuries, even millennia, of earlier work. The long view of cancer history is a tale of intrigue, courage and extraordinary dedication. It's a story of physician scientists who were also keen observers of everyday life and who expressed a rigorous urgency to learn, no matter where their inquiries took them. Many of the basic causes of cancer were identified hundreds of years ago. Mining, painting, smelting, forging, distilling, curing, smoking, grinding, and cleaning were portrayed in literature and excellent medical accounts—some dating from the Middle Ages—as risky enterprises. These observations were confirmed in the first half of the twentieth century by experimental studies of rodents, rabbits and other small mammals. Much of this knowledge deserves to be central to medical education and practice. But modern cancer medicine has a collective amnesia about its own history.

BEFORE THE TWENTIETH CENTURY, physicians and scientists had an expansive view of what it took to be able to say that anything could be considered a cause for cancer. A broad range of natural experiments, some carried out by researchers on themselves, repeatedly showed one simple thing—our health reflects the sum of our life experiences. Most cancer arises not because of who our parents were but because of what happens to us after we are born. Where and how we live and work, what we eat, how we spend our private time, how we move about: all these things affect the kind of health we will have. Heat, cold, dust, dirt, radiation, soot, fumes and myriad natural and synthetic agents combine to affect the chances that anyone will get any disease. Cancer develops not because of one unique circumstance, whether hereditary or environmental, but out of the sum total of the goods and bads of our lives.

Hippocrates was not the first of the ancients to be fascinated with the uncommon and monstrous growth of cancer, nor was he the earliest to describe a sprawling crab-like tumor of the breast, called *Karkinoma*. Nearly four centuries earlier, around 900 BC, one of the first depictions of the disease is found in a collection written on pressed papyrus reeds from Egypt—the world's first preserved paper. The Edwin Smith Papyrus, named for the English surgeon and Egyptologist who translated it in the nineteenth century, describes eight cases of tumors or ulcers of the breast in startlingly modern terms. The author of the papyrus reports only one treatment for these ancient tumors: repeated use of a "fire drill" to burn out those growths that had broken through the skin.

Cancers that could be seen were sometimes removed successfully as long ago as the Middle Ages. Even then a healthy life was considered to lessen the chance the disease would occur. The twelfth-century Jewish polymath Moses Maimonides, who served as the chief rabbi of Cairo as well as chief physician to the sultan of Egypt, carefully described how to excise a cancer and uproot all surrounding tissue. But he warned that this wouldn't work "if the tumor contains large vessels & [or] the tumor happens to be situated in close proximity to any major organ."[8] To prevent the disease, he counseled staying away from dusty cities and dirty air, eating chicken soup and garlic, and getting regular exercise.

In the mid-sixteenth century, the geologist and physician Georgius Agricola spent years preparing a massive report on mining that included detailed information on the ailments of miners. He didn't just rely on what others told him. Agricola went underground into the Erz Mountains of Central Europe to watch boys and men extracting, preparing and processing ore. He was struck by the number of young miners with tumors in their chests.

Agricola's magnum opus, *De re metallica,* appeared in 1556, one year after the author's death, and included some of the earliest reports on the chronic ailments of underground work. Those who entered the mines the youngest, if they did not perish in gruesome accidents, fared the worst and eventually died from lung diseases and tumors. Agricola's work was printed with 289 remarkable woodcuts (see illustration at beginning of this chapter) portraying the brutal work of mining both above and below ground.

Sometimes it takes a few centuries for important news to make the rounds. In 1912, Herbert Hoover, then one of America's top mining engineers, and his wife, Lou, a Latin scholar, published the first English translation of Agricola's work in *Mining* magazine with the four-century-old wood cuts.[9] In their introductory comments, they explained that they made the translation because this sixteenth-century work remained relevant to the lives and deaths of miners into the twentieth century—something we are reminded of today by occasional reports of mining disasters in Russia, China or West Virginia. The Hoovers admitted that harms to workers were regrettable, although the ways they could be avoided were less apparent than the profitability of the materials with which they worked.

By the turn of the eighteenth century, the path-breaking Italian physician Bernardino Ramazzini had documented more than three dozen different cancer-prone professions, including mining of coal, lead, arsenic and iron. At that point the disease was still uncommon and usually lethal. Ramazzini could not tell you which specific part of the job caused which maladies, but he knew that people in many different jobs were subject to risk, including metal gilders, chemists, potters, tinsmiths, glassmakers, painters, tobacco workers, lime workers, tanners, weavers, coppersmiths, mirror makers, painters, sulfur workers, blacksmiths, apothecaries, cleaners of privies and cesspits, farmers, fishermen, soldiers, printers, confectioners, carpenters, midwives, wet-nurses, and corpse carriers. For each of these trades, he explained what particular agents or conditions he thought gave rise to certain classes of illness. Those who worked with dust and fire, like miners, blacksmiths, glass workers, printers, bakers and smelters, tended to suffer from weakened lungs, unstoppable coughs, and occasionally suffocating tumors of the lung.

When he reached his late sixties, itself an achievement at the time, Ramazzini published his major work, *De morbis artificum diatriba (Diseases of workers),* which showed that the things men and women did at work played a major role in determining what ailments they developed.

Ramazzini died at eighty-one in 1714, in an era when most workingmen did not reach forty years of age. In addition to being fairly adventurous, he was an observant doctor with a penchant for record keeping. He noted that nuns tended to be free of cancer of the cervix, then one of the most common fatal tumors of women. At the same time, those who lived celibate lives were more often plagued by breast cancers than other women. Ramazzini speculated that both of these anomalies could be related to the same cause—nuns didn't bear children but underwent a lifetime of menstrual cycles uninterrupted by pregnancy or nursing. His theory that something associated with the failure to bear children affected cancer risk remains a central tenet of cancer research today.

One other thing distinguished Ramazzini's work. He believed that those who learned of workplace hazards had a simple moral duty to warn workers about the risks of their employ and urge them to lower those risks for themselves, their families and their towns. He offered this modification of Hippocrates' ancient advice: "When a doctor visits a working-class home he should be content to sit on a three-legged stool, if there isn't a gilded chair, and he should take time for his examination; and to the questions recommended by Hippocrates, he should add one more—What is your occupation?"[10] Ramazzini based this advice on his own practice. "I for my part have done what I could and have not thought it unbecoming to make my way into the lowliest workshops and study the mysteries of the mechanical arts."[11]

A few decades after Ramazzini, the English surgeon Percival Pott reported a link between working with soot and an unusual tumor that was hard to miss—cancer of the scrotum. In his detailed study of chimney sweeps, he noted that "the disease in these people seems to derive from a lodgment of soot in the . . . scrotum." German and Swedish sweeps, who wore leather trousers or bathed more often, were found to have been less affected by the disease. English sweeps understood the dangers of their work, calling scrotal cancer "soot wart."

This late eighteenth-century example provides four points of some relevance for workplace cancer. First, workers often know the dangers of their work. Second, differences in risk can be associated with different workplace practices. Third, uncommon events, like cancer of a male reproductive organ, get noticed. Finally, even as clear-cut an association as soot and scrotal cancer reflects multiple causes. Like many causes of cancer, coal soot is not one chemical but a mixture of hundreds. Even though Pott's finding was a milestone in medical science, this complex chemistry would mean that his chimney sweeps would have trouble collecting damages in a modern U.S. court. If any of them were to sue the chimney makers or the coal companies for damages, he would be asked to show exactly which of those hundreds of chemicals found in coal tar had caused his cancer. The true answer, "All of them," is no longer considered acceptable.

A TECHNOLOGY THAT would revolutionize our ability to find cancer surfaced at the end of the nineteenth century, long after Ramazzini and Pott had warned of the effects of work on health. Like many modern breakthroughs in cancer research, this miraculous invention turned out to increase the risk of the disease as well. On November 8, 1895, Wilhelm Conrad Röntgen, the physics professor and rector of the Julius Maximilian University of Würzburg, spent an entire evening repeating the same unbelievable experiment. Each time he sent an electric current coursing through an evacuated tube, a paper far across the room mysteriously began to glow. Even when placed behind the walls of the room next door, the paper emitted the same radiance.[12] The professor called these penetrating rays x-rays.[13]

That winter the Röntgen household must have rocked with excitement. On New Year's Day, 1896, Röntgen sent his paper, "Uber eine neue Art von Strahlen" ("On a new type of rays"), to leading physicists in Germany, England, France and Austria. Each package included stunning, ghostly x-ray images of the bones of Frau Bertha Röntgen's right hand against a dark background.[14] She refused to have another such image made of her body, seeing in these skeletal pictures a portent of death. Her premonition would be borne out years later when it became clear that these rays damage the bone marrow and seed many different types of cancer. Frau Röntgen never bore children, a fact that I find hard to separate from the thought of the two of them living in unshielded rooms just below the lab where Röntgen conducted years of experiments with x-rays.

Anyone opening those envelopes in January 1896 knew they were glimpsing something amazing. On the fifth day of the new year, Röntgen's discovery—along with his wife's finger bones—made the front page of the *Vienne Press*.[15] Within weeks x-rays became the first world-

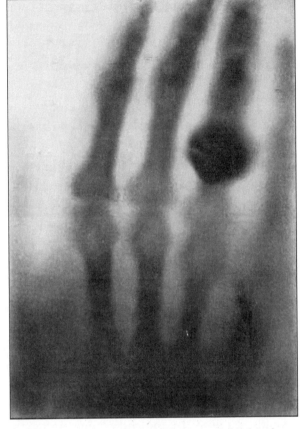

Figure 2-2 *The x-ray of Frau Bertha Röntgen's hand (with a ring on her finger) was featured in newspapers around the world within days of her husband's, Wilhelm Conrad Röntgen, discovery of x-rays on November 8, 1895.*

wide medical vogue. In Paris, London and New York, x-ray machines popped up at public demonstrations, as fashionable party entertainment and in the chicest clinics.[16] A small museum at the headquarters of Genzo Shimadzu in Kyoto, Japan, displays x-ray images that were produced in October 1896, just ten months after Röntgen circulated his report.

As the twentieth century dawned, things that had once occupied the fantasies of lonely laboratory investigators were changing the way the world ran and looked. Electricity, motor cars, telephones and lightbulbs transformed the scale and scope of time and space. Information that had once taken months or years to disseminate got passed around the world within days. The public appetite for science soared. So did the faith that these new discoveries would radically change human life for the better.

That x-rays and other technological and chemical breakthroughs might harm our health was not imagined by those who enthusiastically rushed to put them to work. Thomas Edison would become a notable exception. Clarence Dally, Edison's chief assistant, became one of the first people to die from radiation. Burned in x-ray experiments carried out with Edison at the end of the nineteenth century, Dally's arms had sores that never healed. Cancer spread from these sores over his limbs and into his lymph nodes. Distraught when Dally lost first his ulcerated arms and then his life, Edison refused to have another x-ray.[17]

Researchers' impulse to experiment on people's health, including their own, in the name of a greater good has a long and intermittently respectable history. Some of these self-experimenters, like Edison, escaped dire consequences. Others didn't. Marja Sklodowska, the Nobel Prize–winning Polish chemist, better known as Madame Marie Curie, may have been the second, after Dally, to die from her own research with radiation. Her work and that of her husband, Pierre, detailed the miraculous phenomenon of radioactivity in pitchblende, the ore that yields uranium. By the end of June 1898, they had produced a material that was about three hundred times more radioactive than uranium. The next month they wrote, "We . . . believe that the substance that we have extracted from pitchblende contains a metal never known before, akin to bismuth in its analytic properties. If the existence of this new metal is confirmed, we suggest that it should be called *polonium* after the name of the country of origin of one of us."[18]

The years of shoveling raw radioactive materials into various compartments and repeatedly capturing pictures of their own bones took their toll on the Curies. Cataracts fogged their eyes, and constant ringing dulled their ears. In the spring of 1906, Pierre died when struck by a Parisian carriage he had not been able to see or hear. In 1911 Marie isolated radium, unlocking the capacity of this element's atoms to emit bone-piercing energy from their nuclei. The two Nobel prizes she received for her work didn't protect her from the damaging properties of the agents she had discovered. She died in July 1934 after the marrow of her heavily radiated bones lost its ability to make the right amount of white blood cells. The disease that felled her (and her daughter Eve years later) is nowadays called leukemia, a Greek translation of the German phrase for white blood. For years after her death, the Curie notebooks remained too hot to handle. Their radioactivity was measured in the units to which the couple gave their names and ultimately their lives, the curie.

Just four scientific generations ago, many forms of cancer were so rare that clinical reports were matters of curiosity and autopsies were occasions of great wonder. Late in his life, the pioneering medical researcher Andre Cournand recalled how, during his training as an intern at the Hôpitaux de Paris in the 1920s, he and his colleagues had rushed to see their first case of lung cancer. "You have got to come down here and look at this," said Cournand's medical chief. "You will never see anything like it again!" The residents hurried downstairs and crowded into a circular balcony to see this exceptional death splayed in the operating theatre below. With a shiny steel scalpel, the pathologist expertly carved a V-shaped incision into the grayed skin, starting with a diagonal slash at the front of each shoulder. These two lines converged above the belly into a solitary vertical cut that sliced through the navel down to the pubic bone. Once the muscles, sinews and sacs that hold the organs in place were separated from the skin, the rib cage was cracked and sawed open. Below the rib lay the lungs, blackened and pink, smudged throughout with gray-black stickling. In the center, sprawling in several directions, was a rubbery looking, thick, glutinous glob—that great rarity, a cancer of the lung.

In 1915, when Cournand was a medical student, Paris vibrated with enthusiasm for what x-rays could do for medicine. Like many he found his education interrupted by the First World War. On the battlefield as a surgeon for three years, he learned not to flinch and when to duck. "It had been necessary to develop the attitudes of mind and feeling to face danger and take risks."[19]

For his fortitude during the war, Cournand earned three Bronze Medals. He also earned a keen sense of the value of real world experiences and a skeptical attitude toward received wisdom that rested on theoretical but unproven beliefs. When he returned to medical research in Paris after the First World War ended at the advanced age of twenty-five, most of his colleagues did not believe that the human heart could survive being probed.[20] Decades earlier, the brilliant French experimental physiologist Claude Bernard had inserted a catheter directly into the heart of a living horse to track the way that blood circulated through the body. Yet French researchers feared that what was possible in a horse might prove fatal for a human.

Cournand had a different view. He was intrigued by relatively free-wheeling American research that relied heavily on the use of x-rays and active examinations in contrast to its limited scientific applications in France. In 1930 he emigrated to America to join with the physician Dickinson Richards at Columbia University. They began experimenting with various animal systems, devising small probes for laying bare the remarkable path that blood took as it coursed into and out of the hearts of dogs and horses.

They were astonished to learn that in Germany in 1929, a brave young physician named Werner Forssmann had managed to do what many had thought impossible—take x-rays of his own catheterized heart. Assisted by a lone technician, Forssmann had punctured his left cubital vein with a well-lubricated two-foot-long rubber cannula. With the small cannula in place and a slightly warm sensation in his chest, Forssmann then climbed two flights of stairs to the Radiology Department, where an x-ray showed the tube lying right inside his heart. His German colleagues, some of whom were about to take an entirely different approach to human experimentation, ridiculed this work as both dangerous and pointless.[21]

Cournand and Richards did not attempt Forssmann's procedure on another human until 1941, when they conducted studies on live patients at New York's Bellevue Public Hospital. The fact that their first work involved patients on public assistance probably kept their research outside the limelight.

Forssmann was later captured by the Allies while serving as a physician in Berlin and spent time in a prisoner of war camp until his release in 1945. We know that he had done quite a bit of work on cadavers, taking x-rays of catheters that snaked through their arteries, before he took those stunning images of a tube threaded into his own heart. We do not know all of what he may have done during the war. But we do know that he refused offers from Karl Gebhardt, Himmler's personal physician, to receive human subjects for further research. After the war ended, with many of Germany's collaborating physicians in disgrace, Forssmann was a broken man with limited possibilities. His stature rose in 1956, when the work he had done some three decades earlier, along with the studies of Cournand and Richards at Bellevue Hospital in the 1940s, was awarded the Nobel Prize in medicine.[22]

AMONG THE MANY FACTS about that remarkable cancer congress of 1936 that are not well known (in fact hardly known at all) is that most of the assessments regarding cancerous effects from hormones, arsenic, sunlight, radiation, benzene and other chlorinated hydrocarbons were well accepted by official industrial sources at that time. Ten years earlier, the American National Safety Council issued a final report on the hazards of benzol, the German term used to describe benzene. Their document noted the doses at which it induced narcosis and severe weight loss in animals, and included 125 different references. Highly exposed workers became anemic and sometimes died when overcome by fumes they encountered cleaning out deep tanks. Those without lethal exposures had a range of blood problems that were well studied.

Out of a total of eighty-one workers studied in all plants, the Council reported, "26 gave a blood picture characteristic of benzol [benzene] poisoning; and this ratio of about one man in three affected was maintained even in those workrooms with efficient local ventilation. . . . We were therefore forced to conclude that . . . the use of benzol (except in enclosed mechanical systems) even when the workers are protected by the most complete and effective systems of exhaust ventilation . . . involves a substantial hazard."[23]

In response to these reports of serious health problems in men working with benzene, a series of studies were conducted in cats, dogs, rabbits, guinea pigs and rats. These studies, like many carried out in toxicology at that time, chiefly asked how much benzene was needed to anesthetize or kill the animal and how quickly this happened. Animals were observed for minutes, hours or days to see when they developed jerky tremors, weakness and muscle contraction, and at what point they simply dropped dead. Their blood was drained after death and examined for evidence of what benzene did. Those animals that recovered from these exposures looked normal within days. One study that injected much smaller amounts of benzene in rats found that it induced an array of symptoms including loss of appetite, reduction in infection-fighting cells of the blood, and tremors.[24]

Based on this work, the safety council decided that benzene was a highly problematic material in the industrial workforce: "We are forced to conclude that the control of the benzol hazard (except where the substance is used in completely closed systems) is exceedingly difficult; that in practice, systems of exhaust ventilation capable of keeping the concentration of benzol in the atmosphere below 100 parts per million are extremely rare; and that, even when this is accomplished, there remains a decreased, but substantial hazard of benzol poisoning."[25]

Echoing this work nearly two decades later, the American Petroleum Institute in 1948 conceded that "it is generally considered that the only safe concentration for benzene is zero. . . . Skin contact should be avoided. Acute poisoning by benzene should be considered as an acute emergency. . . . Chronic benzene poisoning is refractory to treatment. Practically all therapeutic measures attempted have failed."[26] The American Petroleum Institute today (as we will see in Chapter 12) takes a radically different position on benzene, actively working to fund research that it expects will overturn national standards in many countries.

In 1949 a report in *Scientific American* by Groff Conklin featured a graphic display of "carcinogens known to be present in human environment."[27] Asbestos was described, along with solar and ionizing radiation, chromates, tar, synthetic dyes and arsenic, as causing cancer by physically damaging the body or chemically inducing malignant growth. The article offered a clear statement: "Scientific and technological progress has exposed man to new physical and chemical agents. Some are believed associated with the rise of cancer as a cause of death." It's worth quoting Conklin's half-century-old views at length because the ideas are remarkably contemporary.

> The growth in the relative importance of cancer as a cause of death is one of the outstanding medical facts of the past fifty years. The disease has moved from eighth to second place in the U.S. since 1900, and to-day only heart ailments surpass it. The reasons customarily given for this rise—improved diagnosis and the aging of the population—do not entirely explain it. They provide no satisfactory answer to the fact that 7.5 percent of the known cancer deaths in 1944 occurred in age-groups under forty. There is evidence, moreover, that the disease is not an inevitable consequence of bodily degeneration due to age, al-though the changes of senescence under certain conditions may be contributing factors. It seems certain that there is a net increase in true cancer deaths, if only because fewer people die from other dis-eases than in the past.
>
> An explanation of this increase and of the causes of the disease is therefore being sought in our environment, so much more complex than it was in 1900. The investigation is focused on carcinogenic agents, as the substances that produce cancer are called, and on the general question of the extent to which the increase in cancer may be caused by agents in the environment that have hitherto been consid-ered comparatively harmless.
>
> It has been established that certain agents to which people are ex-posed through industrial occupations cause cancer if the exposure to them is sufficiently intense and prolonged. As an example, over 75 percent of the miners in the Schneeberg cobalt-uranium mines of Ger-many die of lung cancer; more than 50 percent of those in Joachims-thal uranium mines across the border in Czechoslovakia die of the same cause. In the eighteenth century it was learned that among chim-ney sweeps exposed to intense concentrations of soot, deaths from cancer were between three and four times as high as those in the gen-eral population. It is known also that certain common substances in concentrated doses can produce cancer; for example, mouth cancers

are uncommonly frequent in a tribe in India which smokes cigars with the lighted end in the mouth. This causes them to suffer frequent burns and to receive a concentrated dosage of tobacco tars.

Although these are special cases of intense exposure, they naturally suggest speculation as to whether the average human being's relatively mild but long-continued exposure to new substances in a contaminated atmosphere, in processed foods, in cosmetics and in other elements of our environment may be a contributory cause of cancer. We have as yet no conclusive evidence for or against this possibility. We have no accurate estimate of how many of the artificial substances common to our industrial civilization may be carcinogenic under special circumstances, nor how many seemingly harmless substances, interacting with others that appear to be equally innocuous, may produce carcinogenic results.

Generally speaking, however, employers are no more responsible for the lack of information about industrial cancer than are the many thousands of physicians who have cancer patients in industrial areas or who actually are associated with factories. It is an unquestionable fact that an appreciable number of occupational cancers slip through the hands of doctors unidentified. This is due in a great degree to a general ignorance of the occupational aspects of cancer. Physicians have never been adequately informed of the basic symptomatic and sociological factors involved in identifying occupational carcinogenesis.

The medical profession should be better educated about the need for exhaustive case histories which carry the individual's jobs record in detail back as far as twenty-five years, about the urgency of checking medical suspicions of industrial cancer hazards against careful epidemiological studies of all workers in a plant, and about the paramount importance of impressing plant management with the seriousness of the problem.

The standard protective and hygienic measures currently used in industry to combat industrial poisons and other health hazards are not always adequate for the control of occupational cancer. The following case history is a compelling illustration. Some thirty years ago workers in one of the newer metal industries began to develop lung cancers. At that time the cancers were found to remain latent from ten to fifteen years. The incidence was unusually high in certain operations where the carcinogenic substance was present in particularly high concentration as an airborne dust. An effort was made to safeguard these operations. Up-to-date equipment for removing dust and fumes was installed, and a standard industrial hygiene program was inaugurated, included protective clothing. But the outcome was exactly the opposite of what had been expected. The incidence of lung cancer did not diminish; in fact soon afterward cancers began to appear among workers who had been exposed for less than six years, a much shorter period than had been previously observed. What had happened was that at the time when the protective measures were adopted the factory had also begun to use a more finely ground material to improve production. The finer dust, though present in a much lower concentration than the original material, penetrated farther within the bronchi of the lungs. Thus despite the installation of the latest in protective devices and procedures, the cancer hazard was actually increased.

It is obvious, therefore, that the control of occupational carcinogenesis—and even to a great extent of cancers stemming from more indefinite environmental agents, if and when they are discovered—is a public health problem of considerable magnitude. This is made even more apparent by the scope of a control program that has been proposed by Dr. Wilhelm C. Hueper, head of the new environmental cancer section of the National Cancer Institute. Dr. Hueper, one of the world's leading experts on occupational cancer, has studied the problem in the U.S. and elsewhere for many years. Several of the elements of his program have already been put into effect in European countries. The program proposes that carcinogenic agents be eliminated from industrial military and civilian use as far as is possible and practical; that manufacturing processes which must use such materials be enclosed; that the community be protected by the prevention of the discharge of carcinogenic wastes; that factories be licensed and inspected; that workers be provided with protective clothing, equipment and medical supervision, including frequent and thorough physical examinations.

Four centuries ago some observant physicians laid down the basic foundations of public health research. By the 1950s, some scientists had developed a program aimed at training physicians to recognize and reduce risks from workplace and environmental hazards. How do scientists today go about figuring out the hazards of work or the world around us for our health? We do pretty much what Agricola, Ramazzini, Pott and Roffo did. We look around. We visit and talk to people who are going through natural experiments of their own sort to learn about the goods and bads in their life histories that could account for their health. In classic scientific experiments, results are contrasted between two groups that ideally differ in only one way.

In public health research, we rely on our ability to compare groups that may differ in many ways, in order to conclude whether or not some of those differences account for why some are healthier or sicker than others. For workplace hazards, such as those that first fascinated Ramazzini, we compare various measures of the well-being of those in some jobs to those in others. We ask and count what types and amounts of illness arise in those who work directly with certain agents. Where we can, we measure residues in air and water, or even in blood and urine. We then contrast this information with what happens to those who lack such experiences.

By 1938, the world's top scientists from Italy, France, Germany, Argentina, America, England, Japan and Russia understood that much cancer came from workplace, nutrition, hormones, sunlight, radiation and other external sources. One is left to wonder if the U.S. National Cancer Institute had begun a program to train doctors to look out for signs of these health risks and to promote their reduction by 1949, why were these efforts stymied? What happened to derail programs to reduce the burden of cancer? Why have we spent so much effort treating cancer and so little understanding how to keep the disease from happening?

In the run up to World War II and its aftermath, science could not remain an abstract matter carried out solely because of the inherent curiosity of lone geniuses. Instead, scientific investigations became part and parcel of vital national efforts to conduct and carry out warfare. During the various early-nineteenth-century French revolutions, the philosophes had boasted—at least until some of them were beheaded for doing so—of the value of pursuing cross-national exchanges.

For humanity, the specter of death and national conflict that began to course around the world in the second quarter of the twentieth century concentrated the imagination wonderfully. But it seldom did this in a way that inspired clear thinking about the future. The future got shorted by those who looked solely at the present.

As we have come to know, the mid-1930s, when this august cancer congress was held, was an era of mounting hostilities and widespread militarization of the most common aspects of life. As a committed Unitarian, the biologist Walter B. Cannon saw international scientific collaboration as a moral duty. He resisted nationalistic impulses to pull back from working and meeting with scientists from other nations. He journeyed to Leningrad, Russia—then in the grips of its own revolutionary violence—to meet his colleague Ivan Pavlov, the pioneering behavioral psychologist, in 1935. His address to this congress foretold the lapse of long term interest in any scientific matters, including the ability of chemicals and radiation to damage human life:

> During the last few years how profoundly and unexpectedly the world has changed. Nationalism has become violently intensified until it is tainted with bitter feeling. The world-wide economic depression has greatly reduced the material support for scholarly efforts. What is the social value of the physiologist or biochemist?[28]

Cannon is known today for having coined the term "fight or flight" to describe the physical response of living beings to life-threatening terrors. A chance finding of what made cats get their backs up under duress led him to a lifework examining the complex physical ways that bodies deal with danger, collaborating across oceans and national borders to do so.[29] When facing danger, the body mobilizes. A surge of hormones turns on the ability to fight or run harder, faster, longer. The heart beats more powerfully. Energy surges throughout the body. The hair stands on end, every organ system marshaled in defense against the perceived threat.

Nations do much the same. The prospect of massive, unrestrained global conflict fundamentally changed public priorities and altered the way science was supported and used by those who underwrote its efforts. The immediate need to defend against threats of Axis conquest trumped consideration of the longer term results of living crisis-driven lives. To be concerned with preventing cancer requires planning for and thinking about what will happen in a few decades. A world facing highly uncertain, potentially cataclysmic, prospects was not inclined to ponder the future.

And then, once the war was over (and a slower, colder war took its place), the old knowledge about cancer hazards fell victim to enthusiasm for modern industrial advances and the social and economic forces that lay behind them. A combination of optimism about the industrial future, bona fide improvements in the ability to see and grasp the basic biology of disease, and darker forces fueling that optimism guaranteed that the burden of proving any modern activity caused cancer would become impossibly heavy. The search for more scientific information easily morphed into a reason to reject what had once been known.

THE REASON I KNOW Percival Pott's young chimney sweeps would have had a hard time in a modern courtroom is that in 2000 I participated in a case that was eerily equivalent. Alicia Fernandez was a proud, hard-working Mexican-American immigrant who spent more than ten years in the "clean rooms" of IBM's chip factory in San Jose, California, working with more than a dozen different chemicals known to cause cancer in lab animals. She and her coworkers, all women, wore caps and gloves to keep human hair, skin flakes and perspiration from contaminating the delicate wafers onto which computer chips were etched. But their lungs got no protection from the fumes and dust that filled the room.

Fernandez was thrilled to have what she was told was "a job for life." This turned out to be true in the sense that after ten years at IBM it was impossible for her to continue to work anywhere else. She learned she had breast cancer in 1994. She was the first member of her family in four generations to be diagnosed with cancer. Her eighty-year-old parents in Mexico were alive and well at the time. Fernandez and her colleagues filed a joint lawsuit against IBM in 1998, alleging that their cancer had resulted from their work in an unsafe environment. I agreed to serve as an expert witness for the plaintiff.

Before a case makes it to a courtroom there is a process called "discovery," in which lawyers get to question opposing witnesses in order to gather pertinent information. For this case, the defendants relied on teams of lawyers whose fees for a single day exceeded the plaintiffs' annual salaries. We sat in a paneled conference room while they grilled me endlessly about anything that seemingly crossed their minds. The atmosphere was tense. Every word was taken down by an official stenographer. Eventually the written transcript of my testimony would take up two large boxes. The documents supporting my views filled an entire filing cabinet. Once sworn in, I could not eat or leave the room without permission. Coffee was served freely but bathroom breaks had to be negotiated.

Around day ten of this ordeal, an expensively dressed, leather-tanned lawyer leaned in across the glistening mahogany table and asked, "Now, Dr. Davis, are you telling me that you know which *specific* chemical caused which *specific* cancer in Alicia Fernandez at what *specific* time?"

The repetition was hard, heavy and unmistakable. Dictionaries define *specific* as precise, particular, definitive, unique—all qualities that the causes of cancer rarely display.

I squirmed. "Of course I can't. As you know very well, cancer comes about from all the things we encounter in life, not just one exposure," I replied. "But if I were to take a two-by-four and hit you over the head and knock you out, would there be any question about what had caused you to drop?"

The lawyer reddened and recoiled. "Dr. Davis!" he snapped. "You are out of order. I will have you cited for contempt if you pursue this further."

I knew as soon as I spoke those words that I had been had. Over the intense war of words that had gone on for twenty days, the lawyer had baited me repeatedly. This time I had bit. While I have never struck another person in my life, I intended my words to hit him as hard as he had been hitting me with his repeated challenges. His shocked response was of course an act: its purpose was to make sure I didn't get a chance to explain that many of the things that combined to knock

Fernandez out had been known to cause cancer early in the last century. We couldn't say which distinct one had decked her, or if perhaps it was several in combination, but the lawyer and I both knew one thing. IBM was well aware of all the chemicals that Fernandez worked with, and they knew that many of them caused cancer. At the time, however, that information was a trade secret.

Today, Fernandez's job is done in the dark by robots in what is called "lights out" manufacturing.

IBM won that lawsuit in 2004. The judge ruled that the plaintiffs had not proved a causative link between each of the many different chemicals IBM used and the cancers that had occurred in each individual woman. The increased rate of breast cancer for the women who created the guts of computers became a matter of public record only after the company's legal efforts to prevent publication had been exhausted.[30] By then it was too late for Fernandez. In lawsuits as in much of life, timing is everything.

Brilliant, very well-paid lawyering has produced a highly restrictive set of rules regarding what sorts of evidence provide proof that any one thing causes another. Much of the real world is shut out. Centuries-old knowledge about the causes of cancer, accepted by scientists and businesses alike in 1936, is now inadmissible in court—not because it has been disproved, but because it has been superseded by legal maneuvers and disqualified by a subtle shift in what is considered proof of harm. The kinds of scientific evidence that can be used today in these situations has been carefully culled. Animal research is usually ruled out. That work may be suitable to make drugs, but not for showing harm to people. To establish in law that harm has occurred from a toxic exposure, first you have to show that a group of people with precisely the same exposures has come down with the same health problem. Then you have to prove that this same problem is occurring in another group to which the person you are representing belongs. Then you also have to find and identify a responsible party that is still in business who should be held accountable. Basically, before you can collect damages, you must get cancer or some other awful disease, show that someone else already got it from the same things you did, prove that you had specific exposures to a particular agent, find the firm that caused your harm and can now pay for it, and prove that they knew the exposure was harmful.

In Fernandez's case, these specifics proved her undoing. I was not allowed to testify about the fact that Fernandez had worked with dozens of widely known cancer-causing agents. Instead, I was challenged to show which specific chemical, like benzene or asbestos, had actually caused her particular cancer. By insisting on this tortured and highly artificial idea of a single cause of a single case of cancer—and because many U.S. courts now go along with it—IBM's lawyers won the case before it even started.

The Nazis ran the first public advertisements that warned of the dangers of smoking. They also banned smoking in public places, especially for women of reproductive age.

3

A Broad Enough Principle

It's a very comforting thought that this is a German peculiarity
and nobody we know would ever dream of doing any such
things. That's nonsense, and very dangerous nonsense.
—GERHARD WEINBERG

IN 1907, having flunked out of both boarding and art school, unemployed and drifting, eighteen-year-old Adolf Hitler returned home to
his ailing mother, Klara. Her unrelenting chest pain had turned out to
be a spreading breast cancer. The surgeon removed her entire breast,
all the tissue underneath up to her collar bone, part of the chest wall,
and the muscles of one side of her chest. At forty-seven, Hitler's long
suffering, adored mother was completely bedridden with a gaping,
unhealed wound. His abusive father having died three years earlier,
Adolf alone was left to care for her.

The teenager attended to his weakened mother in a coal-fire heated
cold-water flat in a suburb near Linz, Austria. Their small, damp space
reeked of iodoform—a pale yellow, crystalline substance, with the
penetrating smell of an old hospital. The Office of Strategic Services
interviewed her physician, Dr. Eduard Bloch, who reported that when
he told Adolf his mother was fatally ill, "Tears flowed from his eyes.
'Did his mother,' he asked, 'have no chance?' Only then did I realize
the magnitude of the attachment that existed between mother and

son. I explained that she did have a chance, but a small one. Even this shred of hope gave him some comfort."[1]

Every day, Hitler dutifully poured five grams of liquid iodine onto a yard of gauze and then packed it directly into Klara's open chest lesion. It was a costly and excruciatingly painful treatment, and completely ineffective. There is an extensive literature debating whether the doctor was providing what he thought was a last shot at a cure, or whether Klara was forced to endure this regimen just to satisfy Hitler's demand that something be done.[2]

Klara died shortly before Christmas on December 21, 1907. Hitler was inconsolable. Bloch, who visited the home regularly, wrote of the two, "I have never witnessed a closer attachment."[3]

All madmen differ. But they all warp some segment of reality to suit their particular madness.[4] Jews traded, sold, and profited from tobacco, alcohol, and other cancer-causing products. In Hitler's demented thoughts, if the world had fewer Jews, it would have less disease. Obsessed with ridding himself and the German nation of malignant terrors that grew from within, Hitler took on a mission: Germany would become *Krebs frei,* cancer-free. And it would become *Juden frei,* free of Jews. "The Jews," Hitler proclaimed, "are a cancer on the breast of Germany."[5]

The Nazi master race must be purged of unhealthy elements in its environment, its habits, its gene pool. Hitler wrote, "How many diseases have their origin in the Jewish virus! . . . [The Jews are] poisonous abscesses eating into the nation . . . an endless stream of poison . . . being driven by a mysterious power into the outermost blood vessels of the body politic."[6] These ideas found fertile soil in the minds of other anti-Semites, who seized on the charge that tobacco and many other modern evils had been spread throughout the world by Jews.[7]

It is of some interest that Hitler showed the Jewish doctor who had tried to save Klara extraordinary consideration. When Germany annexed Austria, Hitler personally arranged for Bloch's safe passage out.[8] While he may have felt indebted to this particular Jew, his feelings about Jews in general were unrelenting and unrepentant. Hitler's social program took as its core the inarguably sensible need to extend *Gesundheit*, health, by preventing rather than treating disease. To achieve these goals, the system relied on *Gesundheitsführen,* health leaders, specially designated doctors whose mission was to find and eliminate defectives. These health leaders pressed for restricting environmental agents believed to harm the German "germ plasm," moving to limit tobacco smoking and the use of white flour and sugar as well as aniline dyes and other industrial toxins. They also promoted organic agriculture and natural medicines. Every German citizen had a duty to engage in *Gesundheit über Alles*—healthfulness over all—to ensure the well-being of the nation.

Despite their deranged application, a number of Nazi ideas about how to rid the world of cancer and build a healthier human had respectable scientific pedigrees. To the consternation of the Church of England, Charles Darwin's idea that the fittest survived in the natural world became broadly accepted.[9] Darwin understood that social and economic conditions could completely undermine what he had found in the animal world. Slavery, which he opposed, could not be ascribed to natural selection. Many of Darwin's early followers missed this subtlety. They naively believed that they could create a better human society based on natural selection, through what was dubbed eugenics—literally meaning well-born.

Unlike Darwin, his cousin Francis Galton was not a man of nuance. Like many staunch advocates of eugenics, he opposed philanthropy on the ground that it propped up those who would naturally perish. In a tract written in 1869 he urged that marriage be regarded as an opportunity to promote a better race: "As it is easy . . . to obtain by careful selection a permanent breed of dogs or horses gifted with peculiar powers of running, or of doing anything else, so it would be quite practicable to produce a highly-gifted race of men by judicious marriages during several consecutive generations."[10]

At the turn of the century, the geneticists Thomas Hunt Morgan and H. J. Müller showed that fruit flies can be bred to determine eye color and wing shape. The concept of controlling genes in flies easily morphed into efforts to improve horses, food and flowers. It was a short step from there to an array of social programs aimed at breeding better humans.[11] For the first three decades of the twentieth century, the scientific movement for eugenics captivated scientists and politicians throughout Europe and America, eager to see science applied in arenas never before imagined.[12]

Social Darwinism forged ahead without the endorsement of the man whose name it appropriated. In Germany in 1895, Alfred Ploetz established the grounds for racial hygiene. He contended that in order for the most fit to prevail, the weak had to be denied medical care and the ability to pass on their defective genes. British Social Darwinist John Haycraft echoed these ideas, suggesting that tuberculosis and leprosy be deemed "racial friends" because they chiefly attacked those who weren't meant to survive. In 1904 Plotz and Haycraft joined forces to create the Society for Racial Hygiene and established a journal, the *Archives of Racial and Social Biology*. By the end of World War I, racial hygiene was a respected field of medical science in Germany, England, France and America.[13]

In 1910 the United States created a eugenics record office with Harry Laughlin as its first superintendent.[14] His job was to be sure that deficient persons didn't get a chance to transmit their defects. A decade later, as the "expert eugenics witness" to the House Committee on Immigration and Naturalization, Laughlin provided scientific "evidence" on the damaging impact of race mixing. He contended that populations of southern and eastern Europeans, Mediterraneans and Russian Jews were rife with defects and should be kept out of the American gene pool. Northern Europeans provided better genetic fodder. Based on this analysis, Congress passed the Johnson-Reed Act in 1924, banning immigrants from "weaker" stock and forcibly sterilizing citizens deemed deficient.[15] For the next sixty years, America would set genetic limits on those who wished to enter the country. My grandfather, Sam Langer, six feet four at fourteen years of age, came to Ellis Island from Romania in 1911. He was put in the queue to be turned back because he had been blinded in one eye, and only got into the country with the help of a small bribe given to an immigration officer. As a rule, those who appeared fit, had good eyesight and hearing, loved the sea, and had the lightest skin were allowed to enter.[16] America, like France and many other nations at the time, would protect its future population by keeping the weak ones out. By letting my future grandfather in, the country got sixty years of direct labor, five tax-paying children, fourteen college-educated grandchildren and eighty-four more years of life from Zadie Sam.

The English philosopher Bertrand Russell was not perturbed by the fundamentally undemocratic aspects of eugenics:[17]

> The ideas of eugenics are based on the assumption that men are basically unequal, while democracy is based on the assumption that they are essentially equal. It is politically awkward to advance eugenic ideas in a democratic community when those ideas take the form, not of suggesting that there is a minority of inferior people, such as imbeciles, but of admitting that there is a minority of superior people. Measures embodying the former fact can therefore win the support of the majority, while measures embodying the latter cannot.[18]

Britain, perhaps because it never faced the numbers of immigrants who headed to America, never adopted broad eugenics laws. But the British were not slackers in this respect and found other ways to promote eugenics.[19] The founder of biostatistics, Karl Pearson, nowadays known for a number of statistical tests that bear his name, took a Galtonian view of war as a purifying means of ensuring the survival of the fittest. "This dependence of progress on the survival of the fittest race," he wrote, "terribly black as it may seem to some of you, gives the struggle for existence its redeeming features; it is the fiery crucible out of which comes the finer metal."[20]

War and disease were not the only ways to rid the human race of less hardy souls. Sterilization was carried out on a large scale. So that people could not bear children, surgery, chemicals or radiation were used to destroy the ovaries or testes. Such treatments sometimes left the patients severely disabled or dead. Just before the stock market crash of 1929, one Canadian province, Denmark and the Swiss canton of Waadt passed laws allowing the genetically infirm to be sterilized.[21] A German cartoon from 1933 boasted that Germany was one of more than a dozen nations at the time that were carrying out programs to rid themselves of "defectives." "Wir sind nicht allein" [We are not alone]. The drawing featured the flags of the U.S., Denmark, Norway, Switzerland, Sweden, France, Russia, China and Japan— nations that had adopted eugenic laws at the time. In fact more than two dozen U.S. states mandated sterilization of tuberculars, alcoholics, and the feebleminded.[22] Several hundred thousand people in more than a dozen countries lost the capacity to reproduce. By 1933, when Britain passed laws allowing the sterilization of drunkards and other undesirables, more than 400,000 presumably inferior Americans had been sterilized.[23] Sweden sterilized more than 60,000 "unfit" citizens in four decades.

Opinion leaders throughout America and Europe firmly believed better populations could be bred. They were not perturbed by indications that intelligence and success in life come about from factors that may not be evident at birth. Leading senators and governors, along with reformers such as Margaret Sanger, the founder of Planned Parenthood, and business leaders like Henry Ford, all championed the idea that society's leaders were equipped to decide who merited the chance to pass their genes on to the future.[24] An ardent American advocate of government action against the feebleminded was Leon Whitney, executive secretary of the American Eugenics Society. In 1934 he published *The Case for Sterilization*, contending that the state was obliged "to weed out defective persons from society, just as a farmer would clear a field."[25] Whitney was hardly a lone crackpot. The editors of the *New England Journal of Medicine* in the early 1930s lamented the "dangerous" increase in the rate of American feeblemindedness, calling the economic burden of supporting the mentally feeble "appalling." In 1934 the *Journal of the American Medical Association*'s editor, Morris Fishbein, wrote that "Germany is perhaps the most progressive nation in restricting fecundity among the unfit" and argued that the "individual must give way to the greater good."[26] Presumably Fishbein was unaware that under the programs then taking shape in Germany, he would have been one of those giving way.

Among the individuals sacrificed for the greater good were thousands of men and women whose chief defect was being in the wrong place at the wrong time. Ten years earlier, in 1924, support for sterilization was not simply a matter of a few states, but was endorsed by the U.S. Supreme Court. A Virginia court had ruled that the state could sterilize both an institutionalized seventeen-year-old girl, Carrie Buck, and her seven-month-old infant daughter, Vivian. Carrie Buck's mother, Emma, an epileptic, resided in the State Colony for Epileptics and Feeble-Minded, along with her daughter and young granddaughter. U.S. Supreme Court Justices Louis Brandeis, William Howard Taft and Oliver Wendell Holmes in 1927 endorsed the decision in *Buck v. Bell* to sterilize all of them.[27]

> It is better for all the world, if instead of waiting to execute degenerate offspring for crime, or to let them starve for their imbecility, society can prevent those who are manifestly unfit from continuing their kind. The principle that sustains compulsory vaccination is broad enough to cover cutting the Fallopian tubes. Three generations of imbeciles are enough.[28]

Later it was found out that Carrie Buck's pregnancy had not resulted from the simple-minded promiscuity that her custodians had alleged. In fact she had been raped. As a way to cover up this inconvenient fact, Carrie was institutionalized when her pregnancy by her rapist—the nephew of her adoptive mother, Alice Dobbs—became evident.

Described as an intelligent woman and an avid reader until her death in 1983, Buck worked managing other people's households until her death. Her young daughter Vivian did very well in school before dying at the age of eight from what was termed a stomach infection, probably arising from the sort of preventable diarrheal diseases that killed many children at the time.[29]

In his book *The Mismeasure of Man,* Stephen Jay Gould notes the case against the Buck women rested entirely on bogus descriptions masquerading as science:

> At the original trial in late 1924, when Vivian Buck was seven months old, a Miss Wilhelm, social worker for the Red Cross, appeared before the court. She began by stating honestly the true reason for Carrie Buck's commitment:
>
> "Mrs. Dobbs, who had charge of the girl, had taken her when a small child, and had reported to Miss Duke, the temporary secretary of Public Welfare for Albemarle County, that the girl was pregnant and that she wanted to have her committed somewhere—to have her sent to some institution."
>
> Miss Wilhelm then rendered her judgment of Vivian Buck by comparing her with the normal granddaughter of Mrs. Dobbs, born just three days earlier:

"It is difficult to judge probabilities of a child as young as that, but it seems to me not quite a normal baby. In its appearance—I should say that perhaps my knowledge of the mother may prejudice me in that regard, but I saw the child at the same time as Mrs. Dobbs' daughter's baby, which is only three days older than this one, and there is a very decided difference in the development of the babies. That was about two weeks ago. There is a look about it that is not quite normal, but just what it is, I can't tell."

This short testimony, and nothing else, formed all the evidence for the crucial third generation of imbeciles. Cross-examination revealed that neither Vivian nor the Dobbs grandchild could walk or talk, and that "Mrs. Dobbs' daughter's baby is a very responsive baby. When you play with it or try to attract its attention—it is a baby that you can play with. The other baby is not. It seems very apathetic and not responsive." Miss Wilhelm then urged Carrie Buck's sterilization.[30]

As support for eugenics mushroomed throughout the world, Nazi Germany's *Führer* didn't need to invent arguments for racial hygiene. He simply borrowed them from the Americans, whose efforts he followed and admired. Hitler's promise that the German state would tap science to forge a better race relied on heredity. He told a fellow Nazi, "it is possible to a large extent to prevent unhealthy and severely handicapped beings from coming into the world. I have studied with interest the laws of several American states concerning prevention of reproduction by people whose progeny would, in all probability, be of no value or be injurious to the racial stock."[31]

On the wall of the Reichstag hung a signed photo of Henry Ford to his good friend, Adolf. German translations of leading American genetics researchers, such as Madison Grant, sat on Hitler's bookshelves, along with Ford's book *The International Jew: The World's Foremost Problem*. Mementos of Hitler decorated Ford's office as well.[32] In July 1938, Ford got a very special present for his seventy-fifth birthday: the Grand Cross of the Supreme Order of the German Eagle—Germany's highest civilian award. He was the first American to receive it. America's Whitney envied the evolving German system and believed that Germany's system of racial hygiene gave the Nazi *Reich* a competitive advantage: "While we were pussy-footing around, reluctant to admit even that insanity of certain sorts runs in families, the Germans were calling a spade a spade . . . by this action Germany is going to make herself a stronger nation."[33]

In his unpublished *Zweiter Buch* (Second Book), written in 1928, Hitler declared his great admiration for the American approach to racial purity. He contended that America was a dynamic, "racially successful" society that practiced eugenics and segregation and kept out "racially degenerate" immigrants from eastern and southern Europe. Hitler believed that the majority of Americans were true "Aryans" who felt, as he did, keenly threatened by a Jewish plutocracy.[34]

By the 1930s, the country of Kant, Goethe and Beethoven, the country that had trained many of the world's physicians and captured half of all Nobel Prizes,[35] had the highest rate of cancer in the world.[36] Lung cancer, which in 1910 had accounted for less than half of one percent of all autopsies, had risen fivefold by 1933 and claimed one of every eight cancer victims. The *Gesundheitsführen* and the *Führer* himself were convinced they could change these patterns. As they had done with eugenics, they used the most highly regarded science of the day, not only clinical reports by physicians and pathologists but experimental studies using animals from America and elsewhere. The Jena Institute of Pathology kept meticulous accounts of autopsies carried out between 1910 and 1939.

Those running Jena argued that tobacco had to be behind the dramatic surge in cancer. The *Reich* could hardly create the master race if so many Germans had stinking habits that led to miserable deaths. Hitler himself was rabidly opposed to smoking. Other Nazi generals did not share his aversion. Hitler was not a vegetarian, despite rumors to the contrary. He did advocate that women of childbearing age, on whom the fate of the German nation depended, should eat whole grains and vegetarian diets and be prevented from indulging in tobacco or alcohol.[37] This was not mere rhetoric.

In 1939 the *Gesundheitsführer,* Leonardo Conti, set up the Bureau Against the Dangers of Alcohol and Tobacco, two of the most deadly consumer poisons. The bureau's studies showed that men who smoked had ten times as much lung cancer compared to those who did not smoke. Conti addressed these problems aggressively. Those missing more than a month of work due to conditions tied to smoking were sometimes forced to go to nicotine withdrawal clinics.[38] Pregnant women were barred from buying cigarettes. Smoking was banned on trains and in many public spaces.

The desire to restore Germany to its former prominence, as well as the belief that genetic improvements had to play a key part in that effort, were not limited to the Nazi party or to Germany. In the early 1930s, many of Germany's doctors and scientists were Jews—a matter about which Hitler often ranted. Some of these Jewish doctors had no idea what was coming. Many of them directly supported the idea that the state should determine who would be allowed to bear children. By 1933, Jewish supporters of eugenics would find the tables turned; they themselves were targeted as undesirables by German scientists using eugenics arguments. The widow of one of the Jewish founders of the Arbeitsbemeinscheift für Judische Erbforschung und Eugenik (Society for Jewish Research and Eugenics) in Berlin, told her son many years later, "The Jews would have been better Nazis than the Germans if the Nazis had given them the chance." Jewish advocacy of racial hygiene vaporized when it became clear that eugenic policies would be applied against all who had any Jewish blood, not merely those deemed deficient by some purportedly rational standard.

Jewish opinions became irrelevant anyway, as Jews were driven from academic posts and forbidden to practice medicine. After 1933, the drive to rid the nation of cancer merged with the drive to get rid of defective genes. Non-Jewish doctors, many of whom were out of work and eager to replace the more successful Jewish doctors, ardently supported the effort to become free of the Jews in all quarters.[39]

From the earliest days of the Third Reich, doctors were among Hitler's most fervent supporters. Half of Germany's physicians eventually became members of the Nazi party, serving as key parts of Hitler's exclusionary program.[40] A chief *Gesundheitsführer* urged, "It is not enough for the National Socialist health policymaker to eradicate already existing diseases; he must avoid and prevent them. The healthiest people is not that which possesses the best or the greatest number of hospitals, but rather that which needs the fewest."[41] The next step for Nazi research was straightforward. Not only were those who were allowed to breed limited to the best, but the health of the breeders was to be protected against damage to their "germ plasm."

Enthusiastic about promoting a better breed of man and well versed in eugenic arguments, in 1935 the French Nobel Laureate in medicine, Alexis Carrel, published a popular book, *L'Homme, cet inconnu* (*Man, This Unknown*) that played right into Nazi notions of race hygiene. He argued for a platonic society governed by a hereditary elite. Enforced eugenics was to rely on gas chambers to get rid of inferior genetic stock. When the Nazis invaded France, Carrel had the opportunity to put this scheme into action. In June 1940, as 8 million French citizens fled their homes, Louis Darquier, the Vichy government's commissioner of Jewish affairs, put restrictions on Jews that were more stringent than those in place in Germany at the time.[42] Carrel argued that the obligation to ensure the continuation of only the best elements of an elite race required deporting 100,000 Jews. Those who wished to avoid this fate had to obtain "certificates of nonbelonging to the Jewish race" by proving they had other origins. In 1941, with full support from the Vichy government, Carrel set up the *Fondation française pour l'étude des problèmes humains*, an institute dedicated to studying "all suitable measures for safeguarding, improving and developing the French population."[43]

Medical scientists trained to heal the sick underwent a shocking transformation in France and Germany. With chilling enthusiasm, physicians eagerly took charge of devising a system that would decide who would live or die or be allowed to reproduce. How they went about doing this, what criteria were used, how mistakes were to be avoided—all this was up to the individuals charged with the task. Under the sway accorded medical science at the time, clinical credentials per se proved that those who held such positions merited trust. The Germans thought of themselves as standing against the phony egalitarian ideas of the French Revolution, with its concepts of universal brotherhood, freedom and equality. July 14 is the date celebrated in France as the liberation of the Bastille, the beginning of democracy. The German cabinet chose the same date in 1933 to enact a number of major laws, including one allowing sterilization of those deemed defectives.

In 1934 a system of German health courts, sometimes taking only a few seconds to decide a case, reviewed some 65,000 petitions filed by medical experts and ordered the sterilization of five out of every six persons they reviewed.[44] While no one can be certain of the numbers, one expert on this dark period of German life believes that close to half a million people were selected to have their ability to reproduce ended. About 20,000 may have died from the surgery.[45]

According to one historian, physicians had proposed to Hitler in the mid-1930s that they should begin a program to get rid of defective people even earlier. Hitler put them off, indicating that this could not be done until there was a war going on. By 1939, as military activity expanded, German health administrators set up a centralized system of what was called euthanasia. They began to kill babies with major defects and moved quickly to others, including mentally ill persons and military veterans who had been disabled by fighting in the First World War. By the summer of 1941, about 100,000 people from various industrialized nations had been murdered in Germany and its occupied territories.

This effort to end the lives of defectives was drastically curtailed because of uproar from the families of war veterans. At this time, war casualties on the eastern front were mounting, and it would not have been good for morale if soldiers at the front thought that receiving serious injuries would put them in line to be killed back home through state-ordered euthanasia. Nevertheless, throughout the country hospital patients continued to be killed by German physicians and nurses as part of this overall scheme to rid the population of the ill.

The state killing machine did not always run smoothly. Nuns who worked in the numerous Catholic hospitals had dedicated themselves to a life of healing and other Christian ideals. Perplexed by being ordered to murder sick patients, they asked the bishops what they were supposed to do. Catholic doctrine forbids euthanasia but also requires obedience to the government. At this time, many women in slave labor who had been raped and become pregnant were brought in and forced to have abortions. This too put the nuns in a quandary.

By mid-August 1941, euthanasia had become decentralized in various hospitals throughout Germany and the occupied lands of Austria, Poland and the Sudetenland. These widespread medical murders could not have taken place without the full cooperation of people in the medical profession throughout Germany and elsewhere.

A story from an old German colleague told to the distinguished diplomatic historian and Holocaust survivor Gerhard Weinberg discloses how extensive the killing program became. "My colleague had a sickly younger brother who had been repeatedly hospitalized for a variety of ills. The last time that her mother brought her weak brother to the hospital, the kindly family doctor warned her, 'Don't ever bring him back.' She realized he was warning her that if this sick child had returned to the hospital, he would have been killed."[46]

In 1937 the head of the health office in Berlin, Dr. Hans Reiter, formally declared that the master race must remain free from environmental damage and from inferior genetic material carried by Gypsies, homosexuals, socialists, Jews and communists. Reiter's personal contribution to bringing about a stronger nation was to sterilize hundreds of those judged enfeebled, criminally insane, tubercular, alcoholic, drug addicted, or psychiatrically or politically undesirable.

Germany did not adopt these broad eugenics policies based merely on the rantings of a solitary, insane, grief-maddened politician. If Hitler had not existed, given the tremendous and widespread enthusiasm for protecting the Aryan race and elites, someone else—quite possibly a physician—would have spearheaded the effort for racial hygiene. Germany was not the only place where enthusiasm for these policies existed at the time. France, Austria and Japan were among the nations with deeply shared eugenic views that became the grounds for mass murder.

Kurt Gerstein was one of the few Germans to infiltrate the Nazi system and to have left a record of having done so. A devout young Christian, he was stunned by the murder of his mentally ill sister-in-law, Bertha Ebeling, at Hadamar, a supposed asylum from which many mentally ill and other so-called defectives never returned. Determined to bear witness as Christian duty required, Gerstein became a member of the elite SS storm troopers. With an unrealistic sense of what one man could do, he set out to sabotage, document and expose the killing machines of the death camps. At one point, he buried canisters of poisonous cyanide containing Zyklon B gas rather than deliver them to the camps.

His detailed reports on the operations of the death camps stand out as exceptions in a nation that widely acceded to the barbaric demands for racial purity. In the context of the world in which he operated, of course, Gerstein would have been seen to be insane. As the war progressed, he took bigger and bolder risks. On August 21, 1942, on a train from Warsaw to Berlin, Gerstein encountered the Secretary to the Swedish Legation in Berlin, Baron Göran von Otter. The Swedish diplomat related what Gerstein desperately told him when the train stopped midway and they both got out to catch some fresh air.

Von Otter later described the encounter:

> There were beads of sweat on his forehead. There were tears in his eyes. And his voice was hoarse when he said at once, "I saw something awful yesterday—can I come and see you at the Legation?" I suggested that we restrict our conversation to the train. "Is it the Jews?" I asked. "Yes, it is," Gerstein replied. "I saw more than ten thousand die today."
>
> It was hard to get Gerstein to keep his voice down. We stood there together, all night, some six hours or maybe eight. And again and again, Gerstein kept on recalling what he had seen. He sobbed and hid his face in his hands. From the very beginning as Gerstein described the atrocities, weeping and broken hearted, I had no doubt as to the sincerity of his humanitarian intentions.

Whatever Von Otter may have done with Gerstein's information, it had no appreciable impact on the conduct of the war at that point. Imprisoned after the war ended, Gerstein provided French and German reports detailing the operations of the death camps of Belzec and Treblinka. This information became critical to the prosecutions at Nuremberg. Gerstein's failed efforts to warn the Vatican and other embassies of the death camps while the war was underway have recently been the subject of an extraordinary play, *Either/Or*, by Tom Keneally, the author of *Schindler's List*. We learn of this man and wonder why there were so few others. His life becomes the subject of a stunning play precisely because he was so exceptional.

In 2003 Weinberg told the British newspaper *The Telegraph* that it would be a grave error to think that what happened in Germany could not have occurred elsewhere:

> I have come to think it important for people not to look on the Nazi experience in two ways that are very dangerous and very bad. One of them is to look on it as a kind of freak show. It's not a freak show: it's a coherent, horrible system. Equally dangerous is the opposite: that this is some kind of a German genetic defect. It isn't. Not only were there

decent people in Germany, but these Nazis were people like other people. Human beings can do awful things, and can turn away from awful things and reform themselves.[47]

Another expert on this period, Jeremiah A. Barondess, has noted that

The aim of generating pure Aryans had taken precedence over the most fundamental ethical issues in medicine. Healing acquired a new, sociopolitical definition that swept aside the vulnerability and suffering of large numbers of individual persons, who, in a sense, lost their humanity and were transmuted into pollutants of the state. Physicians in Germany did not simply acquiesce; rather, they accepted, supported, and were instrumental in the application of the racist policies of the Third Reich. They made selections in the concentration camps; they dispatched prisoners who became ill to the gas chambers; they engaged in medicalized killings for political purposes, injecting cultures of live tubercle bacilli or other organisms into party officials and others whose deaths, it was thought, should appear to be from natural causes.[48]

Robert Proctor and Robert Jay Lifton have each detailed efforts by Nazi physicians and researchers to craft scientific grounds for reversing Germany's growing burden of chronic diseases. In the 1930s, German scientists developed epidemiological methods to show that tobacco use increased cancers and other diseases in humans. Not all Nazi leaders shared enthusiasm for what became the first national campaign to combat smoking. Josef Goebbels, the cigar-smoking Nazi propaganda minister, boasted that he never smoked more than one cigar at a time.

Goebbels notwithstanding, Germany officially discouraged civilian use of tobacco throughout the war years, insisting that a better race could not be created if smoking were allowed. Like its enemies, however, the German government supplied small amounts of tobacco to its soldiers, believing that the soothing effects of nicotine might enhance their ability to stand and fight. Supporting evidence for efforts to reduce smoking came from many international experts.

In 1936 the renowned American physician Alton Ochsner reported that he had seen nine cases of lung cancer in six months, after not seeing a single one for close to twenty years. Noting that all of his recent lung cancer patients had started smoking in World War I, he suggested that cigarette smoking lay behind this sudden surge in the disease.[49] Two years later, in 1938, Raymond Pearl, a statistician with Johns Hopkins University, prepared an impressive analysis of the increased death rates of smokers by distilling information from insurance company records.[50] In 1940 the Argentine researcher Angel H. Roffo, publishing in German journals, identified tobacco tars, rather than nicotine, as the principal cancer-causing part of cigarette smoke.[51] Two decades earlier, he had founded one of the world's first cancer institutes in Buenos Aires, where many of his patients were smokers.[52] Roffo was well aware that as early as the nineteenth century mice doused with tobacco tars developed skin cancer, and he certainly knew of Percival Pott's work showing the capacity of coal tars to create cancer of the scrotum. In several hundred papers detailing his own experiments with animals, he described how deliberate exposure to tobacco tars and fumes created tumors throughout what he termed the "smoking highway" of the respiratory tract, from the nose through the esophagus and larynx to the lung.[53] Even when nicotine was taken out of tobacco in his experiments on animals, tumors occurred. This made it clear that tars in tobacco were the chief culprits.

German research on tobacco and other environmental hazards had begun under Kaiser Wilhelm, long before the Nazis came to power, and continued during the Weimar Republic. Under Hitler this work reached an unprecedented level of sophistication and prominence. As a matter of basic policy, the German health officials did not think it prudent to wait for definitive evidence of cancer to develop in humans before acting to prevent it. They were willing to ignore the claims of the tobacco companies and restrict smoking based on what was known at the time from animal studies.

Drawing on research conducted in America, England, Italy and elsewhere at the time, Germany created the first institute for basic research on tobacco. The Scientific Institute for Research into the Hazards of Tobacco was founded on April 5, 1941, at the Friedrich-Schiller University in Jena, Germany.[54] The institute's short life ended with the war, but during that time it managed to produce seven doctoral dissertations on tobacco.

Table 3-1 Dissertations from the German National Scientific Institute for Research on Tobacco, Friedrich-Schiller University

1942	Die Zigarettenraucherin (The woman smoker)	Gabriele Schulz, Käte Dischner
1943	Nikotin-Todesfälle der letzten hundert Jahre (Nicotine deaths over the past 100 years)	Werner Feuerstein
1944	Die Einwirkung des Nikotins auf das Ionenverhältnis von K:Ca im menschlichen Körper (The effects of nicotine on the ratio of potassium to calcium in the human body)	Heinz Held
1944	Die neurologischen Schäden durch den Tabak. Zusammenstellung der Fälle, die darüber bekannt geworden sind (The neurological damage inflicted by tobacco. Compilation of known cases)	Rolf Schroder
1944	Ueber den Einfluss des Nikotins auf Würmer (The effect of nicotine on worms)	Lore Wenzel
1944	Ueber den Einfluss des Tabakrauches auf Tiere (On the effect of tobacco smoke on animals)	Maria Schumann
1944	Lungenkrebs und Tabakverbrauch (Lung cancer and tobacco consumption)	Erich Schöniger

In the 1930s, researchers like Dr. Franz H. Müller at the Jena Tobacco Institute had come up with better ways to study how the world affects our health and ultimately predicts from what we will die.[55] Little is known about Müller. An enigmatic physician from Cologne, the records about his life, like those of many German scientists, end after the war. There can be no doubt that Müller created the first irrefutable modern proof that smoking causes lung cancer in humans. This work began with the shocking fact that in three decades the disease had gone from a rare occurrence to the second leading cause of cancer deaths in Germany.

In 1939 Müller came up with an ingenious approach. He assembled detailed information on smokers and compared their lives to those who had not smoked. He was one of the first to use a technique that looks backward into the natural experiments of people's lives. Like many scientific breakthroughs, his method was deceptively simple. Starting with the German habit of keeping meticulous records on all

Table 3-2 Comparative Incidence of Cancer in Smokers and General Population

Smoking Category	Lung Cancer in Smokers Compared to General Population	All Other Cancers in Smokers Compared to General Population
Very Heavy	16.6	8.8
Heavy	5.8	5.6
Medium	7.8	7.0
Moderate	1.6	1.4
Non-Smoker	1	1

Tested for trends (P<0.00001 in both comparisons)
Source: Müller, 1939.

sorts of things, including deaths, he began asking family members about the lives of those who had succumbed from lung cancer when he conducted autopsies.[56]

The guts of his statistical analysis were clear. Müller contrasted the good and bad habits of healthy people without lung cancer—the controls—with those reported by the surviving family members of ninety-six cases of lung cancer, eighty-six men and ten women. This simple table shows what Müller found: the heavier the smoking, the greater the risk of lung cancer.[57] Those who smoked the most had sixteen times more lung cancer, and eight times more cancer overall, than those who did not smoke at all; those who smoked what was deemed a heavy or medium amount had five to seven times more lung cancer. He also found that the smokers among his control group tended to develop other cancers at much higher rates than those who did not smoke at all. Müller calculated that the chance of these results occurring without there being a direct link between smoking and cancer was one in a million.

The number at the end of this table tells us that there was less than one chance out of a hundred thousand that the results he got were solely due to chance. Just as epidemiologists today would do, he pulled together experimental evidence of the past, clinical reports and the trends in lung cancer of the time, and his own analysis to reach a firm, strong and clear conclusion. Tobacco was an "important cause" of lung cancer. He added that "the extraordinary rise in tobacco use" was "the single most important cause of the rising incidence of lung cancer" in recent decades.[58] Shortly after it appeared in German in 1939, Müller's dissertation was printed as an abstract in English in the *Journal of the American Medical Association*. In an ideal world, this study would have profoundly altered the way people thought about the problem of tobacco. The real world, however, was on the brink of war. An addictive product that drives several major economies is not something to be trifled with.

Four years later, working at the Jena Tobacco Institute, two other German scientists produced a more sophisticated and powerful analysis on the hazards of tobacco, looking at nearly twice as many deaths.[59] Erich Schoniger and Eberhard Schairer surveyed 195 lung cancer cases and 555 others with different cancers, and compared their habits to those of 700 men with no evident disease.[60] The results were stunning. Out of 195 people with lung cancer only 3 had not smoked. Many of them had smoked and worked in dusty trades, with exposure to agents like asbestos and chromates.

As a bright young medical student with an interest in statistics, Richard Doll traveled to Germany in 1936 to attend lectures that laid the groundwork for systematic analysis of health information. The language of medicine in Germany was rife with metaphors of political struggle. During a talk by the radiologist and SS officer Hans Holfelder at Frankfurt, Doll recalled that a slide was shown in which "storm trooper x-rays" attacked "Jewish cancer cells."[61] Other images equating political radicals, Gypsies and Jews with cancer or tobacco regularly appeared in German newspapers and magazines at the time. Müller's work, however, was notable for an absence of such rhetoric and for referring positively to earlier work by Jewish scientists.[62]

Doll's visit to Germany would later cause controversy. In his work with Bradford Hill in the late 1940s, he used the same comparative methods as the Germans, contrasting rates of lung cancer in physicians who had been smokers with those who had never smoked—a study later credited as one of the first rigorous demonstrations of the connection between smoking and cancer, and part of the work for which Doll was knighted. In a 1997 interview with Proctor, Doll did not recall having ever seen or heard of the work of Shoniger and Schairer, though he did remember that of Müller. In a commentary on all this work published in

2001, Doll added that from the perspective of modern epidemiology, even if he had happened to read their study, Schairer and Schoniger's work failed. The numbers they studied were too small; they relied on surviving family members to describe habits of the deceased; the healthy controls came from a narrow age-group.[63] Doll did not mention that the German researchers had reached clear conclusions about the risks of to-bacco in 1943—something his study with Hill would not do when it ap-peared in 1950. History will have to judge whether it was unfair for Proctor to observe, regarding Doll's failure to recall the early German work, that "science can be a forgetful enterprise."[64]

In truth, neither Doll nor the Germans were the first to come up with the idea of assembling and comparing people who are similar in most ways, excepting that one group has the disease. What is today called the case-control design was first used by Janet Elizabeth Lane-Claypon, the first woman to be granted a research scholarship by the British Medical Society. Asked by British health minister Neville Chamberlain in 1923 to tell the government what could be done to study and reduce the burden of cancer, Lane-Claypon came up with a breakthrough approach. She compared equal numbers and types of women who had breast cancer with their healthy counterparts. This was not easy work in the days be-fore computers and dedicated high-speed lines for data transmission.

Working with women physicians from London and Glasgow hospi-tals, Lane-Claypon tracked down five hundred women with breast cancer—the "cases"—and compared them with five hundred others who were free of disease but otherwise broadly similar, known as "controls." No large-scale review of this kind had ever been under-taken. The detailed survey that emerged constituted, as far as we know, the first published epidemiological questionnaire.[65] Being well versed in the scientific literature in many languages, German re-searchers may well have capitalized on this work in creating their own approaches to the study of tobacco in the 1930s and 1940s.

Based partly on these scientific studies and partly on the personal con-victions of Hitler and some of his close advisers, tobacco was singled out as a hazard to be avoided at all costs by the mothers of the master race. This effort to stifle tobacco use never quite took hold. Nazi Germany, like most industrial countries at the time, had come to depend heavily on tax revenues from cigarette sales.[66] Still, German magazines regularly featured ads for products free of tobacco and other "genetic poisons" such as alcohol, along with products like vitamins and nutritional compounds that were touted for their ability to strengthen the gene pool.[67] News stories warned that nicotine passed from inhaled smoke through the lungs into the blood and fat, ending up in breast milk. Tobacco was thought to destroy the ability to make healthy babies or even limit the capacity to have children. In the 1930s and 1940s, leading medical journals in Germany and a number of other industrializing countries frequently warned about the dangers of food preservatives, industrial toxins and coal-tar based artificial colorings while arguing for "natural" products to be employed in drugs, cosmetics and foods.[68]

Just outside of Munich, the lovely suburb of Dachau in 1933 was the site of the world's first concentration camp. News of its creation that year made the front page of the *New York Times,* where it was depicted as a place for political prisoners and degenerates. It was also the site of horrific medical experiments conducted to find out how long it takes a man to freeze to death.[69] Dachau has another, little-known claim to fame. Surrounded by richly forested areas, it housed the world's first large organic garden to grow botanical grade pharmaceuticals, foods and honey for Germany's elite soldiers and their families.

I've been asked not to tell you the name of the physician who ran the botanical gardens of Dachau. In one of those bizarre coincidences that kept happening to me as I was researching this book, I found myself drinking tea with my friend Sophia in Jackson, Wyoming, a few years ago. A lifelong organic gardener and former owner of one of the town's first organic bakeries and restaurants, Sophia hails from the beautiful mountain area near Munich.

"Did you know that Dachau had been the site of the world's first and largest organic botanical gardens?" I asked her one wintry afternoon, shortly after I'd learned this myself.

Sophia has large, piercingly beautiful blue eyes, a strong jaw and a confident look, even when chatting about ordinary matters. She looked at me gravely. "Of course, I know all about that," she replied, looking down as she sipped her tea.

"Why? I was astonished to find that Hitler really promoted organic agriculture myself. Why did you know about this? Did you study it at school in Germany?"

Sophia looked straight at me and shook her head. "No. It was my grandfather," she said softly. "My grandfather was the doctor they brought in to run those gardens."

I was speechless. Sophia hadn't been born until more than a decade after the war ended. I'd never dreamed that her life might have been so directly touched by Nazi horrors.

"My grandmother told me all about it, because her husband had died, so of course I never knew him. He was from the countryside, where he learned to love growing green things. The only way I came to know him was from the stories that were passed on. He was a country physician with what we call here a green thumb. So, when he got the chance to run these really big gardens at Dachau, at first he was delighted. As soon as he found out what else went on there, he tried not to be involved. Each time they would ask him to work inside on the prisoners, he would not come in; he would beg off sick.

"When the war ended, he must have figured what would happen to him. He took the entire family to a lake for a swim. He drowned. Killed himself actually. My grandmother told me that she found the empty morphine vials he had injected before entering the lake."

Sophia assures me that her grandfather refused to work on the inhumane experiments at Dachau, for which physicians were hanged at Nuremberg. A dedicated botanist who had studied medicine, her grandfather had come to the camp believing he had a mandate to produce the largest array of organic foods and natural plant drugs ever cultivated in the world up to that time. Those gardens were in full bloom at the time that he left in 1940 to spend the rest of the war in an obscure region growing medicinal plants.

His case never made it to Nuremberg. When he heard the war was over, he knew what was coming. He joined thousands of officials and other collaborators who took their own lives in the war's aftermath. It's lucky for Sophia that unlike many German soldiers and officials, he did not take his entire family with him.[70]

THE NUREMBERG TRIALS of Nazi physicians, based in no small part on the detailed records of mass killing provided by Gerstein, exposed the simmering barbarism of otherwise ordinary lives. The charges against the German doctors lodged at Nuremberg involved crimes against humanity: murder, extermination, enslavement, deportation and other inhumane acts committed against any civilian population, before or during the war, or persecutions on political, racial or religious grounds in execution of or in connection with any crime within the jurisdiction of the tribunal, whether or not in violation of domestic law in the country where perpetrated.[71]

The physicians' trials, conducted in 1949, laid bare the evil impulses that had infested German medicine and research under the National Socialists. These doctors were not exceptional deviants from an otherwise civil society. Rather, they were ordinary men who accepted the job of controlling, eradicating and extinguishing people deemed politically and religiously unacceptable, sometimes using them as fodder in horrendous experiments. The banality of evil—to use Hannah Arendt's apt metaphor—was everywhere evident. The capacity to do bad things to innocent people was hardly limited to a small band of despotic individuals.

While a few of the Nuremberg doctors were frankly weird, the majority of the twenty brought to trial had been outstanding scientists, just like Alexis Carrel of France. Paul Rostock, the former dean of the University of Berlin medical school, had headed the Department of Surgery, where major advances had been developed in life-saving techniques. Abounding paradoxes have never been explained. This was a society that in 1933 passed some of the world's first laws protecting animals from cruel treatment in scientific research, intending to awaken sympathy for other creatures as "one of the highest moral values of a people." A cartoon featured Hermann Göring receiving the Nazi one-armed salute from rabbits and dogs, presumably acknowledging his efforts on their behalf to ban animal experiments. This same government that protected animals excluded Jews, Gypsies, homosexuals, Catholics and socialists from such protections, deeming them nonhuman and subjecting them to industrial-scale depravities and barbaric experiments.

The physicians' trial perpetuated a convenient myth that guilty Nazi parties were being hunted down and prosecuted and laid the groundwork for a wholesale rejection of any science that came from the Nazi machine. It would be easy to attribute the indifferent response to

Figure 3-1 *A cartoon from a German magazine featured rabbits and dogs giving the Nazi one-armed salute to Lieutenant General Herman Göring in gratitude for his efforts to ban experiments on animals*

tobacco hazards described by the Nazis and elaborated by researchers in other countries to a highly principled moral reluctance to rely on any science forged out of fascism. Yet, sometimes what is easy is also wrong.

After the war ended, not all Nazi science was banished from use. Many innovators from Nazi Germany, such as rocket scientist Werner von Braun, were absolved of their Nazi pasts and quickly put to use building better missiles for the Allies.[72] More than 2,000 German V-2 rockets, with their one ton bomb payloads, had rained down on London, terrorizing the population, killing nearly 3,000 people and injuring thousands more during massive bombing raids.[73] Von Braun and his fellow missile engineers were actively engaged in rocket production. The production facility he directed was an "archetypal Nazi factory" that employed foreign slave labor from the Dora-Nordhausen concentration camps where 20,000 died. The few workers needed for fine assembly of the V-2 rockets—some former physicists and engineers—were treated specially. Most others at Dora perished.

In the summer of 1943, just before a newly improved rocket, the A-4, was scheduled to appear, Hitler personally signed the documents awarding von Braun the title of Professor.[74] Although van Braun headed up the production of instruments of mass death for Germany and had been a member of the Nazi party, he was never tried for these activities. Instead, he and many former German rocket scientists were transplanted to the Redstone arsenal in Huntsville, Alabama. So many German émigrés set up store there that the base was dubbed Peenemunde South, in honor of the site where Germans had produced the V-2 rockets.[75] In a nearly immaculate conversion, the technologies for rockets, data sorting, key punches, protocomputers, voice recording, synthetic rubber and microscopic thin metal films quickly shed their Nazi roots and moved into the British and U.S. mainstream. The scientists who had overseen rocket building in Germany were exempt from prosecution as war criminals in the United States and elsewhere because their work was judged to be critical to the cold war.

In truth not all medical matters with Nazi origins were rejected, nor were all scientists who participated in research at camps brought to Nuremberg for trial. The Nazis had developed some good synthetic antimalarial drugs, which were rapidly put into use in the 1950s.[76] One of the world's most widely read and highly praised medical anatomy atlases to this day turns out to have been the work of a man whose elegant signature shrouded a series of small swastikas with thunderbolts through them. The virulently anti-Semitic president of the University of Vienna in the 1930s was the anatomist Eduard Pernkopf, famed for his extraordinarily fine drawings of human body parts. The embedded Nazi insignia within the signatures of three of the Pernkopf artists (Eric Lepier, Karl Entresser and Franz Batke) were air-brushed away in subsequent editions of the book, which appeared under the imprint of the original publisher, Urban & Schwarzenberg.[77]

The tradition of dissecting executed prisoners in Vienna dated from the beginning of the fifteenth century. By 1742, Austrian law allowed the dissected to include dead paupers. By a decree of February 18, 1939, all executed prisoners were to be dissected for the purpose of instruction. The official report of the University of Vienna on

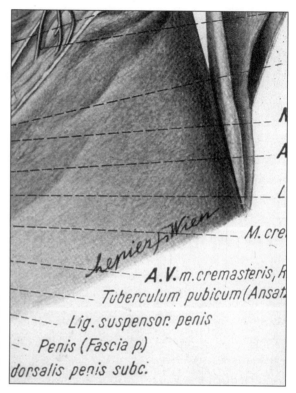

Figure 3-2 A Nazi swastika can be seen between the first and last name of the artist of what was the world's most famous Pernkopf anatomy atlas until the 1990s.

Pernkopf's atlas notes that most of the criminal dissections that could be identified had been political prisoners, pickpockets, petty thieves or those who committed the crime of listening to enemy broadcasts.[78] We will never know the identities of all those whose anatomies are portrayed in this handsome volume. Several commentators have noted that the age, appearance and crude haircut of one of the subjects bears a striking resemblance to a concentration camp inmate; another was a man who had been circumcised.[79] Others have suggested that Pernkopf arranged the deaths of those he wanted to draw in the camps.[80]

An official investigation by the University of Vienna issued in October 1998 found that the institute dissected bodies of beheaded political victims from the Gestapo execution chamber. It confirmed that the cadavers of 1,377 persons executed at the Vienna regional court had been delivered to the anatomical institute. The investigators could not figure out if any of those executed were portrayed in the atlas. In the days before fingerprinting, identifying bodies without heads was not possible.[81]

Stripped of his position as the head of the university, Pernkopf was never put on trial for his work. He was held in prison for two years as an "incriminated person," then became a "lesser incriminated person," and finally was rehabilitated through a formal process of denazification. He died in 1955, having spent the last years of his life working at the university in pathology and overseeing the production of yet another edition of his atlas in 1950.[82]

Only two years ago, the pickled body parts that Pernkopf had sketched were formally buried by the University of Vienna. In truth, the Nazis did not just conduct horrendously unethical experiments or foist controlling policies upon their own people. They also produced important work, inventing methods that are still used today, proving the hazards of tobacco for human health. If Nazi inventions like rocketry were easily adapted by the Allies and the work of Pernkopf continues to be used as a classic reference in anatomy departments around the world, why then were the findings of the cancer hazards of synthetic organic chemicals and tobacco treated so differently?

Why did so little happen to see public health protected against tobacco among the countries that won the war? The answers to these questions are not straightforward, but can best be determined by taking a careful look at the controls under which public health research began to evolve just after these trials ended. While science prides itself on being open and free and vital to democratic societies, scientific studies on matters of tobacco and industrial hazards have been anything but.

Under the guise of protecting trade secrets and in the name of national security, public access to scientific research has been a guarded enterprise. As a Chinese proverb notes, a way of looking is a way of not looking.

U.S. Army Captain Robert Kehoe, on assignment for the Office of Strategic Services, June 1944, to interview leading German scientists, accompanied by a senior German researcher, possibly Ferdinand Flury. A leader in industrial hygiene and toxicology at Würzburg, Flury directed department "E," which developed war gases, pesticides and other toxic chemicals and their antidotes through studies of animals and humans. During and before World War II, Kehoe served as Medical Director of the Ethyl Corporation; the firm provided the formula for leaded gasoline to the Germans, in violation of War Department orders.

4

Phantom
Collaborators

The right to search for truth implies also a duty; one must not
conceal any part of what one has recognized to be true.

—ALBERT EINSTEIN

WAR CRIMES, the saying goes, are what the losers get charged with.
What if, just before the outbreak of World War II, an official in an international, American-based firm had overseen the production of critically important war materials in German, Italian and Japanese
affiliated plants, some of which depended on slave labor? What if this
same man had gone to Germany in June 1945 for U.S. intelligence,
right after the European theater of World War II ended, and had come
back with evidence that chemicals his American firm was producing
caused cancer in workers? Were these dealings criminal? None of them
directly broke the law.

What passes for normal activity in any society is whatever most
people do much of the time. Seen through the lens of what we know
today, some ordinary practices of the past appear barbaric. How can
we explain the fact that in many countries, knowledge of cancer hazards in factories was withheld from those who were most at risk until well into the 1980s? The law protects businesses from revealing
what they deem to be trade secrets. But what if your trade secret
costs my father his life? The line where protected secrets end and

criminal negligence starts is not bright and clear but murky and moveable.

The democratic tendencies of modern science have always been muted by economic and political forces that determine who gets to know what, where and when. To understand the constraints placed on research and public understanding of the environmental causes of cancer, there is no better place to start than with two major figures—Wilhelm Hueper and Robert Kehoe.

Hueper, a German émigré pathologist with an encyclopedic grasp of workplace causes of cancer and a command of four languages, was a founder of the field of occupational and environmental carcinogenesis. Kehoe, a pivotal figure in several major chemical firms, also laid the groundwork for this field and headed one of the first university laboratories dedicated to researching the industrial hazards of industrial toxicology and workplace epidemiology. His work established the basic rules that would guide analysis of worker health and safety for the first half of the twentieth century. Two people could not have had more fundamentally different views of the role of science and scientists in the crafting of public policy. Their oddly parallel careers raise the question, Whose purposes should science serve?

Born into an impoverished family, Hueper was drafted into the German army prior to World War I. He dodged residues of poison gas that wafted back onto German troops during that conflict. After his discharge, he became a pacifist and a critic of religion and reactionary politics. In 1923, while in his mid-twenties he arrived in the United States as a married physician. In 1930 he left his position as a pathologist at Loyola Medical School in Chicago to work at the cancer research laboratory of the University of Pennsylvania in Philadelphia—a facility that was heavily financed by Irénée du Pont, head of the DuPont Company.

Hueper's first contacts with DuPont were encouraging. They came about through the recommendation of the DuPont family physician, who had been impressed with Hueper's research at Penn. Encouraged by what he thought was du Pont's open and frank attitude, in 1932 Hueper sent him a letter. German researchers, he wrote, had found that men manufacturing some synthetic dyes, just like those being made at DuPont's Delaware facilities, often developed blood in their urine and blockages in their bladders—early signs of tumors. Hueper candidly warned that the DuPont Company could expect to see many cases of bladder cancer in the workers at its dye plants in Delaware.

At first the letter got no response. His boss at the university lab told him that DuPont workers had no bladder cancer at the time. Hueper replied, "Well, that may be, but they will get them."[1]

In 1933, as the economic depression spread, Hueper was broke and feared that he was about to become unemployed. At the same time, Germany had begun investing heavily in institutions to study and prevent cancer. In desperation, Hueper put his furniture in storage in Philadelphia, cashed in his life insurance policy and traveled with his young wife back to Germany to try to find a job. The Nazis had just seized power, and he signed his letter seeking work with "Heil Hitler." Science historian Robert Proctor says he cannot tell whether Hueper's journey was a matter of conviction or opportunism.[2]

In any case, Hueper was stunned by what he found. Once the world's leader in scientific research, the Germany that he toured in 1933 had become warped by politics and was losing its scientific edge. Hueper was appalled to see that Germans banned experiments like those he had learned to conduct there on rabbits and dogs but continued to carry out studies on humans. Gypsies, homosexuals and Jews were not accorded the legal protections granted laboratory animals.

With no obvious job prospects at hand and disheartened by the chaotic state of Nazi Germany, Hueper and his wife returned to the States. In 1934 he got one of those once in a lifetime chances. His prediction had come true: the DuPont plants making industrial dyes now had twenty-three cases of bladder cancer. The company brought him in to tackle the problem. Appointed pathologist at its newly created Haskell Laboratory of Industrial Toxicology in Wilmington, Delaware, in 1934, Hueper was granted a scientific platform in one of the world's leading firms to study the hazards of which he had warned.

Coming up with methods for studying a danger is one thing, but trying to keep that danger from affecting people is quite another.

With a spirited start, Hueper set up novel experimental systems to determine precisely how animals respond to precise amounts of some of the dyes and solvents being manufactured at DuPont at the time, including beta-Naphthylamine and benzidine, two compounds that had

been well studied in Germany. From his review of the German literature, Hueper knew others had already documented these compounds'
damaging properties in humans. In 1895, less than two decades after
production of synthetic dyes had begun in Frankfurt, Germany, the
surgeon Ludwig Wilhelm Carl Rehn reported that one of every ten industrial dye factory workers had bladder cancer. By 1906 physicians
from every nation where such production had begun up to twenty
years earlier reported dozens of additional cases. The International
Labour Office had published a technical report in 1921 declaring 2-
naphthylamine and benzidine to be human carcinogens.

By 1930 the high risk of bladder cancer among those who worked
regularly with such dyes was clear enough that Germany and Switzerland officially agreed to pay dye workers who developed such cancers—making this one of the first formally compensable occupational
illnesses.[3] Germany also devised workplace rules allowing only older
workers to work with certain cancer-causing materials—on the theory that these older men would die soon enough after their employ-
ment ended that cancer was not an issue.

Convinced that the German experience with dyes foretold major
health problems in America, Hueper tried to monitor the health of
DuPont workers handling these same agents. This did not go well. Like
physicians throughout the world at the time, Hueper wanted to believe that doctors would be held in high esteem and their advice
heeded by the governmental and corporate worlds, just because they
were medical men. In his unpublished autobiography, Hueper explains
that his honeymoon with Haskell laboratory proved short-lived. He
tracked how the bladders of dogs first became reddened, then scarred
over and finally developed tumors after repeated exposures to some of
the same chemicals then being produced at the Chambers Works. Eager to see how aromatic amines were being made firsthand, one day
Hueper set out to look inside the factories. Here's how he described
his first and last visit to the beta-Naphthylamine manufacturing plant
just across the Wilmington River at Deepwater:

> The manager and some of his associates brought us first to the building
> housing this operation, which was located in a part of a much larger
> building. It was separated from other operations by a large sliding-door

allowing the ready spread of vapors, fumes and dust from the be-tanaphthylamine operation into the adjacent work rooms. Being im-pressed during the visit by the surprising cleanliness of the naphthlamine operation, which at that occasion was not actively work-ing, I dropped back in the process of visitors, until I caught up with the foreman at its end. When I told him "Your place is surprisingly clean," he looked at me and commented, "Doctor, you should have seen it last night; we worked all night to clean it up for you."[4]

Hueper decided he needed to see what a real plant looked like. He took his team up the road to a benzidine operation that had not been expecting them. The small building was caked with white powder. Residue—the dried deposits of the chemicals being made there—cov-ered the road, the loading platform and the window sills.

The response of Haskell's directors to Hueper's foray into this regu-lar operating factory was swift. He would never set foot in a DuPont factory again. Hueper ordered him to stop studying humans and re-strict his research to the labs. In November 1937 he was threatened with legal action if he tried to talk about or publish any of his findings regarding worker health dangers.[5]

By 1938 Hueper had come up with innovative environmental proof that what had happened in Europe would take place in America. Beta-Naphthylamine—then being widely produced as an industrial dye by DuPont and leaving chalky remainders in its wake—caused bladder tumors in dogs that looked much like those found in men in Europe. He predicted similar results in humans. This didn't fit the company man-agers' definition of useful research; it was certainly nothing they wanted others to know. After continuing disputes with his bosses about what could be published by whom and where, Hueper was fired in 1938.

Hueper went on to hold a number of other research posts, develop-ing experimental studies of suspect materials and amassing a library of papers on workplace hazards. He ended up in what seemed to be a dream job at the U.S. National Cancer Institute. From 1948 until his retirement in 1968, Hueper led its first section on environmental can-cer, where he provided original experimental research and synthe-sized the world's literature on avoidable causes of the disease. He is the person Rachel Carson credited in her path-breaking 1963 book *Silent*

Spring for exposing the connections between the environment and cancer. What began as a dream for Hueper slowly became a nightmare.

ROBERT KEHOE took a different path. Trained in medicine at the University of Cincinnati in 1920, just as Hueper was completing his medical studies in Germany, he became comfortable in a world of relative privilege. Mark Twain quipped that he wanted to be in Cincinnati when the world ended because it would take ten years for the news to get there, but Kehoe found the town well suited to his ambitions. The wealthy were expected to help those less fortunate, and they were assumed to be better able to do so than ordinary people. Kehoe was recognized early on for his brilliance and innovation in coming up with ways to improve workplace conditions. He also displayed considerable business acumen and made a career of advising corporations under fire about how to protect themselves from accusations that their products were dangerous—an early expert in what might be called defensive research. For more than forty years, Kehoe remained a central figure in public health circles, at various times serving as president of the American Academy of Occupational Medicine and as director and president of the American Industrial Hygiene Association—groups charged with setting standards for exposure to workplace hazards.

Kehoe's views on the value of industrial hygiene research were not developed as abstract scientific principles; they reflected the corporate crucible out of which they emerged.[6] Consider the position he took on leaded gasoline. On February 2, 1923, the world's first leaded gasoline was introduced, touted as a salve for engine knocks and pings caused by incomplete combustion. That same year, in a small General Motors plant in Dayton, Ohio, the two workers responsible for bottling liquid lead died. The production line was shut down in April 1924. Charles Kettering, chief of GM's effort to develop leaded fuels, blamed the workers.

"We could not get this across to the boys," he said. "We put watchmen in at the plant, and they used to snap the stuff [pure tetraethyl lead] at each other, and throw it at each other, and they were saying that they were sissies. They did not realize what they were working with."[7]

Kehoe, then a young assistant professor of pathology at the University of Cincinnati, was brought in to advise on how to prevent such mishaps in the future. A methodical man, he figured out what had killed the workers in a single day. According to a newspaper account of his early work for the companies, Kehoe deduced that the danger lay in the bottom or ground level where heavy lead fumes sank. Putting in fans to blow the toxic fumes outside and giving men hip boots to prevent the liquids from being absorbed through their skin would allow them to work without keeling over.

Kehoe backed Kettering's view—if only workers could be trained to be more careful, their health would not be endangered. This was one of the early breakthroughs in industrial hygiene. Workplace training in safety saves lives. Other developments spearheaded by Kehoe proved less enduring. Measuring lead in the blood of all workers in the plant, Kehoe decided that background levels of lead were high in all workers, not just in those directly handling lead. Kehoe confidently declared that lead was a natural compound found in all humans—an essential micromineral. Years later, when scientists showed that people living on remote mountaintops in Nepal had no lead in their bodies at all, Kehoe's earlier conclusions were understood as reflecting the fact that all the samples had been contaminated by the industrial workplace. But at the time, Kehoe's view that lead was a natural contaminant was music to the ears of this growing, highly profitable industry.

When it became known that men who worked with liquid lead were going insane, Kehoe provided assurance that setting up less sloppy industrial processes would eliminate the problem. Standard Oil, the company that owned the Ethyl Corporation at the time, went to great lengths to keep the fatal health costs of producing leaded gasoline well-hidden. Joseph G. Leslie was one of those whose life and death were kept secret. He worked making liquid lead at the company's Bayway, New Jersey, plant. After an explosion of leaded gasoline in 1923, he spent the last forty years of his life locked away in a psychiatric hospital, unable to speak or communicate with others. To the rest of his family, he had simply died that year. Only his wife and son knew he was alive. His grandchildren and other family members did not learn what had happened to him until historian Bill Kovarik put all the pieces together in 2005.[8]

As one of the top specialists on how to train workers and set up plants so that men did not die working with this liquid brain poison, Kehoe fit easily into the corporate world. Realizing the value of being able to tap such advice on a regular basis, the Ethyl Corporation, General Motors, DuPont, Frigidaire and others promised to give Kehoe $100,000 every year starting in 1929 (equivalent to several million today) to run the industrial toxicology laboratories on the University of Cincinnati campus—named for one of his main benefactors, Dr. Kettering. Just thirty years old, despite the global economic depression, Kehoe had hit the scientific jackpot. A university official told the *Detroit Free Press* in 1936 that the companies "would meet all salaries and expenses. . . . Dr. Kehoe's expense account for research was unlimited."

As part of his duties, Kehoe spent a fair amount of time consulting with lawyers and other doctors looking at records of injured workers. Convinced of the value of making industry cleaner and more efficient, he came up with important methods for tracking the health and well-being of workers in order to craft better ways to organize production. Many of Kehoe's students at Kettering would go on to distinguished careers in industrial hygiene, including Eula Bingham, who became director of the Occupational Safety and Health Administration in 1976, and Paul Kotin, who became the first director of the National Institutes of Environmental Health Sciences in 1968 and later worked for the asbestos industry. The impressive new facilities at the University of Cincinnati became a major center for studies requested by industry, providing private advice to companies and never releasing results unless given permission to do so. Kehoe's reports were often used to defend these companies against lawsuits and claims for compensation. From 1930 until a stroke ended Kehoe's active involvement in 1963, much of this secret research was never published.

At the same time that he served as professor at Cincinnati and director of the Kettering laboratory, Kehoe was also handsomely paid as medical director of the Ethyl Corporation—a company with extensive and growing global holdings. These overlapping positions say volumes about the close relationships between business, government and the university that continued throughout his leadership at Kettering. In a candid admission of the industrial cast of the enterprise, Roy Albert, one of the more recent directors of the Kettering labs, notes in his memoir that Kehoe ran the facility as the "Medical Department of the Ethyl Corporation."[9] Subsequent leaders would change that, making Kettering the locus of first-rate scientific work published in leading academic journals.

But for Kehoe the labs were an industrial outpost, conducting studies of rodents and "human organisms" under contracts signed with Monsanto, DuPont, General Motors, Stauffer Chemical Company, the Tennessee Valley Authority, U.S. Steel, Mobil Oil, the Ethyl Corporation and others. Each contract stipulated that "the investigative work shall be planned and carried out by the University, and the University shall have the right to disseminate for the public good, any information obtained. However, before issuance of public reports or scientific publications, the manuscripts thereof will be submitted to the Donor for criticism and suggestion."[10]

Such wording was standard. Businesses routinely funded research at universities and even within the federal government under similar agreements. The same businesses that produced the materials Kettering tested also decided what findings could and could not be made public.

As late as March 25, 1965, after he had retired from active management of Kettering, Kehoe sent a memo to the staff regarding papers prepared for publication, reminding people not to refer to private reports in their public talks or papers. "It is undesirable, as a rule," he wrote, "to refer to reports of the Laboratory made to Sponsors in papers prepared for publication, since such references bring requests for these reports. As these reports often contain confidential information, they cannot be supplied, except confidentially, to other interested persons, and unless one knows that they are suitable for issuance to others . . . they should not be mentioned in public."

When information on workplace cancer hazards emerged from the Kettering labs, it was passed on to those who ran the factories but not to those who worked inside them. Kehoe was a kind man and ahead of his time in many ways, according to Bingham, who is now a professor at the University of Cincinnati. When she was pregnant and conducting research, Kehoe saw no reason for her to disappear, as was usual for women at that time. She was encouraged to work so long as she wished.

Still, when it came to dealing with corporate heads on matters regarding the safety of their products, Kehoe took an old-fashioned view. He believed, like many doctors who worked for industry, that businessmen would listen to him. And he believed that details of workplace hazards were best kept confidential.

These were not the only points on which he would be proved wrong. Within two months after the war in Germany ended, intelligence agencies from Britain and the United States sent agents there to retrieve critical information. One member of the U.S. team was Kehoe, then a captain in the U.S. Army. He interviewed key German scientists and brought back critical studies on topics ranging from chemical warfare to pesticides, pharmaceuticals and industrial materials.

Throughout its history, Ethyl Corporation had maintained close ties with many foreign companies—much closer than the U.S. government realized at the time. Germany was one of its most important clients. In March 1942, Thurman Arnold, a U.S. assistant attorney general, told a Senate committee investigating war profiteering that without the unique formula for making leaded gasoline—Ethyl's main product—the Nazis could not have flown their planes or fueled their land vehicles. Committee Chairman Harry Truman termed the alliance between some American companies and German national company I.G. Farben "treason."[11]

I.G. Farben was a firm of such excess and extravagance that it inspired novels, movies and myriad doctoral dissertations. Formed after World War I, its name was an amalgamation that stood for *Interessengemeinschaft,* "Association of Common Interests," of fabric dyes *(Farben).* Among the group's common interests was the election of Adolf Hitler. I.G. Farben was the single largest donor to the Nazi party and throughout its history was well funded by many Wall Street investment firms. Its members included BASF, Bayer, Hoechst and other German chemical and pharmaceutical companies.[12]

Ethyl itself was a truly global firm, and had independently joined forces with Hitler's staunchest corporate supporters, forming the firm Ethyl Gemeinschaft in 1934. Regarding the value of Ethyl's work for the Nazis, a report from I.G. Farben credits the firm and its major partner, Standard Oil, with directly fueling the German war machine. Standard Oil gave Germany one thing it did not have—a way to boost the power of its gasoline—through the production of ethyl gasoline. A German memo from 1940 found after the war gave credit where it was due:

> The closing of an agreement with Standard [Oil] was necessary for technical, commercial, and financial reasons: technically, because the specialized experience which was available only in a big oil company was necessary to the further development of our process, and no such industry existed in Germany; commercially, because in the absence of state economic control in Germany at that time, IG had to avoid a competitive struggle with the great oil powers, who always sold the best gasoline at the lowest price in contested markets; financially, because IG, which had already spent extraordinarily large sums for the development of the process, had to seek financial relief in order to be able to continue development in other new technical fields, such as buna [synthetic rubber].[13]

The company's official history claimed that the deal between the Americans and the Germans was not entirely one-sided. Standard Oil of New Jersey and General Motors—the companies then owning Ethyl—learned quite a bit about relevant war technologies of interest to them as well.[14] Having acquired the capacity to create powerful fuels for cars, tanks and airplanes from the Americans, the German war effort got an advantage without which it would have folded much sooner:

> By reason of their decades of work on motor fuels, the Americans were ahead of us in their knowledge of the quality requirements that are called for by the different uses of motor fuels. In particular they had developed, at great expense, a large number of methods of testing gasoline for different uses. On the basis of their experiments they had recognized the good antiknock quality of iso-octane long before they had any knowledge of our hydrogenation process. This is proved by the single fact that in America fuels are graded in octane numbers, and iso-octane was entered as the best fuel with the number 100. All this knowledge naturally became ours as a result of the agreement, which saved us much effort and protected us against many errors.

As a consequence of our contracts with the Americans, we received from them, above and beyond the agreement, many very valuable contributions for the synthesis and improvement of motor fuels and lubricating oils, which just now during the war are most useful to us; and we also received other advantages from them. Primarily, the following may be mentioned:

(1) Above all, improvement of fuels through the addition of tetraethyl-lead and the manufacture of this product. *It need not be especially mentioned that without tetraethyl-lead the present methods of warfare would be impossible.* The fact that since the beginning of the war we could produce tetraethyl-lead is entirely due to the circumstances that, shortly before, the Americans had presented us with the production plans, complete with their know-how. It was, moreover, the first time that the Americans decided to give a license on this process in a foreign country (besides communication of unprotected secrets) and this only on our urgent requests to Standard Oil to fulfill our wish. Contractually we could not demand it, and we found out later that the War Department in Washington gave its permission only after long deliberation. [Emphasis added.][15]

On this last point the memo's author was mistaken. The War Department never agreed to give the Nazis the formula for enriching gasoline with lead. Quite the contrary, when the American firms' intent to create a German subsidiary became apparent, the War Department specifically ordered the companies *not* to let the Germans have the methods for enriching aviation or motor fuels. The War Department's official, unsigned letter on the matter, sent to the heads of Standard Oil, General Motors and Ethyl on December 15, 1934, was clear:

I am writing you this to say that in my opinion under no conditions should you or the Board of Directors of the Ethyl Gasoline Corporation disclose any secrets or "know how" in connection with the manufacture of tetraethyl lead to Germany.[16]

In response to these concerns Ethyl lied to the government. On January 12, 1935, Earl W. Webb, the president of Ethyl, sent a letter to the chief of the Army Air Corps promising that no critical technical knowledge would be given to the Germans. Later that year, Ethyl secretly signed joint production agreements to produce leaded fuel with I.G. Farben in Germany as Ethyl Gemeinschaft, and separately with Montecatini in fascist Italy.

Files captured from I.G. Farben after the war confirm just how precious this technology was to the German military effort:

> Since the beginning of the war we have been in a position to produce lead tetraethyl solely because, a short time before the outbreak of the war, the Americans had established plants for us ready for production and supplied us with all available experience. In this manner we did not need to perform the difficult work of development because we could start production right away on the basis of all the experience that the Americans had had for years.

Big American firms collaborated extensively with the Nazis. A letter of October 19, 1936, from the American ambassador to Germany, William Dodd, to President Franklin Delano Roosevelt explained the complex ties between American and German firms.

> At the present moment more than a hundred American corporations have subsidiaries here or cooperative understandings. The DuPonts have three allies in Germany that are aiding in the armament business. Their chief ally is the I. G. Farben Company, a part of the Government which gives 200,000 marks a year to one propaganda organization operating on American opinion. Standard Oil Company (New York sub-company) sent $2,000,000 here in December 1933 and has made $500,000 a year helping Germans make Ersatz gas for war purposes; but Standard Oil cannot take any of its earnings out of the country except in goods. . . . Even our airplanes people have secret arrangement with Krupps. General Motor Company and Ford do enormous businesses [sic] here through their subsidiaries and take no profits out. I mention these facts because they complicate things and add to war dangers.[17]

As ETHYL'S MEDICAL DIRECTOR, Kehoe was responsible for monitoring the toxic exposures of the company's workers around the world, including those in Germany. His archived papers at the University of Cincinnati contain boxes of reports from plants making leaded gasoline in New Zealand, France and Argentina, along with Germany, Japan, Italy and other foreign facilities. His files also include technical reports from that period, long, touching letters to and from his wife and the sorts of printed things that people keep long after they are no longer useful, either out of sentiment or because they are too important to throw away.

In Kehoe's case, these mementos include postcards, souvenirs, snapshots of him in an army uniform and some stranger items. We know that he visited the Colosseum in Rome because someone took his picture there. He also kept disturbing images of piles of corpses at Buchenwald, Tekla, and unnamed concentration camps. One of the photos shows the body of a man in the foreground lying with his back on the ground. His hands and feet are frozen into the position of someone who died in a chair and was tossed off after rigor mortis set in. Another shows emaciated dead adults arrayed like wooden logs, along with a lifeless plump toddler. It is unclear who took them. An expert in photographic history advises me that they are snapshots taken with a common Kodak Brownie camera. They tell us nothing about the thoughts of the person taking them, other than that they were judged of sufficient import to capture on film and for Kehoe to keep them his entire life.

In 1950 James Stewart Martin, who had been chief of the economic warfare section of the Department of Justice investigating Nazi business interests, charged that American and British businessmen got themselves appointed to key investigative positions after the war as a way to keep the lid on what might have become known about what their corporations had done to build the German war machine.[18] Was Captain Kehoe, like other American industrial experts, engaged in efforts to cover the trail of ties between Ethyl, Standard Oil and the Nazi companies, many of whose factories used slave labor from the camps?

We can't be certain how well Kehoe spoke German, but we do know that fluency in the language was required of all medical or science students at the time. His files include reports in German from German chemical companies with which he was apparently working, but nothing written by him in that language.

There is a bond among scientists that typically cuts across nation and culture. A typewritten report from the Field Investigations Unit—the intelligence group under which Kehoe went to Germany—written in the third person and signed in his strong, clear handwriting tells us that on July 21, 1945, Kehoe took custody of the distinguished sixty-eight-year-old Ferdinand Flury, one of Germany's leading toxicologists, whose release he secured afterward.

Ferdinand Flury was known to have perfected antidotes to poison gas because he had also created the poison in the first place. He knew all about how they worked through studies he had directed at the nation's top chemical toxicology laboratories.[19] He reportedly advised against the use of war gases toward the close of World War II and was dismissed from the institute he had directed shortly after it was bombed to bits in 1945. Kehoe learned a lot from the man who became his temporary prisoner. The fact that he worked to obtain his release suggests one of two possibilities. Either Flury was exempt from prosecution as a war criminal because he had no direct engagement in war crimes, or his research was judged too important to be lost to prosecution. The latter category of exemption from prosecution was one that many experts managed to fall into. Flury died in 1948, a few months after returning to the university laboratories. He was never prosecuted.

Kehoe's report to the field investigations unit on Flury relays information that may have been obtained through reading documents. In the crisp language of a scientific observer, Kehoe explained how wartime conditions impaired efficiency. Slave labor was inexperienced, underfed and unclean, he noted. As a result, productivity suffered.

The general level of health of industrial workers was influenced adversely by the war in a number of important respects. In the first place the increased rate of production brought in many inexperienced workmen, and as the war progressed many foreigners ("slave labor") were brought into the plants. The rates of occurrence of injury and illness among these increased. There were many new chemicals with more or less unknown properties, and a number of highly dangerous war products—explosives, chemical warfare agents etc.—, to be dealt with and the toxicological investigation of these compounds inadequate. By far, the more important influences, however, were the disruptions caused

by inadequate housing, improper food, uncleanliness, vermin infesta-
tions, and the loss of sleep and rest occasioned by frequent enemy
attacks.

About those photos of corpses from the concentration camps, we
can only infer from the words of others the circumstances under
which they were taken. We know from records of SS Storm Trooper
infiltrator Kurt Gerstein how the deaths were engineered. We know
Kehoe visited some of the camps, where he interviewed those who
had worked in factories and found the stories not believable. Kehoe's
own descriptions suggest the normal privations of wartime, nothing
more.

> Thousands of foreign workers and their families were crowded into
> camps adjoining or near to plants, in barracks and buildings that varied
> from filthy hovels to reasonably clean and passable quarters. As houses
> were destroyed by bombing, German workmen and their families
> were put into barracks also. The crowding of these quarters led to mi-
> nor epidemics of communicable diseases, including typhus in quite a
> number of localities into which Polish and Russian civilians and war
> prisoners were brought. Sleep was disturbed during bombing attacks,
> and water and sewage lines were broken. All water had to be boiled,
> and apparently was, since little or no typhoid fever was reported in
> any of the plants in which specific inquiry was made.
>
> Nutrition was not especially good in any of the labor camps, as at-
> tested by statements of laborers interviewed (which could not be
> wholly accepted), and by the fact of an upward trend in pulmonary tu-
> berculosis which began to appear in 1941 and continued upward with
> a corresponding increase in the frequency of the more severe type of
> case. During the last year of the war there was also a fairly general loss
> of weight among workers.

"A fairly general loss of weight!" Healthy people naturally fend off
various bacteria and viruses that live on the skin. The skin of the mal-
nourished becomes a breeding ground for blood-sucking flies or
worms, scabies, and wounds that cannot heal. Without commenting
about the starvation imposed on slave labor, or the fact that the Ger-

mans had enslaved others, Kehoe attributed their skin infections to poor hygiene.

> One of the striking phenomena was a tremendous increase in the oc-
> currence of dermatoses, especially furunculous [caused by tiny
> worms]. Scabies was also very common and caused great difficulty,
> since it was necessary to disinfect the entire family group that lived to-
> gether. The furunculus was the result, apparently, of the general im-
> possibility of personal cleanliness among the workers. Soap and other
> detergents were not available, water was scanty and often non-existent
> because of broken lines following air raids. Therefore neither skin nor
> clothing could be cleaned effectively or with sufficient frequency. In-
> festation of quarters with vermin was common, so that frequent fumi-
> gation was necessary.

Surely Kehoe saw the ovens. By the time he arrived on the scene, we know from reports by others that the camps were full of dead and skeletal refugees. He kept photos of their corpses. Yet his report makes no mention of the industrial-scale genocide that was carried out. His soulless recollections give no indication that he grasped the depravity of the camps. His written account depicts starvation and disease— evidence of mass murder was not escapable.

Contrast Kehoe's detached tone with that of the report relayed on the radio by Lt. Colonel Richard Seibel after he entered the camp at Mauthausen: "We've come across this big camp, we don't know what it is. People are dying everywhere. . . . People don't do this to peo- ple." Or listen to the disgust in the words that Fred Friendly wrote to his mother of the same scene: "I saw their emaciated bodies in piles like cords of wood. I saw the living skeletons, some of whom, regard- less of our medical corps work, will die. I saw where they lived; I saw where the sick died, three and four in a bed, no toilets, no nothing. I saw the look in their eyes."[20]

If Kehoe's depictions of the camps were clinically detached, his de- scriptions of research developed by the Nazis suggest professional ad- miration. As the center of chemical research on many topics, the Germans had come up with an innovative solution to a special prob- lem. At the time, diethylstilbestrol (DES), which was the first synthetic estrogen ever created, was added to animal feed to fatten cows, pigs and chickens for the German nation. The problem was that young boys who worked in the factory that made DES developed painful, swollen breasts.

Today we might wonder how such a product could have been considered appropriate for general use even then. DES was easily made from coal tar and had not been patented because it was first produced in the laboratory by publicly funded British researchers. It is now banned in industrial countries, but back then its ability to make a scrawny cow put on several hundred pounds in a few months was especially valuable to a wartime economy, breast bulges or no. Kehoe reported the straightforward German solution to this problem—let women be the production workers, since they already had breasts. Presumably they would not mind having fuller ones. Kehoe wrote:

> A drug effect of interest in relation to industrial hygiene is that of DES, in the manufacture of which only female workers are employed, because of the untoward effects induced in males by the absorption of this material in the course of a day's work. Boys develop a mammary swelling with such severe pain the pressure of a shirt cannot be endured. On discontinuance of exposure the condition subsides spontaneously within a week or two. No sequelae have been reported and no abnormalities of the testes have been seen. On the other hand older males develop some atrophy of the testes and some apparently temporary loss of sexual potency. This condition is said to have been cured by the administration of androsterone. Exposed women had no nausea or menstrual abnormalities.

Today we know that pregnant women who were given DES, in the mistaken notion that it would prevent miscarriages, from 1948 up to 1972 are at risk of breast cancer years later. Their children, whether boys or girls, tend to have a host of difficulties reproducing and serious health problems, including a higher risk of cancer and deformities of the reproductive organs. But this should not be surprising information. The increased risk of cancer from DES did not surprise those who understood, when it was first synthesized in 1938, that hormones could cause cancer, a matter that was discussed at the second

International Congress of Scientific and Social Campaign Against Cancer in Brussels in 1936.[21] Kehoe makes no mention of this in his report.

In amassing information on industrial hazards, Kehoe had lots to draw on. The Germans had understood much about the ways various toxic agents kill people quickly and cause permanent damage that leads to death more slowly. Just as Hueper had warned Irénée du Pont in 1932, bladder cancer was rampant in dye workers. Men who worked with 2-naphthylamine and benzidine died from bladder cancer twenty years younger than men without such exposure, according to information provided to Kehoe by Professor Heinrich Oettel of Ludwigshafen am Rhein, Germany, in 1939.

The introduction to one report Kehoe prepared for the U.S. and British Intelligence Objectives Subcommittee on I.G. Farbenindustrie, Leverkusen, in January 1947, was straightforward: "The objective was to study incidence of and methods of prevention of bladder tumor among workers in the benzidine plant." The I.G. Farben researchers had developed precise methods for correlating the amount of chemical residue found in workers' urine, the percentage of all those working who had developed bladder cancer, and how many years they had been employed at the plant. The bladder is a kind of natural storage system that collects all sorts of poisons before they flow out of the body. But things that move through the bladder can leave traces behind.

To carry out their studies, German researchers snaked crude rubber tubing into the penises of workers going through urethras up to their bladders, checking for residue of blood or early signs of tumors. Researchers complained that some workers were not especially cooperative and needed days off after the test.

The information that Kehoe collected for the Army Field Investigations Unit can be turned into a simple bar graph. Each solid block (shown on page 92) represents the proportion of workers who had worked for various time periods at Leverkusen and found themselves urinating blood, having leaks and spasms, or dying after the tumor spread from their bladder to their bones. There are of course many problems with producing such analyses. The graph only shows those whose deaths were recorded or who were still alive at the time of the study.

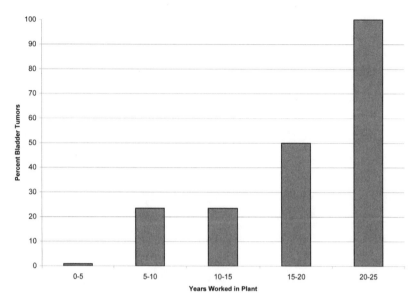

Figure 4-1 *Percent Bladder Tumors in Benzidine Plant Workers, Leverkusen, Germany, 1946*

Still, we learn some remarkable things from this picture. The German findings were ·quite simple: the longer a man worked at Leverkusen, the greater his chance of developing bladder tumors. Of those who had survived working for twenty years or more, every one had bladder cancer. Half those working fifteen to twenty years had such tumors, as did one-fourth of those working fifteen years or less. This was an early and clear illustration of a basic public health concept called a dose-response relationship. The higher the dose, or in this case the longer time that exposure has taken place, the stronger the response. We cannot say what happened to everyone who ever worked in this plant. But we can be sure that those who made it through twenty years of factory work all developed bladder tumors.

The mimeographed report with this information that Kehoe provided to the Office of Strategic Services (later the Central Intelligence Agency) came with a stricture: "This report is issued with the warning that if the subject matter should be protected by U.S. Patents or patent application, this publication cannot be held to give any protection against action for infringement." In other words, "trade secrets" were not to be compromised by the release of the report. The state of employees' health was apparently a trade secret. Not until many years later, when lawsuits were filed alleging earlier knowledge of some of these hazards, were these forgotten reports unearthed from moldy boxes where they had been stored. That was when it became clear just how strangely connected the lives of Hueper and Kehoe had been.

So, what happened to the evidence of workplace and other chemical sources of cancer the German researchers had produced? Where did it go? Right after World War II ended, the rich array of scientific information on industrial cancer hazards developed by the Nazis was translated into English by various field investigation offices, such as the one Kehoe manned.[22] Detailed summaries of Nazi research on the hazards of tobacco and various chemicals made their way to executives of many of the U.S. corporations then producing these materials.

We do not know all of what this unpublished material contained. But we do know that Kehoe resumed his highly profitable and distinguished career, working with some of the same firms that had employed him in the 1930s. DuPont, one of the principal funders of Kehoe's Kettering laboratories, would in due course become one of the largest chemical companies in the world.

BUILT DURING WORLD WAR I, DuPont's Chambers Works in Deepwater, New Jersey, produced some of the most profitable and toxic compounds in the world: synthetic dyes like beta-Naphthylamine and benzidine, and tetraethyl lead. To those who warned that these compounds had been shown to damage animals, Kehoe insisted that when considering the safety of lead or dyes, what mattered was not experimental findings produced in lab rats under controlled conditions but the real-world experiences of those working with any given compound.

He argued, successfully, that lead should be permitted in gasoline because natural residues of lead are already found everywhere, and the additional amount from such a small source would be trivial. When challenged on this view by Sen. Edmund Muskie at a hearing in 1966, Kehoe offered this immodest rejoinder: "It so happens that I have more experience in this field than anyone else alive."[23]

Like most of my colleagues, I had thought there was little new to learn about lead. After all, this compound has been studied for more than a century and is now believed to be controlled in modern societies. I was stunned when I opened some of Kehoe's files.

In 1926, when debate first surfaced about plans to put a liquid poison into gasoline, Kehoe knew a lot more about what lead did to the brain than he disclosed at the time. His files include dozens of autopsies that he conducted on dead newborn babies born with lead throughout their bodies. One report noted that a particular baby in St. Louis had been the mother's third child to die in infancy from lead poisoning. Another case, from Waynesboro, Mississippi, in 1926, described three infants born to the same mother, the last when she was thirty-four. All three died of lead poisoning; their autopsies showed elevated levels of lead in their blood, liver, and bone.[24]

Kehoe's dozens of autopsies of dead babies are the work of a very meticulous man, showing precise amounts of lead measured in the brains, livers, hearts and kidneys of poor black and white infants. For toddlers with lead poisoning, it's always been assumed that their exposure comes from the habit of putting things into their mouths. But there was only one source from which newborn infants could ingest lead: through their mother's blood while in utero or through their mother's milk, after tiny residues of lead dust brought home by their father got into their mother's body. Where this lead came from was not speculated on; the fathers' occupations are not even noted in Kehoe's reports. Today we understand that workers can bring invisible residues of metals, asbestos and other toxics home on their shoes and clothing. Fine particles, dozens of times smaller than a human hair, can be laden with heavy metals and accumulate in the homes and bodies of those who work with such compounds. This ability of the dust men worked with to affect the health of their children was a matter on which Kehoe remained silent throughout his career.

Another matter on which we can only speculate regards the circumstances that led Wilhelm Hueper to contribute information on the swollen, bloody brain of a two-and-a-half-year-old girl on whom Kehoe was asked to perform an autopsy in 1936. It was not uncommon for pathologists to pack up and ship various body parts to others for examination. From Hueper's letter of referral for this case, we learn that the Haskell toxicology lab had asked Kehoe on May 14, 1936, to review the collected tissues that had originally been examined at the

Children's Hospital of Philadelphia. The typewritten reports from Kehoe's files are all we have. At the time, his laboratory was certainly the only place where lead could be studied in this manner. The report that Kehoe signed on this toddler said that her bones were nearly 1 percent lead. He concluded that she had probably eaten lead paint shortly before her death.

Beginning in 1928 and frequently afterward, Kehoe published a number of reports purporting to show that lead was an unavoidable natural contaminant.[25] In reaching this view he consistently made a fundamental error. He compared men with high exposures to lead in smelters and factories to others who worked in comparable factory jobs that he believed had no exposure. There was one big problem with this approach. If you worked in a factory where leaded fuels and paints and batteries were essential parts of industrial life, you couldn't escape lead. In 1966, the Cal Tech geologist Clair Patterson revealed this fallacy in Kehoe's work. Famed as the man who had come up with techniques that showed the earth was four and a half billion years old, Patterson invented many of the basic methods of geochemistry. Using rigorous laboratory tests, he found that bones from prehistoric humans had almost zero lead; it was not a "natural" contaminant at all. Ethyl, in response, tried to get Cal Tech to fire Patterson. When that failed, they sought unsuccessfully to discredit his work.[26] Patterson's findings meant that much of Kehoe's analyses had missed this essential fact: rather than being a ubiquitous part of the natural world, lead is an indicator of the industrial world.

Other than this single autopsy on a poisoned toddler, Hueper and Kehoe seemingly had little direct contact. When they finally connected, it was not as scientific collaborators but as shadowboxing adversaries in legal wranglings.

Behind the scenes the two men were joined at the hip. In 1960, as part of his work on a legal matter involving an injured worker, Hueper learned that Kehoe had been reviewing his work for years, conducting secret studies on matters Hueper had been ordered not to pursue. In 1937 Hueper was directed to stop studying or talking about the cancers he had found at increased rates in men who handled synthetic dyes and solvents at DuPont. Eventually he was also instructed not to disclose evidence that animals exposed to these same compounds, or to coal tar derivatives, also developed tumors. But research on these matters did not end when Hueper left Haskell laboratories.

Archival materials at the University of Cincinnati show that much of Hueper's halted work continued under Kehoe's secret direction. Benzidine was one of many compounds that Hueper had originally studied and on which the Kettering laboratories carried out private research starting in the 1940s. Long known to afflict rats with bladder tumors, benzidine dyes remained a mainstay of chemical production at DuPont. Hueper was well aware that the first studies on the cancer-causing properties of coal tar dyes, and those on the carcinogenic nature of butter yellow food coloring—an amino azo dye—had been published in German in 1914 and 1935 respectively.[27]

For twenty years after Hueper left DuPont, the Chambers Works did not report on any new cases of bladder cancer in its workers. No national or state cancer registries existed at that time; the only reports of such cases were what executives chose to reveal to public authorities. Despite the company's silence, however, cases kept accumulating. In 1980 it became known that 364 cases of bladder cancer had occurred in this one factory since its beginning.

Under a private contract between DuPont and the Kettering laboratories, Kehoe and his colleagues looked into the cancer-causing properties of synthetic dyes and coal tar contaminants of paraffin in workers and studied the response of lab rats to these same agents. One confidential agreement for research identified thirteen different formulations of paraffin oils that Kettering labs would study in animals, ranging from naphthenic oil, solvent refined, to chlorinated sperm oil in sulfurized fat base. Keeping any findings on the cancer-causing hazards of these and other workplace chemicals secret was perfectly legal at the time. Kettering's clandestine work on coal tars and dyes remained unpublished and only surfaced years later, as part of a defensive strategy in a lawsuit.

Kehoe's archives contain numerous detailed private reports showing that animals develop tumors when exposed to many of the ingredients tested in the labs. They also contain memos explaining that compounds shown to induce cancer in animals shouldn't be assumed to represent a similar hazard in humans. In fact there are lots of differences between experimental rodents and humans, though not quite as many as some might suppose. These distinctions proved fertile grounds for constructing a useful argument when findings of increased risk in animals were eventually made public: if a chemical causes cancer or some other poor health outcome in purebred rats or dogs, that finding may not be pertinent to humans. We don't bark, chew through wood, or make babies after three weeks, so we should not assume that what happens to other mammals when exposed to chemicals necessarily happens to us.

Kehoe did more than argue for the limited value of experimental work when it came to the hazards of lead, synthetic dyes or other materials. He also insisted that those who would emphasize lab findings on the dangers of these agents were engaging in irresponsible speculation. What counted most, he claimed, were plain, hard facts, the sort that only come from looking at the health of real live—or sick or dead—workers. In response to the concerns of leading public health researchers about the potential impact of toxic exposure, Kehoe confidently argued that only such "facts" could be used to balance the sometimes conflicting values of industrial progress and human health. Yet he spent decades gathering precisely these sorts of facts and routinely kept them from the public and other scientists.

For Hueper, who had charted the ruined lives of workers exposed to lead and aromatic amines, this line of reasoning was anathema. He knew that in America, Germany, Switzerland, England, France, Italy, Austria, Czechoslovakia and Poland—basically every country where aromatic amines were used—within two decades outbreaks of bladder cancer would hit.

In his autobiography, Hueper notes four basic ways industry can avoid focusing attention on cases of occupational cancer. First, they can feign blindness by refusing to see or record cases, as DuPont did with bladder cancer at Chambers Works after 1936. Second, they can create negative evidence by only counting disease in workers who have been employed a short period of time and excluding from their records those with long-term exposure who are neither working nor alive. Or they can pack the study population with many workers who have no exposure to the agent being examined, thereby diluting evidence of an effect. Finally, they can suppress or delay publishing the results.

Sometimes legal proceedings reveal that a company did all three. Here is how Hueper finally learned that Kehoe had secretly continued the studies he himself was ordered to drop. As an expert witness in a 1960 lawsuit brought by three men who alleged that their cancers came from working with waxes contaminated by petrochemicals, Hueper offered the opinion that the men developed skin cancer because of their daily contact with oily hydrocarbon residues. To counter Hueper's testimony, the defense revealed that its own expert witness had also conducted private research on these compounds for many years. Hueper writes:

> The Director of this organization in Cincinnati [Kehoe], testifying as a consultant of the oil company, had to confess that none of the results of his institute's studies with these oils had been published or had been made available to the medical profession in general or to labor organizations, because the data were considered by the oil company as 'privileged' information, i.e., the property of the oil company. When after more than a year's time, the final information became available to the court and the plaintiff, there was no longer any doubt that even in the hands of the members of the Kettering Laboratory the incriminated oils had carcinogenic properties, although its Director had found it proper at the first hearing to make some snide remarks about my scientific reliability.[28]

The studies in the Kettering laboratories on aromatic amines remain so secretive that few have even heard about them even to this day. Their existence is confirmed by memos in the Kehoe archives describing the work, but even now I have not seen the reports themselves.[29]

Hueper found out that Kettering laboratories had also conducted secret research on other coal tar products. At one point industry reps claimed that Kettering had shown that while foreign coal tar might cause cancer, good old American tar did not. In his autobiography Hueper tells of one tragic result of this yahoo attitude. An executive of one of the companies making coal tar–based paraffin waxes came to Hueper's National Cancer Institute office to explain the better properties of coal tar. As he extolled the benefits of American products, Hueper saw the unmistakable mark of two small irregularly shaped cancers on the company man's neck—precisely the kind of tumor created by work with coal tar.[30]

A more protracted example of denial arose regarding what looked like the elimination of scrotal tumors in petrochemical workers. In his autobiography, Hueper explains that for a while what had once been a big problem for refinery workers seemed to have disappeared. Prior to World War I, a number of workers were found to have increased rates of scrotal tumors. But after the industry reported cleaning up its manufacturing processes in 1920, not a single case was recorded. This was ballyhooed as a triumph of industrial hygiene. It turned out to be no such thing. Whenever a scrotal tumor had appeared after 1920, industry health officials simply called it a venereal tumor, chalking it up to something not talked about much at all—sexually transmitted diseases. When Hueper examined the medical records of these alleged venereal tumors, every one turned out to have been a scrotal tumor. The epidemic of scrotal cancer had not ended but was literally defined out of existence.

Hueper's career after DuPont was not a happy one. In a remarkable case of bad timing, he published his magnum opus just weeks after the Japanese attacked Pearl Harbor. His book—a massive tome covering laboratory and public health studies of workplace hazards in several countries, entitled *Occupational Tumors and Allied Diseases*—brought together epidemiologic and experimental studies from four continents over more than a century to assert that workplace factors were important and controllable causes of cancer and other illness.[31] Hueper carefully explained that despite the limits of available statistics on cancer in humans, there was much evidence that rich people's tumors tended to differ from poor people's, and those who worked in dirtier, dustier trades developed distinct ailments. The book was intended as a public health call to arms. But with the war finally under way, the only arms of concern to most people were those that could provide for national defense.

Despite his run-ins with DuPont, Hueper found some receptive colleagues in mainstream medicine for a while. During the 1940s, he wrote editorials for the *Journal of the American Medical Association* on the cancer risks of solar radiation, aromatic amines, estrogens, coal tar products, arsenic, asbestos and other environmental hazards. In 1948 he began work with the fledgling National Cancer Institute, heading its first environmental cancer section. At first this must have seemed kismet. Under Hueper's leadership, in 1950 the NCI issued a blunt pamphlet for the general public, depicting a number of avoidable causes of cancer.

The twenty-page pamphlet began by pointing out that while cancer has been a recognized medical problem since eight centuries before Christ, in the modern era patterns of the disease had shifted radically. Medical x-rays, dietary deficiencies, tobacco and drinking habits, sunlight, and toxic chemicals all played a role. When it came to cigarettes, Hueper took the view that while smoking could be a cause of cancer, its role was easily exaggerated by those eager to downplay the impact of toxic chemicals.

This NCI public report fingered major workplace causes of cancer, ranging from radiation to specific toxic chemicals such as asbestos, aniline dyes, aromatic amines, paraffin oil, shale oil, crude oils, benzene, chromates, and nickel carbonyl. Figure 4-2 reproduces a small illustration from this report showing the various ways that people could be exposed to aromatic amines. The chance of developing bladder cancer varied with the extent of exposure. Workers making this product who inhaled it directly or absorbed it through their skin had the highest risk of all—with nine out of every ten developing bladder cancer. Risks did not stop at the factory gates. Those who worked or lived nearby also had more disease than others without such exposure.

This small diagram turned out to be more prophetic than Hueper could have realized. A quarter century later the NCI found that citizens of Salem County, home to the Chambers Works, had the highest rate of bladder cancer in the nation.[32]

Hueper's pamphlet ended with a clear set of proposals for a program of cancer control that, except for his views on tobacco, remains pertinent today. The program includes eliminating carcinogenic agents from industrial, civilian and military use whenever practical; instituting safety procedures for the handling of suspect materials; and providing careful medical monitoring of exposed workers for early signs

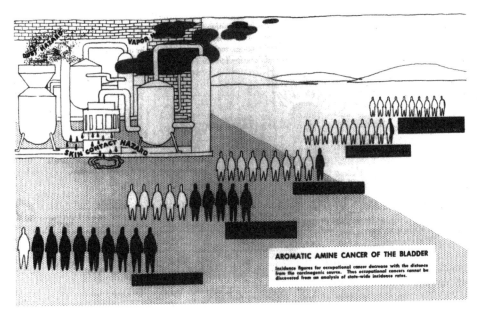

Figure 4-2 *A diagram by Wilhelm Hueper, from the National Cancer Institute, 1950, depicted incidence of bladder cancer among industry workers and predicted community risks.*

of cancerous and precancerous conditions. More than half a century later, the pamphlet's conclusion bears repeating:

> Environmental carcinogenesis is the newest and one of the most ominous of the end-products of our industrial environment. Though its full scope and extent are still unknown, because it is so new and because the facts are so extremely difficult to obtain, enough is known to make it obvious that extrinsic carcinogens present a very immediate and pressing problem in public and individual health. It should become one of the most urgent tasks of all medical men, public health officials, labor and management leaders, and members of legislatures, to become familiar with the problems of environmental cancer. They must all work together to combat its causes at the source, before the dread disease spreads to more and more of our people."[33]

But the political pendulum was swinging in a different direction. In 1959, when Hueper submitted an update of this pamphlet for publication, the editorial board of the NCI delayed so long that it never appeared. Later he failed to get official approval from the NCI to publish an expanded edition of *Occupational Tumors and Allied Diseases*.

Throughout the 1950s and for the rest of his career, Hueper's work was dogged in many ways. He undertook a long study comparing patterns of lung cancer in various workforces, which found that chromate workers and uranium miners had ten to twenty times more lung cancer than other workers, and that blacks had double the rates of whites. But as his studies of worker cancer risk mounted, his ability to release and discuss that work began to contract.

A memo Hueper wrote in 1961 to labor unions interested in his experience provides a sobering account of official corruption, obstruction of science and suppression of information at the NCI during the 1950s. Even his ability to speak to medical students as an NCI official was restricted "under the pretext that 'my time as a scientist at the agency is too valuable for engaging in such educational activities.'" As his efforts to study industrial hazards increasingly came to be seen as antibusiness and thus potentially pro-communist, Hueper found that he could no longer work at the NCI freely. He had to submit all his papers for reviews that seemed interminable.

Years later he learned what was behind this: "In 1960, I was told by a former Member of the Haskell Laboratory of Industrial Toxicology, a startling tale. . . . Rather unexpectedly my colleague, whom I had met before at various occasions, said, 'I am well acquainted with your recent publications. They are being handed to me as carbon copies before publication, for comments and appraisal after having been forwarded to the management of the DuPont Company, from someone in Washington.'"[34] Just as at DuPont, Hueper was soon ordered at the NCI not to publish his findings or talk about subjects he could not write about. Convinced that the government's never-ending editorial reviews were just thinly veiled censorship, he stopped trying to revise his book on occupational causes of cancer.

When he set out to show the cancer-causing properties of mining chromate ore and uranium, he ran straight into a fan blade of scientific criticism from which he would never recover. Uranium and chromate are critical to the production of nuclear weapons. Hueper learned the hard way that inquiries into their dangers were not welcomed. "One nice day when I submitted that manuscript for clearance, I was called later on in to the office of my director. He said, 'Now the high medical officials of the Atomic Energy Commission object against that. They said there are quite other reasons why the uranium miners develop cancer; it's not the radioactivity. You shall omit that from your presentation.' I said to the director, 'I will tell you something. I did not join the Public Health Service to be made a liar!'"[35]

Nonplussed by the orders of his superior, Hueper sent his manuscript warning of the dangers of uranium mining to the Colorado Medical Society and to one of the senators from Colorado. That senator, perhaps concerned about losing jobs or increasing production costs, apparently complained to the NCI about its renegade scientist.

When he first landed at the NCI, Hueper had hoped that DuPont would ignore him. For a while, it did. When he started working for the government, Hueper seemingly had the capacity to pursue some of the same research questions DuPont had ordered him to stay away from. Hueper's explanation of how his work at the NCI ended frankly smacks of paranoia. He wrote, "I soon found out that I had underestimated its unrelenting spirit of personal persecution."

When people talk in this way, of course, we have to wonder whether they are perhaps delusional. But sometimes a person believing himself to be under assault really is. In Hueper's case his unpublished autobiography provides ample evidence that people really were out to get him. Industries had been given extraordinary access to his papers prior to their being submitted for publication when he worked at the NCI.

Several weeks after meeting with a plant physician in Denver regarding aromatic amines and bladder cancer cases, Heuper received a letter from the Federal Loyalty Commission "informing me that I was under investigation for disloyalty." A colleague confided that he had been asked whether Hueper was a Nazi. A few weeks later, a letter from George H. Gehrmann, Hueper's former boss at DuPont, alleged that "I had shown communistic tendencies." After these attacks, Hueper was officially barred by the NCI from conducting studies of human risk of bladder cancer from aromatic amines, and he was forbidden to finish the work he had begun with other researchers on chromates, beryllium and other cancer-causing agents in the workplace.

His fate at Haskell played out again at the NCI. Hueper was told to focus solely on experimental animals and stay away from workers. The argument continued to be that there is no convincing evidence of human harm from these exposures. So long as Hueper and the state health departments with which he was working were prevented from studying the effects of chemical exposures on workers, of course, no such evidence would exist—at least publicly. Studies of sick or dead workers remained secret or were not tallied. Using statistical evidence on cancer rates, Hueper began to amass more evidence at the NCI that validated his little sketch from 1950: environmental exposure to synthetic dyes well outside the workplace affected cancer patterns in the community. He was forced to withdraw papers from consideration under threat of legal action.

There is one matter on which it is important to fault Hueper. His single-minded focus on chemical hazards led him to minimize the dangers of cigarette smoking. In his autobiography, he explained that he had never doubted these dangers. He himself had stopped smoking in 1938. But he insisted that if smoking were demonized as the only cause of cancer, then other causes of the disease could easily be ignored. "Fluctuations in concentrations of industries, their types of products, and their environmental wastes, the intensity of solar radiation, the variety of occupations, the condensation of populations, the types of fuel used, the concentration of automobile traffic in regions and roads," along with the growing use of synthetic additives and contaminants in food and water, combined to shape patterns of cancer.

All these modern hazards, Hueper feared, might be obscured by the smoking bandwagon. "It became clear that the acceptance of the cigarette theory would require the extermination of a great deal of factual medical knowledge . . . which was painfully and tediously acquired over many decades of medical science, and to replace it with an ill-documented, simple, unitarian theory which appears to offer no plausible solution." Hueper worried that industrial dangers would be dismissed as unimportant, because so many people smoked. He also noted a number of paradoxes regarding cancer and smoking. Women and some men with lung cancer were not always smokers. Lung and other cancers occurred in people with no smoking history whatsoever. What especially vexed Hueper was the fact that only about one in ten smokers would come down with lung cancer, whereas ten out of ten workers exposed to synthetic dyes for twenty years would develop bladder cancer.

Until he left the NCI in 1968, Hueper continued to publish basic scientific investigations on experimental animals. Having learned, late in life, some political survival skills, he provided expert testimony in a number of major lawsuits, but only when subpoenaed. In his final months, after his wife of more than fifty years had died, Hueper typed out his memoirs with some resignation:

> It is not surprising that the promotion and development and the factual evidence supporting the concept of environmental carcinogenesis as the dominant cause of human cancers has aroused more disbelief as well as objections and condemnations than any other concept advanced regarding the causes of cancer in man. [36]

By 1950 knowledge of the ways that sundry physical and chemical agents affect cancer had been around for about 175 years. While Hueper and Kehoe shared a naive confidence that careful, methodical assembly of the facts would lead directly to a better world, they held fundamentally different views of what that world might look like and who should be in charge of making it right. As a result, Hueper ended his career as an alienated outsider. Kehoe remained at the top of his profession, hailed as the founder of modern industrial toxicology. Neither fully appreciated that what passes for facts depends on when and whether those facts are ever permitted to surface, and to whom they are ultimately given.

TIME

The Weekly Newsmagazine

R. H. Hoffmann

DR. CLARENCE COOK ("PETE") LITTLE
"If we can get all the women talking about Cancer. . . ."
(See MEDICINE)

Number 12

Featured on the cover of Time *magazine in 1937, Clarence Cook Little headed the American Society for Cancer Control in 1929, and the American Cancer Society in 1944. In 1954, the pipe-smoking geneticist became the founding director of the Scientific Advisory Board of the Tobacco Industry Research Council, a group that dispensed millions of dollars to leading university researchers for four decades.*

5

Fear Sells

If you want to go fast, go alone;
if you want to go far, go together.
—AFRICAN PROVERB

ONE DAY IN 1991 I received a phone call in my office at the U.S. National Academy of Sciences from a gentleman named Emil Bizub. He had read newspaper accounts of my work showing that cancer rates had increased above and beyond those which could be explained by smoking or aging alone. He thought I should have something he had been saving for more than thirty years.

When his package arrived, it contained a well-preserved copy of *Life* magazine from May 5, 1958. The cover (see page 2) showed a young, dark-haired woman, eyes shut, lying on a table. From above, what looks like a massive cannon pointed to her midsection. The decades-old headline read, "Fresh Hope on Cancer." It could have been written yesterday. Hope is still the ultimate drug for cancer patients.

When I called Bizub to thank him for the magazine, he told me that the woman in the photo was his aunt, Isabelle Messinger. "Look at it," he said. "She was young. By the time they picked up this tumor that started in her cervix, it had spread everywhere. I'm a funeral director. I get to see it all. I see people at the end. Here in New Jersey, cases of cancer keep coming in younger and younger."

Dealing with the deceased, Bizub saw firsthand how those with cancer ended up. "People who are desperate to live a little longer will go through anything. I see people without breasts, or breastbones, or legs or arms. The other day, I had to prepare a woman for burial who had gone through eight operations. She ended up with no legs or breast-bone. This woman would have done anything to live. So would my aunt. And she did. She still didn't make it past forty."

Bizub's aunt Isabelle died of cervical cancer shortly after she made the cover of *Life*. We know now that her death could have been avoided.

At the start of the twentieth century, cervical cancer was a death sentence, and often a secret one. It was not unusual for women to go to their graves never knowing what was wrong. By the time its physical tracks could be felt, the illness had often been brewing for more than a decade. The warning signs of bleeding, shortness of breath and pain—then touted as the indications of cancer—were not at all clear in the disease's early stages. After all, women have been bleeding, sometimes uncomfortably, for as long as the species has existed. By the time the strange discharges, unstoppable cramps, and relentless full-belly feeling were evident, the cancer was unstoppable.

Doctors had long known that cervical cancer did not arise overnight. But it was accepted as an inevitable result of some poorly understood deficiency, perhaps a moral one. Disease of any kind was seen as a result of not trying hard enough to be good enough. Then as now, far more black women got the disease and died of it. The fact that this ailment was uniquely a women's disease that disproportionately affected those then called "coloreds" combined to made cervical cancer less than a high priority for doctors.

More accessible forms of cancer might produce visible or palpable lesions, lumps and bumps that, once found and recognized, gave doctors a fighting chance to cut out the growth and arrest the disease. But the cervix, the most common locus of cancer at the beginning of the last century, was a private place whose means of access was closed to all eyes. At a time when women felt shame at the mere mention of their genitals, the topic of cancer of their private parts was not considered appropriate for public discussion.

The narrow opening to the womb can be examined directly when a light is shone into the open vagina. As early as 1908, the little known

German physician Walther Schauenstein pointed out that the naked eye can discern the difference between a vital cervix and one that would become or already was cancerous. A healthy cervix is shaped like a winking eye, and looks soft, pink and rosy.[1] A sick one can appear gray and full of pus. By 1920 Hans Hinselmann had invented a simple device called a colposcope, basically a low-power microscope on a flexible stick, that could magnify the surface of the cervix. Recognizing the ability to diagnose cancer visually, a group of gynecologists formed the American Society for the Control of Cancer (ASCC) in 1913. They had one simple aim: to persuade physicians to learn how to look at the cervix and persuade women to allow regular exams.

It never has been easy to get women to take their clothes off and lie on their backs with their legs spread apart and their feet in cold metal stirrups. A leading ASCC pathologist, James Ewing, ruefully conceded in 1926 that even if doctors became convinced of its value, the prospects for persuading women to undergo this exam twice a year were slim and daunting.[2] It was not clear at the time whether it was harder to find physicians who understood the value of performing the exam or to persuade women of its value.

In 1928, an enterprising Greek American researcher, George Papanicolaou, found a way to do more than just inspect the womb's opening. Papanicolaou reported that a few cells easily swabbed from the womb's opening could reveal the health of the cervix. "Carcinoma of the fundus [uterus] and carcinoma of the cervix," he wrote, "are to some extent exfoliative lesions, in the sense that cells at the free surface of the growth tend to be dislodged and subsequently find their way into the vagina." Healthy cells come in regular shapes and form orderly patterns, but malignant cells are out of control. When looked at under a typical microscope the difference between healthy and cancerous cells is usually unmistakable.

Papanicolaou was not the first to use the microscope to find malignant cells, nor did he claim to have invented the method. Walter Hayle Walshe, a London physician, had done that nearly a century earlier.[3] The idea of using cervical cells to detect cancer had also been proposed by the Romanian pathologist Aureli Babès in a French journal in 1926.[4] But the Pap smear, as it came to be known, provided an ingenious way to take a cellular snapshot of what was going on inside of the uterus by probing the metabolically active glandular cells that poke into the top of the vagina.

Where cells occur determines not only their shape but the size and density of their nuclei, the command and control centers that house our genetic material. The heart of the cell contains most of the materials that determine how it responds to various assaults. The job of the cell's center or nucleus is to defend the ability of our DNA to repair itself and direct activities that control whether cells remain ordered or go awry. Cells taken from the outermost layer of the cervix, only one or two dozen cells thick, come from just above a basement membrane, beneath which lie deeper tissues that make up the body of the cervix. The surface layer contains large, dense, flat cells that have small centers, while the deeper cells are smaller and more rounded, with fuller nuclei.

At the certain point just before the cervix ends, uterine cells give way to vaginal ones. This zone carries a weighty name: the squamo-columnar junction. Shorter squamous cells populate the vagina, while longer columnar cells line the uterus. The area where the longer cells meet the shorter ones is where the action is. Papanicolaou figured out that cancers tend to arise in this fast-growth zone.

Ordinary cells stay in place and follow well-established orders to stay alive. An exquisite system of cues and cellular signals determines cell growth, size, the nature of their centers and even their time of death. The terms we use to depict cellular derangements reflect the vital importance of order. The path from normal, to dysplastic (meaning literally out of order), to neoplastic (meaning newly disordered cells), to cancer is not simple but appears to be fairly direct. When magnified, cancerous cells, or those likely to become cancerous, present as modern-day equivalents of Hippocrates' disorderly, menacing crab, or *karkinoma*.

Today, when a Pap smear uncovers an abnormality, the deviant cells can simply be taken out, long before cancer has a chance to arise. If the growth has spread, a snippet of tissue can be removed and examined by a pathologist for signs of premalignant changes, including the growth of abnormal blood vessels. If all cells of the cervix have converted to cancer cells, this is called carcinoma in situ and the entire uterus must be removed. But in the 1930s, the medical world was not moved by Papanicolaou's work, and women, black and white, continued to die at alarming rates.

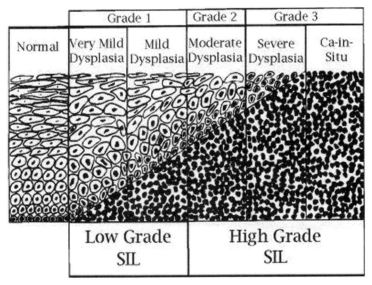

Figure 5-1 *The cells of the cervix send tell-tale signals of their health. Normal cells, seen on the left, are round and regular. Cells that can become or are cancerous take increasingly irregular shapes.*

THE WOMEN'S FIELD ARMY

In 1929, Clarence Cook Little took over the reins of the ASCC part-time. Under his leadership the organization would mount an unprecedented campaign for public awareness of a disease seldom discussed in polite company. A blueblood with lineage going back to Paul Revere, Little was the scion of one of the most prosperous families of Boston. His father, James Lovell Little, was listed in 1882 (six years before Clarence's birth) among the wealthiest citizens of Boston. At the tender age of seven, Clarence showed an unusual knack for studying genetics and bred prize-winning pigeons. Using some of the same techniques of selecting animals with stronger traits that H. J. Müller and others had pioneered with the lowly fruit fly, Little took on the challenge of breeding hardy mice appropriate for laboratory studies. At Harvard he sought to create a uniform strain of mouse that could be tested for response to cancer-causing agents, a research effort he would maintain for the next twenty years. At that point, he espoused the view that studies in laboratory animals would enlighten the study of humans; he once commented that when the history of genetics was better understood, a statue would be erected to honor the lowly mouse.

Little had a meteoric career. After World War I, he helped Margaret Sanger and Lothrop Stoddard found the American Birth Control League, and he became president of the University of Maine at age thirty-three. At thirty-seven, he became president of the University of Michigan, a post he held from 1925 to 1929. While there he attracted controversy over his frankness about sexuality, including the need to teach unmarried men and women how to prevent pregnancy. In his first major address as Michigan's president on November 19, 1925, he aligned himself with those who believed eugenics could be practiced safely. Noting that immigration laws were already protecting the American gene pool from defectives, he urged sterilization of the mentally and criminally deficient.

In the throes of a divorce, he left Michigan in 1929 and founded what is still one of the world's largest colonies of mice for genetics research, at the Jackson laboratory in Bar Harbor, Maine. That same year he became managing director of the ASCC, an appointment that put the pipe-smoking Little on the cover of *Time* magazine.

Within a short time, the ASCC was struggling. During the Depression fund-raising dwindled to less than $50,000 per year. Convinced of the need to broaden the ASCC's reach and realizing the potential of directly engaging women in the effort, Little arranged to meet with the leaders of the General Women's Federation in 1935. By then more than 140,000 Americans were dying of cancer each year, and the disease had become the nation's second leading cause of death after heart disease. The sobering fact was that more women than men were dying of cancer and usually from cancers of the cervix, stomach and breast.

The federation was not a prissy club run by repressed matrons. In the days of racial and religious segregation, its membership included representatives from national councils of Jewish, Catholic and Negro women. Under the leadership of Jane Addams and Julia Ward Howe, the federation tackled some of the most troubling social issues of the twentieth century, including women's suffrage, birth control, child labor, Indian welfare and illiteracy.[5] One can imagine that Little saw them as a group to which he could make his sales pitch with complete frankness.

One of those whom Little impressed was Marjorie B. Illig. Her biography describes her as a "radiologist until marriage." The choices for women in any profession at the time were limited: either be married or have a career. So Illig did each in succession.

The wife of a General Motors executive, Illig chaired the public health division of the General Women's Federation in 1935. As a radiologist, she had seen dozens of young women whose X-rays revealed abdomens riddled with spreading white blotches of disease. Illig explained to her colleagues at the federation that doctors could identify subtle abnormalities of the cervix long before cancer showed up on X-rays. Women's lives could be saved if they would show up for regular physical exams. The suggestion that early detection could prevent deaths provided a radical and hopeful solution for a problem that had come to be fatalistically accepted. If women could be mobilized against this devastating disease, the world would be better off.

Under Little's direction, the ASCC set aside $100,000 to forge a new entity for women only in the struggle against cancer—the Women's Field Army. Illig became its national commander. Organized into paramilitary groups with khaki uniforms, ribbons of rank and ornamental swords, the WFA roared into existence in 1936, giving women a safe and constructive way to express militaristic impulses many perhaps did not know they had.

Practically overnight, ASCC moved from trying to persuade medical societies to take an interest in cancer to commanding battalions against the disease in some thirty states.[6] They targeted not only cervical cancer but breast and throat cancer. In a radio broadcast on February 10, 1936, Rep. Edith Nourse Rogers of Massachusetts and Commander Illig explained the rationale for their military-style campaign.

The title of our program, "Women, Enlist, This Is Your War," may sound a little grim in these days when newspapers and newsreels are filled with stories and pictures of the tragic conflicts in Spain and China. These are wars in which women are indeed playing a part, mostly as victims. But our war is of a very different kind. It is a war to save a human life, a war for health and happiness. We are not using bayonets or tanks or machine guns: our weapons are leaflets and lectures. We are fighting with facts and our military objectives are the putting to rout of fear and ignorance. This war is against one of the greatest enemies of health. It is against cancer.

Under the banner of national honor and raw political power, men had left a legacy of unresolved conflicts throughout the European battlefields of World War I. The Women's Federation bemoaned armed conflict, as did other pacifist groups during this period of growing national and international militancy.

Grace Morrison Poole, adviser to the Women's Field Army and general chairman of the National Federation of Women's Clubs, urged, "Instead of killing each other in military battles as some nations are doing, let us organize to rid humanity of one of its most dreaded diseases. Instead of fighting against humanity, we can war for humanity by enlisting in the cancer drive."[7] The leap from militaristic language and symbolism to a quasi-military battle organization was perhaps inevitable. Rather than passively waiting for the disease to strike and providing solace and compassionate care, the activists behind this army, wrote Poole, would conduct "trench warfare with a vengeance against a ruthless killer."[8]

Recalling the launch of the WFA in his 1939 book *Civilization Against Cancer*, Little wrote:

> To be sure a few overcautious and worried extremists among some of the very valuable groups of pacifists grumbled a bit at the military analogy, but their objections vanished as the movement began to acquire momentum. It was not difficult to make them see that it was far better to turn man's desire for fighting away from his fellow man as an objective and to encourage the development of a war to save lives.[9]
>
> If men would spend their major energy in fighting common enemies instead of undermining each other's stability, we should all be much happier. It seems strange that it must take a ruthless killer like cancer to bring home that fact.[10]

On November 21, 1936, the *New York Times* reported "Cancer Foes Begin Nation-wide Drive." At the time women had few constructive channels outside the home in which to fully participate in politics of any sort. The WFA provided a welcome and probably unique vehicle for their political and social energies. Barred from work outside the home and from military engagement, women willingly joined the WFA for what seemed a safe and vital battle. Little wrote that most of these women had never "possessed an outlet for their desire to fight a disease that they had always hated."[11]

"This is merely the beginning," he told *Time* magazine (March 22, 1937). "There is no longer a need to fight cancer alone. Hundreds of thousands will share the burden, understand the sufferings which too long have seared the very soul of men and women. At a time when our country is inclined to develop class, race or creed consciousness or hatreds the menace of a common enemy and the inspiration of fighting it together may have a sorely needed and deeply significant religious and moral force . . . It is a hard task requiring patience— trench warfare with a vengeance against a ruthless killer. No quarter need be given or asked."

When asked why he felt so deeply about the need to educate the public about cancer, Little admitted that his motivation was not merely scientific, but deeply personal. "Because I have both experienced, understood and, I am afraid, caused too much suffering and hate it. Because my own father died as a result of cancer. Because perhaps whatever ancestral desire I have to explore the unknown is appealed to by the research work and the wish to be a 'crusader,' which almost all of us have, if given a chance to express itself. Finally, because I believe that Americans will be happier and saner if they combine in fighting a scourge like cancer than they will be if they continue to fight each other for money and power."[12]

In this same story *Time* described the launch of the largest social movement ever loosed against the disease, noting that:

Three hundred thousand U.S. women have cancer. Some 80,000 will die of it this year. Some 40,000 need not die of it if they take or have taken advantage of the resources which Medicine has so far marshaled against the nation's second most common cause of death. About six women get cancer to every five men. The most prevalent forms of cancer in women, however, are cancer of the breast and womb, which are the most curable.

The *Time* report included a contemporary-sounding explanation of the ways in which inheritance and environment combine to cause cancer.

> Investigators have at last got a glimmering of what causes cancer. Some people inherit a susceptibility to the disease. But they do not develop cancer unless some susceptible part of the body is unduly irritated by: 1) carcinogenic chemicals, 2) physical agents (X-rays, strong sun light, repeated abrasions as from a jagged tooth), 3) possibly, biological products produced by parasites. Carcinogenic chemicals occur in coal tar, bile acids, female sex hormone. However, no one understands the exact way in which any of these causes cancer in those individuals who are susceptible to cancer.[13]

This view of the numerous environmental causes of cancer and the role of genes quickly faded from public awareness. But within a year, the energized women of the WFA joining forces with the lackluster ASCC proved a masterstroke. Public educational efforts mushroomed. Cancer became a women's issue. Fighting for their lives, the women of the WFA pushed the ASCC to financial and social levels it had never achieved before.

The military metaphors of the enterprise tapped an unexpressed and fierce determination women identified as uniquely their own. Fighting against a disease marshaled militaristic energies that women were not permitted to vent in other arenas. The ferocity of their commitment became legendary. Imagine a bevy of mamma bears on the prowl, alert and primed to attack a hidden enemy. Little reported that women in this army put themselves through daunting conditions to get the message out.

> In one far-western state women snowshoed over mountains and across country for miles to reach small communities with a message of education and hope. In another state ravaged by floods, the captain of a local unit of the army conducted her organization's campaign from a second-story room above the inundated first floor.[14]

At its peak the army numbered over 2 million. Set up as the ASCC's public education arm and fielding battalions in every state, WFA mem-

Figure 5.2 *A Women's Field Army poster from the 1930s appealing to the growing militaristic attitude toward cancer.*

bers marched door-to-door collecting money and handing out pamphlets. They organized public events, canvassed neighbors and friends to raise money and generally spread the word about cancer, with a candor that even today seems refreshing.

The ASCC's motto had been "Fight Cancer with Knowledge." The WFA altered it slightly to "Cancer Thrives on Ignorance. Fight It with Knowledge." Bleeding, lumps and unhealed sores became the subjects of slogans, posters and pamphlets distributed liberally to members and the general public. Posters from the WFA's heyday convey the military imagery and fighting spirit of the enterprise. One of the most popular showed a feminine hand firmly grasping the phallic base of a vertical pointed sword, the tip of which nearly brushes the word *cancer*. If women couldn't have their own weapons of battle, this was one war where they could still grab on to symbols of power.

Making cancer a word people could utter was good but not enough. Once Illig and the WFA got hold of the issue, training physicians and educating the public was no longer a sufficient focus for many of the group's members. They wanted to do something about the disease, not just talk about it. By the end of the decade, WFA interests had expanded to encompass a number of broad and somewhat daring public health concerns. Among their other missions, the WFA sought to provide women with ways to avoid unwanted pregnancies. In an age when public funding for medical care was paltry, the WFA carried out community studies confirming the need for public health programs and advocated group budgeting for hospitals.

The WFA also showed an impressive knack for attracting attention. One exhibit, created by members from the state of Georgia, featured a miniature graveyard with wooden tombstones proportional in size to the number of individuals dying from various causes. A tiny stone marked deaths from infantile paralysis, the subject of a million dollar yearly outpouring in public largesse. The tombstone for cancer was huge, though public funding for research about the disease was considerably lower. Commenting on the exhibit, Little pointed out the importance of crafting compelling messages.

> The public forms its judgment of the relative importance of its health needs on visual and emotional stimuli rather than by the process of thought. Infantile paralysis hits children and leaves very clear evidence of disease. . . . Cancer, too, moves silently and quickly. Its ravages, however, are less obvious because it kills rather than maims.[15]

Through the ASCC and other political and social ties, Little lobbied for the establishment of the U.S. National Advisory Cancer Council in 1937, a group that was headed by the surgeon general, Thomas Parran, and included such luminaries as James B. Conant, president of Harvard, and James Ewing, director of cancer research at the Memorial Hospital in New York City.[16] But creating more national institutions dedicated to cancer came at a price. Folks began to believe the claim that early detection was the key to treating the disease, but such claims raised serious questions. Why were cancer rates increasing when more people were seeing their doctors sooner? Why was so little real progress being made against finding and treating the disease? Why were more and more women dying of cervical cancer? Significant numbers were beginning to question what "cancer control" really meant.

Little was nothing if not confident. He needed no scientific proof that getting large numbers of women to talk about cancer would lead to reducing its toll on society. As he told the *New York Times* in 1937, "We can save many lives, probably upward of forty percent of those who died from cancer each year, if we can educate laymen to visit their doctors at the first indication of cancer." Women, he argued, faced a heavy burden to help save themselves and were rising to the cause.

In another *Times* interview the following year, Little boasted of the WFA's extraordinarily successful membership recruitment. "In 1935 there were 15,000 people active in cancer control throughout the United States. At the close of 1938, there were ten times that number."

In their candor and zeal, Illig and Little were far ahead of their time. A cinema noir–style film produced by the ASCC and the U.S. Public Health Service in 1942, *Enemy X,* features the two of them in a half-hour whodunit.[17] The film opens with smoke drifting upward as a pianist, lit cigarette dangling from his lips, plays a Schumann sonata.

A police inspector interrupts the music, "I'm going nuts! There's so many dying!"

"What are you talking about?" the piano player asks.

The scene shifts. A motionless man sits on a stool slumped forward over a lunch counter. The inspector lifts the man's head back by the hair, exposing an X mark on his forehead. "Would you believe we have fifteen more of these? Nobody knows what's causing it. They all have this X on their foreheads."

Who is the mysterious killer? A disease that is running rampant—cancer. The second reel includes a panel discussion with Illig and Little; each appears physically forceful, a bit overnourished and not easily intimidated. At one point, the male interviewer turns to Illig with a smirk and asks, with an insultingly slow cadence, "Exactly . . . what . . . do . . . you . . . *do,* Mrs. Illig?"

While Illig answers, her questioner peers at her with his hand under his jaw and skeptically pursed lips. [You can imagine that the script directions read, Gaze through the speaker; try to look interested, although she is just a woman.] She replies calmly, "We educate people about how to find cancer. And we press ahead fearlessly to push them to go to their doctors."

Next the interviewer deferentially looks upward and asks Little what doctors can do. Little diffidently replies, "If we find cancer early enough, we can cure it!"

Despite these public displays of conviction, Little was not naive about prospects for a cure. Moreover, the very existence of the ASCC, and much of its early support, derived from the public dread of cancer. ASCC focused on training doctors and providing comfort to those dealing with the disease. Basic research was best left to the presumably natural market of free ideas, unfettered by practical concerns and not shepherded toward specific ends.

By THE 1940s, Little had acquired many hats. He was the managing director of ASCC and director of the Jackson labs. In 1937 he had also become president of the American Birth Control League—a post that allowed him to push some of his ideas about eugenics for the American public. His experience and public profile as one of the nation's best-known cancer experts made Little a perfect candidate to become the first director of the National Cancer Institute. This was also a part-time job. When it started in 1940, the NCI consisted of a few scientists working in universities and set up in labs at the National Institutes of Health campus, then located in downtown District of Columbia. The institute studied a few surgical and pharmaceutical remedies and searched for treatments to help people combat cancer.

One expert meeting on cancer that Little organized garnered this sardonic report in *Time* magazine, October 9, 1944:

> The influence of heredity on cancer was a topic for 40 experts in Bar Harbor, Me. last week. They concluded that they knew very little about it. The meeting was called by Dr. (of Science) Clarence Cook Little, head of the Roscoe B. Jackson Memorial Laboratory where, since its founding in 1929, the devious ways of mouse cancer have been studied. Some things the experts did not know:
>
> How to define the word cancer (they finally agreed that, for the present, a cancer is whatever a qualified pathologist thinks it is).
>
> Why some mice develop cancers when suckled by mice of a cancerous strain.
>
> Whether identical twins are more likely to get identical cancers than ordinary brothers & sisters.
>
> Whether susceptibility to cancer depends chiefly on inherited factors, body chemicals or external environment.
>
> Though the experts emphasized their own ignorance, the meeting brought hope to many cancer researchers: it passed a 14-point resolution outlining a program for coordinated cancer research—the

first time in the U.S. that a concerted attack on the disease has been organized.

The budget of the fledgling NCI never topped half a million dollars under Little's leadership. Meanwhile, the ASCC consistently brought in more funding through its cancer crusade than the NCI managed to get from the federal government.

In medicine as in other fields, chance favors those who are prepared and motivated. Thus it is not surprising to learn that a fundamental shift in the way the world thought about and supported cancer research came not from researchers but from the demands of an irate woman. Before her marriage to Albert Lasker, a man famous in his time as the father of modern advertising, Mary C. Lasker was an accomplished businesswoman. As a young child, she had been deeply upset by a visit she and her mother paid to their family laundress who lay dying of cancer. When her beloved cook, Maria Amosio, developed cancer some forty years later, Lasker was appalled that nothing had changed. She was told nothing could be done to help the poor woman, except place her in a hospital for incurables and the insane.

After her cook died, Lasker took it upon herself to do something about the dismal status of cancer treatment and research. Having learned that the ASCC existed, she marched into Little's office one day in 1943. The two had never met before. Lasker demanded to know how much money the organization spent on efforts to improve treatment of cancer. Little replied, "Nothing."[18] He explained that the ASCC only worked on encouraging people to go to their doctors and teaching doctors to competently look for the disease. His board, consisting mostly of medical experts, didn't really believe that research on curing the disease was a worthwhile investment. They were convinced that finding a tumor early and cutting it out and radiating it would reduce the cancer burden.

Mary Lasker was astonished. Her husband, Albert, had created public relations campaigns that got women to start smoking and consumers to buy cars and radios. She knew that with the right approach, cancer treatment would have to improve. Devra Breslow interviewed Mary Lasker, as part of her study of the early efforts against cancer,

and remembered quite clearly that "Mrs. Lasker was absolutely infuriated that no single institution in the United States had as much as
$500,000 for cancer research, an amount that she said 'wouldn't even
be suitable for an advertising campaign about a toothpaste.'" She gave
the ASCC $5,000 to prepare a pamphlet on the state of existing re-
search and promised she would return. When she did, neither
Clarence Little nor cancer research would ever be the same.

The threat to medical practice

Throughout the 1930s, George Papanicolaou continued to study the
pathology of cancer, preparing slides of cells taken from women's
cervixes. He and his colleagues were achieving what Lasker had asked
of Little—an improved way to keep cancer from killing people. In
1940 Papanicolaou and gynecologist Herbert Traut began training
medical technologists to reliably distinguish healthy cervical cells from
unhealthy ones on a microscope slide. By testing the same women for
two years, they were able to show conclusively that cells from the
cervix signal how much trouble exists and whether surgery is called
for. Changes in the outermost layer of cells, from thin and flat to a
fuller appearance, indicated a quickly spreading cancer. Once detected, the disordered cells could be removed and cervical cancer
avoided. The demonstration that the Pap smear actually worked created a stir in the medical community.[19] The question changed from
whether the procedure should be done to who should do it.

When the Pap smear was first proposed in 1928, medicine was
practiced largely on the basis of belief and tradition. This had hardly
changed by the 1940s. A doctor provided health care through house
calls and private office visits. If he thought a diagnostic test made
sense, he performed it himself. No one conducted scientific evaluations of the effectiveness of physicians' methods.

Thus Papanicolaou and Traut's suggestion that medical technicians—mere laboratory workers—could evaluate Pap smears hit a
major professional roadblock. Surgeons and pathologists were not eager to give up their control over women's wombs. They argued that

only tissue removed from the uterus by a surgeon could reliably indicate potentially cancerous growths. Surgeons, doing a lucrative business removing uteruses, fought bitterly with gynecologists, who thought they should regularly observe the cervix and remove tissue pieces rather than the entire uterus. Big money and big egos were at stake. The private fee-for-service insurance systems advocated by medical societies designated medical doctors as the sole source of testing.

A Pap smear could be performed in a clinic by a technologist and read under a microscope by a lab technician. Neither task required a medical degree. The idea that reading a glass slide might replace the need for surgical removal of a piece of the uterus to determine its health constituted a major threat to surgeons. Furthermore, the notion that public health agencies and nurses could conduct tests, train experts to read them and screen large numbers of people for signs of illness was seen by many physicians as a plot to socialize medicine.

The controversy smoldered throughout the 1950s. On one hand, intransigent, skeptical surgeons sought to maintain their bread and butter biopsy practice. On the other hand, Herbert Traut and a few other gynecologists and researchers felt sure that putting patients with no outward signs of disease through the Pap test would prove life saving.

A number of articles in popular magazines minced no words: if women got cancer, it was their own fault. As one recent historical analysis noted:

> "False modesty," Virginia Gardner reported in 1933, was "in large measure responsible" for the persistently high rates of "cancer of the cervix of the womb." Gardner also blamed the "prudery" of the public for this state of affairs because public lecturers on cancer avoided the topic of female reproductive health even though cancer of the cervix was the greatest "menace" to women. . . . A 1952 *Reader's Digest* image of "false modesty" shows a woman hiding her eyes, overshadowed by the shame of exposure.[20]

From childhood, women had been taught to be ashamed of their reproductive organs and their sexuality, to protect their reputations, and to save themselves sexually for their husbands. Anticancer campaigns told women to overcome these familial and social teachings and to do in doctors' offices what they had been expressly taught not to do anywhere else: lie down with their skirts up, their underwear off and their legs apart, exposed to a man not their husband. It's hardly surprising that women would avoid pelvic examinations. The term "false modesty" belittled female anxieties and denied the sexualized and dangerous meanings of displaying women's bottoms to male eyes.[21]

THE BRESLOW HISTORY carefully tracks the foot dragging and infighting that ensued when various medical specialties began to debate who was best suited to break down this modesty, false or not, and conduct Pap tests. One of the experts Devra Breslow interviewed was Charles Cameron, the research and medical director of the ACS from 1948 until 1958. Cameron questioned whether merely telling people to look for the warning signs of cancer would ever amount to much, arguing that something else had to be done besides shepherding patients into surgeons' offices to have various body parts snipped and radiated.

After Papanicolaou and Traut's demonstration in 1943 that medical technologists could be trained to read Pap smears, Cameron worked directly with Papanicolaou for more than a decade to collect additional evidence and set up studies in several states on the value of the test. This work consistently showed that, contrary to what the surgeons maintained, trained technologists could read the Pap smear effectively and at lower cost. Cameron told Breslow how throughout the late 1940s and early 1950s, universities and public programs in state after state requested the ASCC's and then the ACS's help in training people and developing public health laboratories to carry out cervical cancer evaluations. The requests were always tabled. Believing that physicians alone could read cervix slides reliably, the ACS supported a small number of training programs for doctors only.

By the early 1950s, few physicians had been trained in exfoliative cytology—looking for signs of disorder or cancer in cells scraped from the cervix and placed under a microscope. Efforts to comprehensively institute the Pap smear were further undermined by the slow construction rate of pathology labs. Like many parts of this campaign, this was no accident.

While American physicians were arguing among themselves, ten out of every 100,000 women in the United States died of cervical cancer each year. Farther north, a physician in British Columbia, Arthur Boyes, tackled the problem directly. He screened large numbers of women outside traditional medical offices and proved that these screenings saved lives.[22] Physicians took cervical tissue samples and sent them to a central laboratory for analysis and interpretation, which then provided recommendations about follow-up or treatment.[23] Based on this pilot study, a government program for the early detection of cervical cancer was established in British Columbia in 1949. The soaring cancer rates were stabilized and within three decades had dropped by a third.

British physicians, meanwhile, sided with the American medical establishment throughout the 1930s and 1940s. In the suppressed history he completed in 1977, Lester Breslow wrote,

> The British just did not believe that Pap smears could be carried out by having non-physicians examine cells from the cervix. They felt that having samples leave a doctor's office meant that no reliable information could ever be developed. . . . This physician in British Columbia showed that these fee-for-service advocates were plain wrong. Screening for Pap smears could easily be done by non-doctors, so long as the laboratory was well qualified.

Britain had been hit hard by World War II, physically and economically. In addition, the British approach to public education focused on preserving resources rather than protecting people. As a result, efforts to promote screening and public education regarding cervical cancer were seen as likely to stir up public hysteria and also threaten health care costs. Just after World War II, gynecologist Malcolm Donaldson tried and failed to persuade the British Empire Cancer Campaign (BECC) that its prewar program of lay cancer education should be enlarged to a national campaign. Those in charge of such matters in Britain looked down on their Yankee compatriots who were committed to fighting cancer with public knowledge.

As one scholar of this epoch describes it,

> In Britain, much of the cancer elite—the clinicians, researchers, public health workers, and government officials who made the disease their business—rejected the idea that they should teach the public about cancer symptoms and treatment. . . . The men and women of the British cancer establishment conceived of themselves as managing the frequently irrational demands of a public they characterized as gullible and emotional, a conceptualization of "the public" that they shared with cancer experts abroad and with other medics at home. But many also believed that the public was so irrational about this disease that education—defined as the large-scale mass-media provision of facts about potential symptoms—was counterproductive. What was more, they argued, popular education could only add to the economic and organizational pressures on the NHS, by setting in motion an ill-informed, uncontrollable demand that would overwhelm the services they had labored to establish.[24]

In their refusal to accept Boyes's conclusion that Pap smear screening worked, the British physicians had the support of the man who would soon be knighted for his work on tobacco, Richard Doll. Over his long career, Sir Richard weighed in on nearly all important questions surrounding public health aspects of cancer. In a 1968 report with a physician from Kuala Lumpur, Doll reviewed deaths from cervical cancer in British Columbia and the rest of Canada and found no evidence that the decade-long screening program had saved lives or that medical technologists were as reliable as doctors in providing services.[25] In fact this paper did show a drop in deaths from cervical cancer in younger women, but it was not judged sufficiently convincing at the time. As a result of these views, cervical cancer screening in England was not mandated by the government and was carried out episodically. Deaths from the disease continued to rise.

As evidence mounted from the United States and many different countries of the life-saving value of the Pap smear, England's National

Health Service finally endorsed regular screening for cervix cancer in 1988.[26] In the first ten years of the official British screening program, both deaths and new cases of cervical cancer fell by more than a third. The toll for failing to implement the program earlier—when cervical cancer was often diagnosed at an advanced stage where it cannot be treated effectively—has never been tallied.

BOTH MARY LASKER and her husband, Albert, were at the top of their game as makers of public opinion and knew that modern advertising used proven methods for bolstering sales and markets for all sorts of products, whether radios or motor cars or cigarettes. Why would medicine be any different? Convinced that more money could lead to a cure for the disease, they revolutionized the public discussion of cancer and funding for cancer research.

Mary Lasker would soon learn how right she had been to insist that research would produce fundamental advances in finding and treating the disease. Under Clarence Cook Little's leadership, the ASCC had been funded by thousands of small contributions—typically a few dollars each—raised mostly by the WFA going door-to-door. With commanders and divisions in nearly every state and women physicians in most of the leadership positions, the WFA reached millions of women by appealing directly to their fear of cervical or breast cancer. But the WFA was largely devoted to public education; neither it nor the ASCC had done much to promote research or control the causes of cancer. As the Laskers saw it, this folksy approach would never raise the serious money needed to support a big-time research effort.

After challenging Little in 1943 to come up with ways to treat the affliction that had killed her cook, Mary Lasker became deeply involved with ASCC. She got herself onto the board of directors through her social skills and ready checkbook. Within a year Albert Lasker had placed his protégé Emerson Foote, himself a captain of modern advertising, on the ASCC board alongside Mary. Foote's first major initiative was a name change: ASCC became the American Cancer Society. If the earlier name had signaled an intent to contain or limit cancer, the new name conveyed a more direct and powerful message. The ACS would not seek to control cancer but to obliterate it. Drawing on the militaristic language then coursing around the world, cancer was to be attacked, destroyed, and wiped off the earth. The WFA also got a snappier name, becoming simply the Field Army—a decidedly more masculine entity that incidentally now included men in positions of leadership. Names are not all that was changed. In 1944 the rebranded ACS raised more money than ever before—more than $800,000, equivalent to nearly $9 million in today's dollars. It eventually covered seven national regions and had fifty-nine divisions in forty-eight states and the District of Columbia.

Lasker's price for involvement with the ACS was simple: out of every $4 raised the organization would spend $1 on research. Today's ACS, according to its own reports, devotes only 10 percent of its money to independent research—a fact the organization does not tout.[27] Eric Johnston, a leader in the Hollywood film industry, was brought on board to head up the 1945 ACS fund-raising drive, the first to feature professional fund-raisers. That year's drive netted more than $4 million. Just as Lasker had demanded, about $1 million of this was set aside for research.[28]

Scientists like to think of themselves as free spirits motivated solely by the pursuit of knowledge. Any practical benefits of their findings are either remote or beside the point. Little shared the scientists' perspective. "Few research workers," he wrote, "today expect to discover any startling panacea. They have given up the dream of becoming the dashing leader of a triumphant whirlwind charge to victory. They realize that they are in for a life of grim trench warfare, of hard-earned minor advances and of information slowly gained and painstakingly recorded and analyzed."

But this slow, methodical approach did not play well in the business world. Appalled by Little's passive attitude, Lasker and her friends took him on. Whatever Little's faults, he was regarded as one of the country's top cancer doctors. He had created the field and had survived contentious technical debates. He could not have imagined what was coming.

Before the Laskers became engaged, physicians and scientists had called the shots at the ASCC. Lay leaders, such as there were, played supporting roles to the haphazard management ability of doctors, which is legendary even now. But the ineffective and sporadic efforts made by Little and his colleagues to treat cancer appalled Mary Lasker. She was unconvinced by their reservations about the complexity of the problem and appalled by the meager funding they attracted. Under the Laskers, scientists suddenly had bosses who hailed from a world of money and power. Cameron, the former medical chief of the ACS, described the process to historian Richard Rettig, who profiled the early years of the ACS.

> During the war years, the Board was infiltrated by people who were determined that this organization was a sleeping giant. I suppose it started with one individual, Elmer Bobst, coming on board and saying, 'This has a great potential, let's get my friend, Mr. So and So, like Jim Adams.' . . . They got one of their friends after another on the Board. They were all people of tremendous gusto and enthusiasm and style, and most of them had a good deal of influence.[29]

By 1946, half the ACS board consisted of nonscientists. These were not ordinary folks but pillars of the American corporate world. Elmer Bobst was CEO of the New Jersey pharmaceutical giant Warner Chilcott and a major donor to the Republican party. Known as "Uncle Elmer" to Tricia Nixon Cox, he would go on to engineer the nomination of Spiro Agnew to the vice presidency. Eventually he secured presidential backing for independent funding of the NCI.

As the ACS became a domain for titans of industry, the Field Army fell out of step. A largely volunteer network of amateurs and activists, its members said and did things that the ACS board found embarrassing. Some chapters took on the task of providing health care to those who could not otherwise afford it. These efforts sounded socialistic to some board members (as they probably were) and threatening to medical societies determined to hold on to their monopoly on medical care meted out in doctors' offices. After a particularly boisterous national convention in 1946, a decision was made "high up" in ACS to eliminate the irritant "as conveniently as possible." Within five years the Field Army was history, with some of its leaders being quietly absorbed into the ACS.

The WFA legacy remains part of the ACS and cancer fund-raising.

Nothing can match the singular devotion that comes to those who are fighting for their own lives or those of family members. Volunteers, some of whom are persons of considerable personal wealth, do the bulk of fund-raising for the ACS under various banners.

For his part, Bobst regarded scientists as book smart but terribly ineffective: "I decided that the first priority was to move aside the scientists and physicians who were in administrative control of the organization," he told Rettig. "They were good men, but they were not experienced leaders, and they were not getting results. I wanted majority control to be in the hands of qualified lay leaders."[30]

The chief qualification for a lay leader of the ACS was often check size, not a grasp of cancer. Shortly after the unparalleled success of the 1945 fund-raising drive, Little and Bobst became involved in a bitter fight. Convinced that doctors were at best well-meaning bumblers, Bobst demanded that ACS create a governing board headed up by his lay leaders.

Having led the national effort against cancer for more than twenty years and achieved considerable prominence in the process, Little flatly refused to see the ACS become a bastion of the business community. As Rettig describes it, "The acrimonious debate ended when Bobst said, 'Now, Dr. Little, I would like to conclude by saying that this society is too small to have both you and me in it. I intend to stay.'"[31]

Little resigned. The ACS reconstituted itself with an executive committee largely consisting of businessmen. The Laskers and their allies believed the war on cancer was "too important to be left in the hands of conservative physician-scientists."[32]

Cameron's outrage over delays in using the Pap smear in medicine finally boiled over in 1956, two years after Little's departure. Women were dying while doctors debated who should be in charge and proposed ever more elaborate studies on the issue. At the 1956 annual ACS meeting, Cameron demanded that the board of directors stop stalling and put the Pap test into place nationwide. "I hold that we need not wait for more evidence," he declared; "that there is enough evidence on hand to justify taking the position—women over 40, vaginal

smear twice a year. Can we justify any longer delaying a vigorous campaign to press the use of the smear? . . . My conscience and the opinion of those with the widest experience in its use say no."[33]

Astonishingly, the business-focused ACS board was unmoved.
Group health plans had emerged in Washington State and California,
set up by doctors who shared the radical notion that the private practice of medicine could undermine the goal of keeping people healthy.
Dr. Michael Shadid, the head of Seattle's Group Health Cooperative,
denounced those who opposed group plans. He called the AMA the
"American Meddlers Association" and lamented the fact that they had
forced the government to drop plans to include health care in social
security.[34]

The ACS wanted nothing to do with such programs. Every member
of the board remained aligned with those who defended fee-for-
service medicine conducted solely by doctors in their offices. The notion of cancer clinics or centers was considered socialistic and an
affront to the private practice of medicine.

As Breslow notes, the response to Cameron's plea was hardly encouraging: "Reflective of its bias toward traditional fee-for-service
medicine, the Society adopted standards that 'tissue diagnosis, cyto-
logic and histologic, is a professional medical' function . . . and that
adequate tissue diagnosis should be provided 'in cooperation with
pathologists.' The society shunned active support of public or quasi-
public mass screening initiatives."[35] At every step, pathologists and
surgeons on the ACS board referred to "the poorly understood
changes occurring in the uterine cervix" and the need for more research to improve diagnosis of cervical cancer.[36]

While their benefactors dithered, women with the disease continued to report for radiation treatments and surgery. Occasionally these
techniques saved lives, since radiation kills the fastest growing cells.
Patients who survived the heavy doses of radiation tended to be young
and had cells that could bounce back. Bizub's Aunt Isabelle was not
one of them.

Those who guided people in and out of cobalt radiation found the
work predictable and psychologically hard. The treatment rooms were

fully shielded with lead to prevent radiation from penetrating the walls and people nearby. The heavy machine under which these cancer patients were positioned was considered an improvement over the old fluoroscopes because the beams were more focused. Michael Lotze remembers from his early days in training that sometimes a woman would survive repeated radiation treatments with a hole burned right through her body from the front to the back.

A nurse who worked for years transcribing cancer case reports told me that she could always tell what treatment phase a person was in by how they entered the room. "In the beginning," she recalled, "they would come in by themselves, eager to talk and a little nervous about what would happen. They would leave that first session just a little woozy but excited. The next week, they would be walking with a family member on either arm, a little too weak to get all the way there on their own. Then things would slow down. Every week they would come in a bit whiter and quieter. Eventually they'd have to be wheeled in on a chair. I always knew when the end was coming, because finally they'd be laying on a stretcher, sighing sometimes, waiting while we cleared the room and started up the big motors."

She added how committed people were to these treatments. "We all wanted to believe that we had a chance to beat it, if we could just get people through it." She kept two piles of folders on her desk, one for patients still living and one for those who had died, and transferred the folders from one pile to the other with a sense of grim inevitability. "The thicker the file, the worse a person was doing."

"We all wanted so much for people to stay in the living pile. One woman told me all she wanted was to get to the seaside and have some fresh tomatoes. They didn't allow tomatoes then and they were hard to find. She never got a chance."

Some forty years later, this lady asked me not to use her name or mention the place where she worked because it was considered the best in the world for cancer treatment. It probably was.

The ACS launched an effort to promote the Pap smear in 1957, fifteen years after the test had been shown to save lives and nearly three decades after it was first developed. By that time, the Breslow history makes clear, the evidence in its favor had become overwhelming, at least in America. A number of clinics were reporting great success in finding and eliminating precancerous conditions. The organization now touts the introduction of the Pap smear as proof of its pivotal role in cancer research, and it is true that the ACS paid Papanicolaou directly for his work for many years. But rather than forging new territory in the use of his technique, the ACS bowed to the pressures of surgeons and others on its board to continue studying the problem instead of fixing it. The Pap smear was held hostage years after it had been shown to save lives, while the majority of the medical community resisted its use.

By 1978, more than a decade after Doll published his paper doubting the value of screenings by nonphysicians, the number of deaths from cervical cancer had been cut substantially in many nations that started regular Pap screenings. In the dozen countries that had set up national cervical cancer screening programs earlier, rates had plummeted to one-fourth of what they had been before screening. Japan was an early adopter of the Pap smear, initiating it in 1962. Finland followed in 1963. In Japan, rates dropped from 12.1 to 4.0, and in Finland, they fell from 14.8 to 3.4 per 100,000 women.[37] For younger women, rates declined three times more in these countries than in England and its former colonies.

Between the 1960s and the 1980s, the overall rates of cervical cancer increased in the British Isles from 13.1 to 13.8. In all countries where screening programs had been put into place, deaths dropped by more than 50 percent.[38] See Table 5-1.

Once screening began in the United Kingdom, it experienced declines similar to those in the rest of the world.[39]

Richard Doll eventually admitted his error to Lester Breslow. Breslow told me, "Doll sent me a short note, 'mea culpa.' . . . He admitted he had been wrong. The Pap smear certainly could be administered by others besides doctors." But Breslow wryly noted that Doll never retracted his views publicly. "He never did change his written stance."

POLITICS PLAGUE the issue of cervical cancer even today. Many hold strong views on how to best deal with cancers of the innermost parts of women's anatomy. Computers have proved Janus-faced when applied to industrial-scale processing of Pap smears. Controversies

Table 5-1 International Incidence of Invasive Cervical Cancer in the 1960s Compared to the 1980s

Populations, Sources	Period		Incidence Rate per 100,000		Rate Ratio Follow-up / Baseline
	Baseline	Follow-up	Baseline	Follow-up	
Finland	1962–65	1988–93	14.8	3.4	0.23
Puerto Rico	1964–66	1983–87	32.0	11.5	0.36
Connecticut (USA)	1935–49	1983–87	17.7	7.2	0.41
Canada: Alberta, Manitoba, Newfoundland, Quebec, Saskatchewan	1963–66	1983–87	27.1	11.1	0.41
Sweden	1962–65	1988–92	17.8	8.0	0.45
Denmark	1958–67	1988–92	31.3	15.2	0.49
Slovenia	1956–65	1982–87	25.5	12.7	0.50
Colombia: Cali	1962–66	1982–86	75.6	42.2	0.56
Germany: Hamburg	1960–66	1978–79	35.6	20.2	0.57
German Democratic Republic	1964–66	1983–87	36.0	22.5	0.63
Norway	1968–72	1986–92	18.1	13.3	0.13
Poland: Warsaw city	1968–72	1983–87	21.5	16.3	0.76
India: Bombay	1964–66, 1968–72	1983–87	23.7	19.3	0.82
New Zealand	1960–66	1983–87	14.4	12.8	0.89

Singapore	1968–77	1983–87	17.6	16.3	0.93
England: Birmingham	1960–66	1983–86	13.1	13.8	1.06
Scotland	1963–66	1983–87	12.4	13.2	1.07

Because the proportion of cancer attributed to 'uterus not otherwise specified' was 29.3 percent in 1935–39 but fell to 2.3 percent in 1970–73, the baseline incidence rate may be underestimated.

Source: Leif Gustafsson, Jan Ponten, Matther Sack, and Hand-Olav Adami, "International Incidence Rates of Invasive Cervical Cancer after Introduction of Cytological Screening," *Cancer Causes and Control*, 8 (1995): 755–763.

surround the use of computers to read Pap smears and to sign the names of doctors who may not ever have seen the slides to which their names are attached.

Scientists now understand that most cases of cervical cancer are caused by four common forms of the human papilloma virus. The virus is carried by men, who usually do not develop any symptoms. Heterosexual men with the virus can infect women. If they are not treated, women can die when the virus triggers cancer. In gay men who carry HPV, anal intercourse can lead to anal cancer, which can be lethal. Cancers of the larynx may also be caused by HPV. More than one hundred different types of this virus are easily spread through skin and mucosal contact, particularly the sort of contact that takes place whenever bodily fluids are exchanged.

Nearly half of all college students today have antibodies to HPV, according to studies released by the Centers for Disease Control and Prevention. Antibodies can be thought of as locks that the body sends out to collect keys that are coursing through the bloodstream and keep them from turning on the wrong sorts of cells at the wrong time. For reasons that are not fully understood, several of the commonest forms of the HPV virus place women at greater risk of cervical cancer, and probably a number of other forms of cancer as well. HPV can intertwine with the basic building blocks of amino acids that form the backbone of DNA. As a result, cervical cells can be tested not only for their shape and size but for the presence of HPV in their nuclei. Testing the cervical DNA of young women for exposure to HPV can identify those who risk developing cancer later on. Researchers in England and in Pittsburgh are exploring the use of broccoli-based nutritional therapies to reverse and repair the viral incursion into DNA.

Scientists have now developed a vaccine that prevents the major forms of HPV infection from taking root in the first place. Trials of this vaccine have shown promising results. It was tested on fewer than 20,000 women between the ages of 15 and 25. It is being recommended, and may be required, in pre-teenage girls, very few of whom were in these trials.[40] Despite the lack of experience with this vaccine in young girls, the Centers for Disease Control recommend that eleven- and twelve-year-old girls be vaccinated against HPV, much as they are vaccinated against polio, diphtheria and typhoid. But what about boys? Why are we prepared to vaccinate young girls but not young boys who can spread it? There are good reasons to think that vaccinating boys would prevent anal cancer in those who are gay, could also reduce the heterosexual transmission from men to women, and might also lower the risk of head and neck HPV-related cancers in men and women.

Groups that are organized to protect the purity of young girls are incensed. Not surprisingly, no one is expressing concerns about the health of gay men or about prospects for reducing laryngeal cancer. Instead it is argued that vaccinating against HPV will open the door to wanton sexual behavior among girls, because, among other things, they will not need to think about whether they might contract cancer in thirty years. I've got news for them. With two out of every three girls having sex by age eighteen, and with close to half of them not necessarily willing participants on each occasion, it makes little sense to suppose that decisions about sexual activity are based on the possibility of getting cancer in middle age.[41] It makes even less sense to propose broad vaccination programs without setting up a system to carefully monitor for any problems that might occur.

Oklahoma's Sen. Tom Colburn urged that the only way to protect against HPV and other sexually transmitted diseases is abstinence. In an interview with CNN medical correspondent Elizabeth Cohen, Colburn admitted that the vaccine could lower the risk of HPV. But, he warned, "it will not eliminate that risk. It does not reduce your risk for HIV or any of a number of other sexually transmitted diseases. Abstinence until marriage and fidelity within marriage is the best form of sexual health."[42]

National and international surveys reveal that HPV remains a common sexually transmitted disease. The most common variants of the virus lay behind seven out of every ten cases of cervical cancer in industrial countries and probably even more in the rapidly developing regions of the world. Even today, 4,000 women in America—many more blacks than whites—die each year of a cancer that is believed to be largely avoidable. We do not know why older black women have three times more cervical cancer than white women and die of the disease more often.[43] In North Carolina, there are differences in the disease that tell us the legacies of racism are not over. The State Department of Health reported that from 1995 to 1998, proportionally fewer black women under age forty-five developed cervical cancer, but when they did, they were three times more likely to die of it than their white counterparts.[44] More white women with cervical cancer received surgery, while more black women got radiation treatments.

Racial differences in cervical cancer are unlikely to have anything to do with the biology of race and everything to do with the social and economic experiences of black women in America today. Some detailed genetic studies have indeed found that certain genes are turned on in those who develop advanced disease.[45] But these genes are no more common in Tanzania, Germany, Norway or Japan than they are in Sweden or England. And we still can't be sure whether the genes found to be active in advanced cervical cancer are a result of how the disease has changed the body, or whether they were there all along and led to its development.

In America, for breast and colorectal cancers as well as cervical cancer, black women tend to be diagnosed at a later stage of disease, receive more radiation and less surgery, and die more often. The ability to get medical care of any kind declines for those who live in the rural South. Even in urban areas today, recent cuts in basic medical services mean that fewer women will receive Pap smears. Access to care for all working-class people remains imperiled as a result of ever-shrinking federal funding.

The situation in developing countries is much worse. Cervical cancer remains a death sentence in much of Africa, China and India. In the parts of the world where the most women live, the disease is not found in time for surgery and radiation. Such remedies remain the province of the wealthy. For most women around the world today, providing an inexpensive vaccination may offer the only hope. But if some American legislators have their way, that hope will never materialize.

In a piquant comment on the conflict over HPV vaccine, Katha Pollitt wrote in *The Nation,* "I remember when people rolled their eyeballs if you suggested that opposition to abortion was less about 'life' than about sex, especially sex for women. You have to admit that thesis is looking pretty solid these days. No matter what the consequences of sex—pregnancy, disease, death—abstinence for singles is the only answer. Just as it's better for gays to get AIDS than use condoms, it's better for a woman to get cancer than have sex before marriage. It's honor killing on the installment plan."[46]

6

Making Goods
out of Bads

Great is the power of steady misrepresentation.
—CHARLES DARWIN

DELAYING THE USE of the Pap smear was not the worst legacy of the efforts against cancer in the 1950s. That honor must go to what the ACS, the AMA, and the American and British governments did to prolong and exaggerate scientific disputes about the harms of tobacco.

During the Nazi era, the Germans tried and failed to enact stringent policies to control tobacco. Their conquerors, despite having full access to German research on tobacco, didn't even try. In the United States and England after World War II, radio, television and print revenues depended heavily on tobacco advertising, and seven out of every ten men smoked. For men and eventually for women, smoking was seen as a sign of freedom and even, for a while, as a healthful habit. Like many of the strange stories in this book, the failure of democratic societies to tackle one of the most obvious and dangerous hazards of the modern world was no accident. It was the result of a deliberate strategy to manufacture and magnify public doubt about scientific evidence.

Behind every public health datum lies a sick or dead husband, father, brother, mother, wife, daughter or child. Whether the details of their lives that gave rise to their illness are ever examined depends on a complex set of social conditions and, as we have seen, is strongly influenced by economic interests.

Often you wonder what took so long. For studies of tobacco's impact on our health, an emphatic and much ballyhooed interest in gathering essential information on the subject has often been the last resort of those who in truth couldn't care less about public health. It is always easier to talk about a problem than do something about it. For the tobacco industry, creating a protracted and mostly artificial scientific debate about the dangers of its product was a brilliant business strategy. Public relations is all the more insidious when it masquerades as science and invokes the virtues of "balance" and "keeping an open mind."

Working first with medical experts, like the pipe-smoking Clarence Cook Little and many other heavy smokers, the tobacco strategists counted on their ability to hire leading scientists who did not want to believe that smoking was harmful. With such an impressive front line, tobacco sympathizers carefully crafted doubt about what evidence is required before we can say that a given agent truly is a true threat to human health. Their confidence was well placed, for two reasons. First, the American public has never been especially savvy about science. Second, with seven out of every ten American men addicted to cigarettes, including many scientists themselves, millions supposed that smoking really couldn't be that bad. By cleverly crafting confusion about what scientific research really shows about the dangers of tobacco, the industry created the standard for what constitutes proof of human harm about any public health hazard. It hardly needs saying that this standard serves industry better than it serves the public health.

Modern society depends on doubt. Insecurities about what we look like and how we feel keep people spending money on things they basically know provide neither eternal youth nor beauty. The selling of tobacco as something that smart, sexy people had to have was integral to the birth of modern advertising. By 1937 nearly half of all American families had a radio in their home or car. Entire families sat listening to catchy advertising jingles coming over the radio. People organized their lives around programs like Groucho Marx's comedy quiz show *You Bet Your Life*. Friends and relatives knew better than to call when it was on. At his peak in the 1950s, the comedian Jack Benny regaled half the nation with his dry wit and musical talent, brought to us all by Lucky Strike cigarettes. Benny's renown was so great that when the American Tobacco Company sponsored his violin concert at New York City's Carnegie Hall, the event was covered by the *New York Times* (whose music critic famously noted, "Benny played Mendelssohn last night at Carnegie Hall. Mendelssohn lost").

Whether or not Mendelssohn lost, the tobacco companies made out big. Through programs like Jack Benny's, Luckies were sold as providers of sex appeal to slim, sleek bodies, "so round, so firm, so fully packed"—as the slogan went—just like the cigarettes they smoked supposedly to maintain that vigor.

Democracies rest on an informed public. People can't be informed about things they don't understand. In deciding how to live and what to buy, the public turns to those they believe are experts. The dean of modern public relations, Edward Bernays, pioneered the introduction of tobacco experts, including the otherwise obscure, chain-smoking British statistician R. A. Fisher, who was recruited as a consummate authority who disputed the dangers of smoking.

With a bevy of scientific eminences to invoke, Bernays pushed the use of science-clad ideas in advertisements. One of his first successful forays was to present smoking as a solution to being fat. A keen social observer, in the 1920s he capitalized on the recent achievement of women's right to vote by expanding what they should consider their social rights. He masterminded brilliant efforts to make cigarettes the sign of both freedom and fitness. At one point, he hired young women models, described in a press release as women's rights marchers, to parade down New York City's Fifth Avenue with their "Torches of Freedom." He then choreographed a classic photo op for the press, featuring the attractive young women puffing away on their newfound freedom—the liberty to smoke in public.

One famous ad for Bernays's client Lucky Strike read, "Reach for a Lucky instead of a sweet." The candy companies struck back. They charged that smoking was more dangerous than being chubby. In response, Lucky Strike provided a scientific-sounding riposte that was unsupported by any real science: "The authorities are overwhelming that too many fattening sweets are harmful and that too many such [sweets] are eaten by the American people."[1] Just who were these authorities? Department chairs at prestigious universities, whose eminence in science and medicine gave them the appearance of expertise.

The basic operating plan was straightforward: get people who are trusted as experts in their field to take your position, and the rest will follow. The best public relations work, according to the wisdom of the trade, appears not as PR but as news and expert opinion.[2] Some ad campaigns played to America's pride in being strong, independent and expertly informed. One Lucky Strike radio commercial featured the folksy twang of a tobacco auctioneer, as background music to the discussion of "expert opinions" on the merits of Lucky Strike tobacco.

"Bidabidabye. Bidabidabye. Whatamibid? Bidabidabye? Whatamibid?" So goes the auctioneer's hypnotic refrain as he stokes the price of whatever he is selling. "Bidabidabye. Bidabidabye. Whatamibid?"

While this chant continues as a lull, a confident, soothing male voice enters over it:

> It sure makes me feel at home to tune in on Lucky Strike programs and hear the auctioneer crying out tobacco bids. And I heard something else on a Lucky Strike program the other night to make me feel right at home. That was when the announcer told me how *independent* tobacco men pick Luckies for their cigarette. You see, *I've* been an *independent* buyer for years and *I've* smoked Luckies for years, too.

Then the announcer chimes in:

> Thank-you, Mr. Valentine. It's interesting to get such reactions to the statements we make over the air from the tobacco experts themselves. For these statements are not claims but facts, backed by sworn records. . . . Remember that when you ask for cigarettes. Remember—the men who know tobacco best smoke Luckies 2 to 1!

Half a century ago the investigative journalist George Seldes noted that the entire commercial was based on trumped-up information portrayed as fact. The purported scientific statements about what "independent" tobacco men prefer were completely made up. From making up science for ads, the industry easily moved to influencing real science. Over the next four decades, tobacco money supported a vast and costly series of investigations by some of the world's leading researchers in major academic institutions. By the time Groucho Marx, Edward R. Murrow and many other chain-smoking raconteurs had died of lung cancer, the habit had become deeply ingrained throughout the land.

Having made explicit or implicit health claims about its product for decades, and seeing those claims endangered by widely reported studies coming out in the late 1940s and early 1950s, the tobacco industry hit upon the strategy of fanning both public and scientific skepticism about any reports that cigarettes could be dangerous. In a move of staggering cynicism, the industry secretly funded studies on the hazards of chemicals, hoping to divert attention from tobacco. The chemical industry, in turn, engaged in similar tactics, seeking to focus interest on tobacco as a cause of poor health. It is not surprising to learn that few studies on these two hot-button issues could elude the economic interests that created the impetus for such research in the first place.

Raymond Pearl published a stunning analysis in *Science* magazine in 1938 showing that, in contrast to those who didn't smoke at all, heavy smokers lost about a decade of life. But this statistical gem remained little known in the days before electronic media.[3] One of the earliest North American studies on the hazards of tobacco was produced in 1947 by the Canadian physician Norman Delarue, working in St. Louis, Missouri, with Evarts Graham. A top surgeon of the day, Graham achieved fame in 1933 for the first successful removal of a cancerous lung. Frustrated by his failure to come up with an effective treatment for the disease, he became interested in its prevention. Delarue proposed comparing people with the disease to others lacking it. Looking into the habits of fifty patients with lung cancer and fifty hospitalized with other diseases, he found that nearly all of the lung cancer patients smoked. Only half those without the disease had a history of smoking. Graham was not convinced by this work.

At the time, most men, including most cancer researchers, smoked. Surgical conventions were full of smokers. A chain-smoker who depended on large doses of nicotine, Graham would not accept that cigarettes could be a hazard. But a persistent medical student, Ernst Wynder, would soon change his mind.

Born Ernst Weinberg in Herford, Germany, in 1922, Wynder fled the Nazis and arrived in America in 1938. He served in Germany as a U.S. Army intelligence officer from 1943 until 1945. It is quite likely that his intense interest in smoking developed long before he entered medical school in America, when he lived in his native country. As a young boy in Germany, he could not have escaped the posters and slogans warning Germans of the dangers of smoking. As an army intelligence officer returning to Germany with a keen interest in medical matters, he could certainly have learned of the Nazis' stunning research on smoking and health.

In his American studies, Wynder brought tremendous zeal to his efforts to pinpoint the damaging impact of tobacco smoking. At medical school in St. Louis, he persuaded Graham to let him go through the medical and military histories of lung cancer cases and compare them to those who had died of other diseases. This is precisely the technique that had been developed by German researchers in the 1930s.

Harold Dorn, a U.S. government scientist, took advantage of the voluminous records on military personnel and came up with another approach to clarify the role of tobacco for health. He reviewed files on 198,000 men (and a sprinkling of women) who served in the military between 1917 and 1940. Whenever a death benefit claim was filed, investigators double-checked the primary cause of death with the physician signing the death certificate, and if possible with the results of postmortem examinations. Many veterans had become addicted to cigarettes after receiving them in their regular rations during World War I. It turned out to be a tough habit to break.

Here again, the analysis was clear—those who smoked the most had the greatest risk of lung cancer, as well as higher risks of oral cancers. Men who smoked at least two packs of cigarettes a day had twice the death rate from cancer as those who didn't smoke at all.

Wynder's studies of Graham's patients would change Graham's views and those of many others. Out of more than six hundred men with lung cancer, nearly all had been heavy smokers. Their work, the first large-scale case-comparison study in English linking smoking and lung cancer, was published by the *Journal of the American Medical Association* on May 27, 1950.

An analysis by Morton Levin and others at the New York State Health Department had been withheld from publication because the journal had doubted their results. But when the Graham and Wynder report came in, the journal decided to include Levin's work in the same issue. Their study looked at patterns of lung cancer in men at Roswell Park Memorial Institute and found that the more a man smoked, the greater his risk of lung and oral cancer. That September, writing in the *British Medical Journal*, Richard Doll and Bradford Hill compared British physicians who did not have lung cancer with those who had the disease. Their analysis rested on interviews with the physicians or with their surviving family members and was corroborated by the findings of Graham, Wynder and Levin. The more a man had smoked, the greater his chance of getting lung cancer.

All of these studies used methods strikingly similar to those employed by Müller, Schoniger and Schnairer in Germany more than a decade earlier. Only Doll and Hill acknowledged Müller's work. Unlike the German studies, which urged that action be taken against tobacco, none of the later papers mentioned the matter. Other work soon repeated the German, Japanese and Argentinian animal studies that had produced skin tumors on rodents painted with cigarette tars.

Doll was a formidable presence even as a young scientist. The son of an affluent family, he had the demeanor of someone accustomed to the finer things in life, and the conversational style and voice of a man whose words were often heeded. His work on tobacco made him one of the preeminent public health authorities of the day; because of it he was appointed an Officer of the Order of the British Empire in 1956 and knighted in 1971. A member of the inner circle of British public life for more than half a century, he wielded international influence throughout his professional career.

Doll stopped smoking in 1950. By the time Graham managed to quit, in 1953, it was too late. He died four years later from lung cancer. How did the ACS respond to this growing proof of the dangers of tobacco? Let's just say they were hardly fast on the trigger. This probably had a lot to do with who was running the show.

The Breslow history reveals that the leaders of the ACS in the 1950s included W. B. Lewis, vice president of the tobacco firm Liggett & Myers, and Emerson Foote, a founder of Foote, Cone & Belding, one of Madison Avenue's leading advertising firms. The ACS showed little enthusiasm for British or U.S. studies connecting smoking and cancer. They were even less enthusiastic when important evidence on the dangers of smoking came from within ACS itself, courtesy of its first chief of epidemiology, S. Cuyler Hammond. But behind the scenes, the accumulating scientific evidence and the growing number of public figures who were dying of lung cancer were beginning to persuade other board members.[4]

When Hammond joined the ACS in 1947, the first task he set about addressing was the need for a standard system for classifying and recording cancers. Researchers had long recognized a paradox: the better the medical care in a given region, the higher the recorded rates of cancer deaths. The only way to figure out whether this was due to a real rise in the disease or simply the result of better record keeping was to develop national standards for counting cancers overall.

Once these standards were in place, Hammond and his team turned to the issue of how to understand patterns of cancer in humans. Recognizing the need to gather information on the lives of millions of people and well aware of the limits of funding, Hammond came up with a brilliant way to make use of ACS's legions of earnest volunteers. Over a period of four years, volunteers—most of them women—were carefully trained in how to gather and record information and then sent out to query their neighbors on their occupation, lifestyle and smoking habits. The effort was not without risk.

The ACS board did not object to standards for counting cancer, but some of its members were not happy with what the use of these standards revealed. In 1950, the Breslow history recounts, Hammond presented his program's first findings to the board: military veterans who smoked bore heavy risks of lung cancer. The businessmen were not pleased. According to one of his family members, Hammond was threatened with financial ruin if he released any of this work. Having seen the previous director of the ACS, Clarence Little, summarily canned, he knew this was no idle threat.

Still, Hammond apparently decided that unpublished findings were better than none: he kept going. In 1951 he gave 22,000 of the society's mostly female volunteers something to do that was far more important than asking their neighbors for quarters and half-dollars. Within four years, they had assembled detailed information about the smoking habits of 187,766 men in nine states and confirmed that those who smoked the most were most likely to die of lung cancer and heart disease. Like the 1950 studies, this massive survey provided compelling evidence of what many had suspected—and some, like Pearl, had reported—decades earlier.

Hammond had firsthand experience with the pressure to find ways to make the deadly effects of smoking look incidental. He certainly understood the deeply disturbing implications of this work. Nonetheless, he pleaded with ACS board members to take a stand against tobacco. Finally, in 1954, as many other studies emerged with similar findings, the ACS reluctantly adopted a resolution admitting that "present evidence indicates an association between smoking, particularly cigarette smoking, and lung cancer and to a lesser degree, other forms of cancer."[5] Hammond was finally given permission to publish his findings showing that tobacco caused lung cancer and many other health problems, so long as he included a litany of reservations about how the association might be tempered by air pollution, workplace dust and other things.[6]

Despite this weak admission of the possible cancer hazard of smoking, the ACS's message, as reflected in the highly qualified materials that Hammond was allowed to publish on the subject, remained consistent for years afterward: more data were required on more people in more different situations before any firm conclusion could be reached. Given that more than half of all men at the time were smokers, including businessmen and doctors, the interest in finding flaws in studies of the hazards of smoking—especially those that pointed to other exposures that doctors and businessmen did not have—proved irresistible. The media's passion for nicotine was financial as well: cigarette advertising was an immense business. Tobacco companies were the largest source of advertising revenues for both print and broadcast media throughout the 1950s, surpassing even cars. This discouraged much (though not all) media interest in the story.

B Y 1954, the year of the ACS's irresolute statement on the hazards of tobacco, the pipe-smoking Clarence Cook Little was tired. Shut out of the ACS as it soared to new financial heights, he found himself an outsider in the world of cancer research he had helped build. Then he had an epiphany: the tobacco industry had a growing problem, and he could be the man to solve it.

Dozens of articles had appeared in the *New York Times, Good House-keeping, Reader's Digest* and elsewhere, reporting new findings on the dangers of tobacco in humans. In addition, studies showing that tars created skin cancer in animals also were widely mentioned in popular magazines. The tobacco industry was alarmed. Sales began to dive. They desperately needed to combat the impressions that these reports created. In their exposé on the subject, John Stauber and Sheldon Rampton describe what went on behind the scenes.

> The tobacco czars were in a panic. Internal memos from the industry-funded Tobacco Institute refer to the PR fallout from this scientific discovery as the "1954 emergency." Fighting desperately for its economic life, the tobacco industry launched what must be considered the costliest, longest-running and most successful PR "crisis management" campaign in history. In the words of the industry itself, the campaign was aimed at "promoting cigarettes and protecting them from these and other attacks," by "creating doubt about the health charge without actually denying it, and advocating the public's right to smoke, without actually urging them to take up the practice."
>
> For help, the tobacco industry turned to John Hill, the founder of the PR megafirm, Hill & Knowlton. Hill designed a brilliant and expensive campaign the tobacco industry is still using today in its fight to save itself from public rejection and governmental action.[7]

The new strategy can be summed up very simply: create doubt. Be prepared to buy the best expertise available to insist that more research is needed before conclusions can be reached. Whenever new studies emerged on the hazards of smoking, the tobacco industry would flood reporters' in-boxes with counterarguments asserting that nothing had been proven. It would marshal its own experts to magnify the appearance of a scientific debate long after the science was in fact unequivocal. John Hill's brilliant innovation remains a staple for those who would fight the conclusions of science even today. From the "debate" over global warming to the "debate" over the theory of evolution to the "debate" over occupational causes of cancer, his legacy of selling doubt and using science to undermine any proof of harm is all around us.

Little, who once had impeccable scientific credentials as a leading genetics researcher, proved invaluable to this strategy. The former managing director of the American Society for Cancer Control and the first head of the U.S. National Cancer Institute became the founding mastermind of the Tobacco Industry Research Council (TIRC). The TIRC sent millions of dollars to universities, the ACS and the AMA to study tobacco science. The supply of dollars perpetuated the demand for research. Serious scientists lined up to propose studies that could be funded by the TIRC. So long as scientists could be found to say that research on the hazards of tobacco was still needed—a necessity, of course, in any request for funds—the industry could claim that the issue was not settled. The TIRC directly underwrote many of the world's top scientific institutions, which in turn eagerly endorsed the critical need for continued research. Among the luminaries who served on its scientific advisory board during the ensuing decades were Howard B. Andervont, scientific editor of the *Journal of the National Cancer Institute;* Michael J. Brennan, president and medical director of the Michigan Cancer Foundation; William U. Gardner, E.K. Hunt Professor of Anatomy at Yale University School of Medicine; and Peter M. Howley, of the NCI pathology lab.[8]

An inveterate pipe smoker, Little radically changed his tune from the days when he advocated studies in animals as a vital part of cancer research. He had once written that historians of cancer research would someday erect a statue to the well-bred mouse as the model of study. Now, as chief of the TIRC, he argued that research involving even the pure animal strains he had helped develop at the Jackson laboratory was of little use for understanding human cancer. He noted that "there have been many experiments here and abroad, and none have been able to produce carcinoma of the lung in animals."[9] What he neglected to say was that when exposed to the levels of tars and smoke that humans took in regularly, most mice died. Dead mice don't get cancer.

Laboratory studies under controlled conditions are what we use to make new drugs and understand how they are distributed throughout the body. But identically designed experiments, when applied to the hazards of tobacco or chemicals, were dismissed as irrelevant to humans. After all, so the claim went, there are myriad differences between the rodents in which toxic agents are studied and the people who will ultimately be exposed to them.

Who better than Little, the man who had bred the standard mouse for cancer research, to take up this argument? His dismissal of animal research proved invaluable for the tobacco industry. Little was correct. It was hard to induce lung cancer in dogs or rodents using the crude techniques of the day. Among other things, rats are what are called obligate nasal breathers. This means that unlike people, who inhale deeply through their mouths and can hold their breath for long periods of time, rats take air or smoke into their bodies solely through the nose. This probably gives them some protection from agents that their otherwise tough constitutions cannot handle. Not until a way was found to place tars and smoke directly into a rat's lung, either by placing a hood over its entire breathing zone or by installing a small tube into its lungs, were cigarette residues shown to induce cancers in each and every rat that had such exposures. But in 1954 those experiments were still a decade away.

In the meantime, belittling animal findings paved the way for public acceptance of a very dangerous notion—only human studies could provide proof of harm. Here too there was an ingenious counterargument. In Little's skilled hands, the absence of definitive proof that tobacco and many other substances actually produced illness in humans became evidence that no such harm existed. Even if most men with lung cancer had been smokers, he cleverly pointed out, not all who smoked contracted the disease. Thousands of doctors advocated smoking as a means of controlling weight. Magazine advertisements touted the sex appeal of cigarettes. The AMA and the ACS refused to take any firm position on the dangers of smoking. This symphony of expert skeptics allowed tobacco advocates to perpetuate a culture of doubt about the risks of their product.[10]

Charles Taylor, the former medical director of the ACS, told the Bres-lows that Hammond was nearly broken by pressure from the ACS board to refrain from taking a public position on something he had shown to be a major danger to public health. In meeting after smoke-filled meet-ing with the board, Hammond's pleas were repeatedly dismissed. At one such session in 1957, Taylor recalled, "I can remember Cuyler standing up at a meeting and pleading, 'these lives are on my conscience.'"[11]

The tobacco industry circled its wagons brilliantly. Realizing that the best defense is a good offense, the TIRC eagerly pointed to other culprits for lung cancer, like air pollution and workplace dust and fumes. They were able to enlist some surprising allies in the process. Besides Little there was Wilhelm Hueper, the first director of environ-mental carcinogenesis at the National Cancer Institute and hardly a friend to industry.[12] In 1972 the renamed Tobacco Research Council gave money directly to Hueper and a collaborator, Tom Mancuso, to explore environmental and occupational causes of cancer.[13] As late as 2005 the website of RJR Nabisco—parent company of R.J. Reynolds Tobacco—boasted that:

> Dr. W.C. Hueper, chief of the Environmental Cancer Section of the National Cancer Institute, says the experimental application of tobacco tar to the skin of mice and rabbits has produced contradictory results in the hands of various investigators. He suggests that studies by Wyn-der and others may have involved strains of animals "with an excep-tional receptivity to cigarette tar not possessed by the average member of the species."

The site goes on to note Hueper's reservations about the cigarette smoke theory of lung cancer:

> It is apparent that any final decision concerning the relative role of cig-arette smoking in the causation of cancer of the human lung should be kept in abeyance until a great deal additional study and more valid and, especially, medically conclusive evidence becomes available. . . . The data . . . make it unlikely that cigarette smoking represents a major factor in the induction of lung cancer and in its recent phenomenal rise in frequency.[14]

The text that appeared on this site in 2005 does not mention that Hueper made this statement in 1954: you have to go to Hueper's unpublished autobiography to find that.

The tragedy of many of these arguments is that there were technical merits to them. Air pollution and mining do increase the risk of lung cancer. But that was of no concern to the tobacco strategists. Their goal was simple: confuse people about what scientific evidence shows and keep them wondering whether science can ever answer these questions. Preventing the AMA from taking a position on tobacco throughout the 1950s was a feather in Hill & Knowlton's cap. The creation of the TIRC was another masterstroke. Whatever the technical validity of the TIRC's criticisms, their broad distribution served to foment public skepticism about scientific reports concerning the dangers of smoking. In the first issue of the industry magazine *Tobacco News*, the institute's president wrote, "The Institute and this publication believe that the American people want and are entitled to accurate, factual, interesting information about this business [tobacco] which is so important in the economic bloodstream of the nation and such a tranquilizer in our personal lives."

This demonstrates another aspect of the industry's new strategy: issue loud, frequent, authoritative-looking proclamations of its desire to discover the truth even while suppressing it. Taking a page out of the Laskers' book on public relations, the tobacco companies issued what was published in many newspapers as the "Frank Statement." On January 4, 1954, an impressive and official looking full-page advertisement appeared in the *New York Times* and other major newspapers around the country. In it, the industry attempted to minimize reports that smoking caused lung cancer in humans by invoking evidence from other scientists, and promised that it would directly address concerns about cigarettes. It announced the establishment of a blue-ribbon research committee to explore "all phases of tobacco health and safety." The effect was to prolong public uncertainty about the dangers of tobacco.

RECENT REPORTS on experiments with mice have given wide publicity to a theory that cigarette smoking is in some way linked with lung cancer in human beings.

Although conducted by doctors of professional standing, these experiments are not regarded as conclusive in the field of cancer research. However, we do not believe [the] results are inconclusive, should be disregarded or lightly dismissed. At the same time, we feel it is in the public interest to call attention to the fact that eminent doctors and research scientists have publicly questioned the claimed significance of these experiments.

Distinguished authorities point out:

That medical research of recent years indicates many possible causes of lung cancer.

That there is no agreement among the authorities regarding what the cause is.

That there is no proof that cigarette smoking is one of the causes.

That statistics purporting to link cigarette smoking with the disease could apply with equal force to any one of many other aspects of modern life. Indeed the validity of the statistics themselves is questioned by numerous scientists.

We accept an interest in people's health as a basic responsibility, paramount to every other consideration in our business.

We believe the products we make are not injurious to health.

We always have and always will cooperate closely with those whose task it is to safeguard the public health.

For more than 300 years tobacco has given solace, relaxation, and enjoyment to mankind. At one time or another during those years critics have held it responsible for practically every disease of the human body. One by one these charges have been abandoned for lack of evidence.

Regardless of the record of the past, the fact that cigarette smoking today should even be suspected as a cause of a serious disease is a matter of deep concern to us.

Many people have asked us what we are doing to meet the public's concern aroused by the recent reports. Here is the answer:

We are pledging aid and assistance to the research effort into all phases of tobacco use and health. This joint financial aid will of course be in addition to what is already being contributed by individual companies.

For this purpose we are establishing a joint industry group consisting initially of the undersigned. This group will be known as TOBACCO INDUSTRY RESEARCH COMMITTEE.

In charge of the research activities of the Committee will be a scientist of unimpeachable integrity and national repute. In addition there will be an Advisory Board of scientists disinterested in the cigarette industry. A group of distinguished men from medicine, science, and education will be invited to serve on this Board. These scientists will advise the Committee on its research activities.

This statement is being issued because we believe the people are entitled to know where we stand on this matter and what we intend to do about it.[15]

There is no question that this persuasive advertisement made some thinking people take notice. Many no doubt reasoned that if distinguished scientists are willing to join the board of this group, there must be something to what they are saying. In an affidavit prepared for lawsuits brought (and eventually won) against the tobacco companies for their prolonged deceits, the historian of science Robert Proctor wrote of the Frank Statement:

Most of these claims were either outright lies, or proven disingenuous over the course of time. The "Frank Statement" stated that the claims of a hazard had been abandoned "one by one" when, in fact, evidence for hazards had grown steadily stronger over time. The Statement said that the statistics used in the classic epidemiological studies "could apply with equal force to any one of the many other aspects of modern life," when the most influential studies had been carefully designed to exclude such possibilities.

The Statement pledged cooperation with health authorities, when industry officials and their PR consorts had already privately committed to a strategy of confrontation, obstruction, and obfuscation. The Statement pledged to aid and assist research into "all phases" of tobacco use and health when, in reality, the research funded through the TIRC was deliberately steered into areas that were unlikely to incriminate tobacco. The TIRC was not supposed to operate as a trade association, but it was in fact a PR organization masquerading as a tobacco health research organization. Its existence allowed the industry to say "we're studying the problem," when it was actually doing everything it could to misrepresent the nature and scale of possible harms.

. . . From a historian's point of view, the "Frank Statement" represents the beginning of one of the largest campaigns of deliberate distortion, distraction, and deception the world has ever known. The tobacco industry in effect becomes two industries: a manufacturer and seller of tobacco products, and a manufacturer and distributor of *doubt* about tobacco's hazards. Tobacco Institute Vice President Fred Panzer conceded as much in a private 1972 memo, noting that the industry's strategy involved "creating doubt about the health charge without actually denying it." Many millions of dollars were spent on this part of the industry's activities, the point being to keep the public believing— as Panzer put it—that smoking "may not be a causal factor in diseases such as lung cancer." The industry became a gigantic engine of deceit, utilizing deceptive press releases, "decoy research," deceitful newsletters and pamphlets mailed to physicians, journalists, and stockholders, and many other strategies. Further strategies included misleading word-smithing, duplicitous skepticism (e.g., of research results), false reassurances to consumers, and (eventually) the hiring of historians to misrepresent history.[16]

At the time these ads were published, the average person in America smoked 2,000 cigarettes a year. Those in Britain smoked a few less. In the decade from 1940 to 1950, U.S. tobacco consumption had more than doubled, and it would double again in the following decade. Lung cancer rates began to grow as well, making the connection tough to dispute. At first the industry dismissed the 1950 studies as irrelevant because they had looked back in time to figure out whether smoking played a role in deaths from cancer. Those old cigarettes might have been problematic, but the new ones were alleged to be healthier. The cigarettes people smoked in the 1920s and 1930s, industry officials claimed, were formulated differently and thus not useful in assessing the health effects of those smoked today. They had no basis for saying this, and anyway the argument was made moot by other studies that looked forward, following matched groups of smokers and nonsmokers to see what happened to them. These so-called prospective studies also found increased rates of lung cancer in those who smoked the longest. Still, the public was not eager to give up a product promoted to keep them slim and smart, and also happened to be strongly addictive.

THE INDUSTRY SPENT a great deal of time and money brewing controversy. On April 14, 1954, the TIRC published an authoritatively packaged booklet entitled *A Scientific Perspective on the Cigarette Controversy,* listing authorities in support of the view there was "no proof establishing that cigarette smoking is a cause of lung cancer." More than 200,000 copies were mailed to doctors, members of the press and Congress, medical school deans, and other opinion leaders. Mainstream newspapers such as the *New York Times* and *Wall Street Journal* reported the document's claims as though it were an established fact that there was no firm proof of harm at the time. This position held for decades, while millions more became addicted to nicotine and the profits and legal protections for tobacco grew. It was silently abetted by the ACS and the AMA as well as by respected academic experts, who continued to use TIRC funds to conduct studies about tobacco hazards for these organizations long after the basic issues had been resolved scientifically.

To understand the myopia that surrounded studies of tobacco, it is important to realize that making the connection between tobacco and cancer was never just a matter of scientific inquiry. The burden of proof shifted with the dictates of public relations. Even if the constituents of smoke caused death or disease in animals, Little argued over and over, this did not mean that humans would suffer the same fate. The marketing campaign that masqueraded as scientific inquiry effectively blocked science by appearing to promote it.

The strategy of fomenting doubt about the role of tobacco in lung cancer extended, on occasion, to support for research on the ability of chemicals to cause the same disease. This practice too continued for decades. In 1979, when I served as scientific director of the Environmental Law Institute, the Tobacco Institute's Fred Panzer offered me funds to study lung cancer in the chemical industry, so long as I not consider the role of tobacco. I declined; many others did not.

The chemical industry has also engaged in finger-pointing by supporting groups whose sole focus has been to finger the dangers of tobacco. The American Council on Science and Health sounds like an impeccably independent research group. It's not. Supported heavily by major chemical firms, the ACSH carries out an active, professionally run campaign on the dangers of tobacco. In 1982, the *Washington Post* reported that the ACSH filed a brief on the side of the formaldehyde industry, without disclosing that it was funded by one of the country's largest users of the chemical, the lumber firm Georgia Pacific.[17] The organization's directors deny that their focus on smoking reflects the fact that the ACSH gets much of its money from the oil, gas, lumber and chemical industries. While the group claims to be independent, its financial dependence on these industries and consistent defense of them in various media suggest that there are major limitations on its autonomy.

WHILE THE TOBACCO companies were spending tens of millions of dollars to influence public opinion in the face of mounting reports of the hazards of tobacco, the silent ACS wrapped itself in the defense that it had invested more than $4 million to study lung cancer. The Breslow history reveals that Howard Taylor, who chaired ACS's first committee on tobacco and cancer in the 1950s, argued that research was not enough. He remembers being mightily frustrated as the prosperous, stubborn ACS board blocked any direct action by the organization against tobacco from 1954 to 1960. Ashbel Williams, who chaired the Committee on Tobacco and Cancer and became president of the ACS in 1967, confirmed Taylor's recollections.[18] There were big fights within the ACS, with the staff wanting to publicize its findings and powerful board members refusing to allow it. "Our early efforts were bottled up," Williams told Devra Breslow. "We accomplished nothing. There were two Board members, one from Louisville, Kentucky, who stymied any assertive statements by the Society."[19]

Taylor felt that the ACS should have picketed the offices of the American Tobacco Council. "I still think that the campaign against cigarette smoking ought to have been less proper. . . . I always imagined letters to the stockholders [saying things like] 'You've got blood on your hands.'"[20]

Each time an official group tried to make headway declaring that smoking was harmful, it was met by a well-oiled set of refutations. In 1957 and again in 1959, Surgeon General Leroy E. Burney asserted that the U.S. Public Health Service believed that cigarette smoking caused cancer. Within two weeks of his 1959 declaration, an editorial appeared in the *Journal of the American Medical Association,* calling this view into question. The editorial claimed that there were not yet enough facts to warrant "an all or none authoritative position" about the relationship between smoking and cancer.[21] The promptness of the rebuttal testified to the close ties the AMA maintained with the tobacco industry for years and the deep tentacles of both groups within the government. This important association of physicians was determined not to annoy powerful members of Congress from tobacco states, whose votes were needed on various issues about which the AMA cared deeply, including the looming threat of national health insurance.

Eventually the two factions within the ACS reached a compromise: the organization itself would not take the lead in resolving the issue of smoking and cancer, but it would urge others to do so. In 1961 the ACS wrote to President John F. Kennedy urging that he bring together scientists to review evidence on the health impacts of smoking— something its scientific staff had been seeking for nearly a decade. This decision turned out to be a huge win for the staffers. The tobacco industry had argued for years that without statistically impeccable human evidence, we can't be sure whether or not people are truly harmed. Many of the executives and admen on the ACS board had apparently come to believe TIRC propaganda that this evidence did not exist. In fact, by this time it was overwhelming. By 1961 no one could seriously dispute the existence of studies in humans showing an increased risk of lung cancer and heart disease in those who smoked the longest and the most. Over the next few years, the top experts in public health research pored over data from twenty different countries examining the health hazards of tobacco.

Kennedy distanced himself from the process. When asked in 1962 about his own opinion on the risks of smoking, the charismatic, cigar-smoking president provided a politically adept, completely noncommittal reply: "The matter is sensitive enough, and the stock market is in sufficient difficulty, without my giving you an answer which is not based on complete information, which I don't have."[22]

Kennedy never would get a chance to finesse the situation. Early in

1964, after Kennedy's assassination, the U.S. Surgeon General made a definitive statement about what Kennedy would have had to have regarded as "complete information." But it could not have escaped his attention that the British had weighed in on the matter earlier.

Three big bombs fell on the tobacco industry in the 1960s. In 1962 the Royal College of Physicians issued a report which concluded that smoking was a health hazard and called for sweeping reforms: "Cigarette smoking is a cause of lung cancer and bronchitis, and probably contributes to the development of coronary heart disease and various other less common diseases."[23] In a significant departure from its usual reserved approach, the Royal College recommended that the government not just study the problem but do something about it. That same year, the American Cancer Society overcame the long-standing objections of some of its corporate board members and finally weighed in on the issue. "Clinical, epidemiological, chemical and pathological evidence demonstrate beyond a reasonable doubt that cigarette smoking is a major cause of lung cancer."[24] Two years later the U.S. Surgeon General issued a report *(Smoking and Health: Report of the Advisory Committee to the Surgeon General of the United States)* echoing the evaluation of the Royal College of Physicians: cigarette smoking causes cancer.

Within three months of the surgeon general's report, cigarette consumption in the United States had dropped by 20 percent. The report put Emerson Foote into agony. A colorful man with a penchant for drama, Foote was the manic model for the 1946 novel *The Hucksters,* an exposé of the inside workings of the advertising industry. Written by Frederic Wakeman when he worked at Foote, Cone & Belding, the book was made into a movie starring Clark Gable as the fictional Foote—larger than life, full of himself, a charmer who would stop at nothing to advance the promotion of tobacco. But after having been a mastermind of tobacco advertising for more than two decades, Foote realized in 1964 that he truly did have blood on his hands. He resigned his post as head of one of the country's top advertising firms, McCann-Erickson, declaring that he refused to be associated with an effort to promote a product he now understood was killing people. That same year, he became chairman of the government's National Interagency Council on Smoking and Health, a group charged with developing and implementing plans and programs aimed to combat smoking.

Years later, Foote told a Senate committee looking into tobacco advertising:

> The cigarette industry has been artfully maintaining that cigarette advertising has nothing to do with total sales. This is complete and utter nonsense. The industry knows it is nonsense. I am always amused by the suggestion that advertising, a function that has been shown to increase consumption of virtually every other product, somehow miraculously fails to work for tobacco products.[25]

Foote admitted to the team creating the Breslow report that he had begun having second thoughts from the moment he first heard about the ACS research on the subject. Never a shy man, he said to *Newsday* in 1964 that "I first became interested in the hazards of smoking fourteen years ago when a friend of mine, then President of the ACS and now heading his own clinic in Boston, first showed me reports linking smoking and cancer."[26]

For more than a decade after the surgeon general's report, the AMA continued to accept funding from the tobacco industry, a fact that the industry touted in presenting itself as seeking truth and sound scientific information. In contrast, the ACS from that point onward took on the fight against tobacco as a critical part of its mission. Board members who were directly tied to tobacco either left or, like Foote, went native. Significantly, the AMA, which had not officially adopted the surgeon general's report, refused to join this effort.

With these reports, the formerly arcane field of applied biostatistics, or epidemiology, became a centerpiece of public policy. The dangers of smoking were shown to affect sufficient numbers of human beings as to be undeniable. Research findings from the relatively young field of epidemiology became the centerpiece of public attention. At the same time, experimental research in cell cultures or whole animals became seen as somehow less rigorous, the sort of information on which one should not base policy. This position also provided fertile grounds for other industries to challenge efforts to address and control environmental risks of cancer.

Epidemiology, as the science of looking at patterns of disease in people as they are arrayed in time and space, attempts to discover how past habits or experiences affect present health. Cancer is especially challenging because the disease typically is not detectable until years or even decades after important exposures have taken place. In addition, we have known for some time that many distinct things can contribute to cancer. Science has only a limited capacity to discover which of these things has caused cancer in humans, let alone in any particular human.

If there is one thing the epidemiologic study of cancer cannot have, it is speed. Cancer can take years or decades to develop, making the generation of data equally slow. The debate over smoking within the ACS set the bar for the evaluation of other modern hazards. If the only information that we can trust comes from humans, the feeling was, it's better to wait.

In the view of many health experts in the first half of the twentieth century, animal experiments showing that tars from tobacco smoke produced tumors, as well as autopsies on lung cancer victims who had been tobacco smokers, yielded black-and-white proof of the dangers of tobacco. The tobacco industry was able to count on those who argued, then as now, that when it comes to understanding how humans respond to a given toxicant, the only definite proof arises when researchers can amass sufficient numbers of ill and dead people. Various tools have been crafted to determine the odds that any given finding could be just the luck of the draw. In the early 1950s a number of different research groups all came up with distinct ways of establishing such proof. The tie between lung cancer and smoking in humans was first established by those who pioneered one of these approaches—the case-control or retrospective study—but it was confirmed by all of the others. Still, these confirmations were never sufficient for the tobacco industry, which admitted its product's dangers only under the duress of litigation. Those who confront the tobacco industry's successors and imitators need to remember that when organizations promoting "sound science" have an economic interest in the prevention of science, no amount of proof may be enough.

SOME OF THE most frequently invoked research in both the British and American official statements on smoking was that of Richard Doll,

Bradford Hill and Ernst Wynder. Their work clearly demonstrated the connection between smoking and poor health. Their efforts were also critical to the process of persuading the public to accept these findings. They didn't just publish their results. They became the tobacco industry's worst nightmare—scientists who looked like scientists and were prepared and well able to speak to the public. They engaged frontally, talking to politicians and professional colleagues about the need for action on this issue.

Still, when it comes to crediting scientists for their work on tobacco, it is worth asking why people like Doll and Wynder have been acknowledged as founders of the field while the work of others who came before them has grown more obscure with time. As to why the pioneering research on tobacco developed under the Nazi government has not been recognized, Proctor writes that "the myth that English and American scientists were first to show that smoking causes lung cancer . . . was a convenient one—both for scholars in the victorious nations and for Germans trying to forget the immediate past."[27]

As young men, both Wynder and Doll had direct contact with the German scientists who devised the original methods for studying the dangers of tobacco in the 1930s and 1940s. Wynder lived in Germany until his early teen years and returned there as a military investigator during and right after the war. Doll studied medicine in Germany, taking classes in the German language, which was then required of most medical students, and attended lectures at institutions where much of this work had been carried out shortly before the war began. Neither man was inclined to share credit with the Germans before World War II ended. Afterward, revulsion toward anything of Nazi origin contributed to a general lack of interest in crediting German scientists with figuring out the basic science of epidemiology.

It is also the case, as Robert K. Merton so elegantly showed us, that the human impulse to be seen as the first to do something of obvious importance sometimes inspires less than gallant behavior. From my conversations with each of them over many years, neither Doll nor Wynder was inclined to acknowledge the value of what the other had done. They engaged in sometimes subtle, sometimes not so subtle campaigns for public recognition of their work. Surrounded by admiring colleagues around the world, each continually sought to persuade his listeners that he himself had actually provided the critical and definitive contribution that proved the dangers of smoking. Each maneuvered and failed to get the Nobel Prize for the discovery—a prize they should have received jointly. But neither ever conceded that the German work had laid down critically important precedents.

In 2001, the seminal papers on tobacco hazards by Müller and Schairer and Schoniger, which had originally appeared in 1939 and 1943 respectively, were translated into English in a special issue of the *International Journal of Epidemiology* that acknowledged their pivotal contribution to the field. Schairer's son took strong exception to what Proctor had claimed about the scientists' Nazi commitments. He wrote in this same journal that, contrary to Proctor's assertions, his father had been appalled by the Nazis and had not belonged to the party. Proctor reaffirmed his view that tobacco epidemiology was a vital part of the Nazi effort to promote racial hygiene. In this same issue, Doll explained that he hadn't actually seen the report of Schairer and Schoniger when it was published because the war had begun, and even if he had, it wouldn't have mattered because it wasn't that well done.[28]

In 1994 a team of investigators led by Stan Glantz at the University of California–San Francisco received one of those amazing gifts academic researchers dream about. Nobody knows who sent it. We can only speculate that someone from the British American tobacco company had had too many sleepless nights. What we do know is that sometime in 1994 an unsolicited box arrived at Glantz's office containing thousands of pages of documents from the Brown and Williamson Tobacco Corporation, charting how much and how long tobacco firms had known about the hazards of their products and detailing their strategies to generate confusion and doubt about these dangers. Over the next year, Glantz and his colleagues analyzed the documents. In an act of considerable courage, the *Journal of the American Medical Association* published a series of five papers analyzing these materials. Very few people have read all of them. They can be found on the web.[29]

These documents, known as the Cigarette Papers, confirmed Breslow's assertion that the firms were not alone in behaving badly.

As Allan Brandt makes clear in his authoritative new book on this subject, *The Cigarette Century*, the tobacco firms consistently tapped some powerful and impressive help.[30] Medical societies, including the ACS and the AMA, not only let their prestige be used but actively teamed up with tobacco companies in their efforts to prolong uncertainty about their products throughout the 1950s, and did so far longer than was scientifically defensible.[31] In 1993, Eugene F. Knopf resigned the position he had held for close to a decade as the chief lobbyist for the ACS in Pennsylvania. "The reason for my withdrawal is that I am being retained by the American Tobacco Institute." He left the ACS after having manipulated the state ACS into supporting a law ostensibly aimed at the good-government idea of uniform state standards. The real effect of this law was to prevent localities from acting to limit smoking in public places. This strategy paid off. As of June 2007 Pennsylvania was still the ashtray of the industrial northeast—one of the last states to not have laws banning smoking in public. As a registered tobacco lobbyist Knopf received $85,000 per year up to 1996 for all of his efforts to keep Pennsylvania from enacting laws against smoking in public. Tobacco was not his only client; as of 2002 he also represented the gambling and health groups Greenwood Racing, Inc., and Lehigh Valley Health Network.

The ACS today is a radically different institution from what it once was. National smoke-out days, smoking cessation hotlines, and state tobacco control programs form the bulk of its efforts. There is evidence that these programs are working. ACS researchers reported in 2007 that those states with the most intense antitobacco programs—Arizona and California—had the lowest death rates from lung cancer in young persons, while those with the weakest had the highest rates—including Mississippi, Arkansas and Kentucky.

The ACS is now one of the most aggressive antitobacco forces in the world. But a person reading the ACS website would get no inkling of how this position came about. If she noticed anything amiss, it might be the site's odd silence about the group's stance on smoking throughout the 1950s and much of the 1960s (with the exception of a note in the Milestones about a campaign against tobacco advertising beginning in 1960). There is no mention of the ACS scientists' struggles against the years of delay caused by those within the organization's own inner circle.

The ACS is not alone in its historical reticence. In its official history on the subject, the AMA hardly mentions its work with the tobacco industry. Nor did the heirs of Clarence Cook Little mention his similar connenctions. When he died in 1971 at the age of eighty-three, his obituary in the *New York Times* made no mention of his life as a flack for the TIRC or of his earlier work with the American Cancer Society. He was memorialized as a cancer researcher who founded the Roscoe B. Jackson Memorial Laboratory in 1929 and ran it until his retirement in 1956.[32] The family members who provided the summary of this remarkable man's life let his last twenty-five years, the time he spent masterminding ways to magnify uncertainties about tobacco products, go unreported.

7

Saving Cigarettes

Some things reflect the failure of an entire sensibility.
—SUSAN SONTAG

POLITICS, WHETHER PEOPLED by scientists or elected officials, make for curious bedfellows. Nowhere is this more evident than in the strange alliance that culminated in the 1960s between the chain-smoking head of the U.S. National Cancer Institute, Kenneth Endicott, and the charismatic, anti-tobacco jetsetter, Ernst Wynder. What brought these two incongruous spirits together? They shared the opinion that addicted men (like Endicott himself and nearly half of all physicians at the time) would never stop smoking. It would be better to devise a cigarette that was safe than to try to keep people from smoking altogether. At this point, revenues from tobacco advertising accounted for more than one in every four dollars spent in the booming business of shaping public opinion.

Wynder, one of the most vocal Americans to warn of the dangers of tobacco, had all along urged that it made sense to try to design a safer, cooler, less toxic smoke. The risks of lung cancer up to that time had been demonstrated with plain, unfiltered cigarettes. People wanted to believe that filters would fix that problem. Lots of smart money bet on selling the elegance of phallic filters that could be sold as less harmful, cleaner and easier smokes. Before 1954, only one out of every ten cigarettes sold was filtered. By 1957, the industry projected that close to 75 percent of all cigarettes would be.

Those who play with fire as children or adults, like those who try to take only one dance with the devil, may find they can't let go when they need to. Convinced that there had to be a safer smoke, Endicott led the NCI throughout the decade. Three years after the U.S. Surgeon General's report had declared that smoking caused lung cancer, this chain-smoking chief began a government program that spent more than $30 million of taxpayer money in the United States to create a safe cigarette in 1967. Similar efforts were mounted in Britain.

The concept of making cigarettes safer has an inherent logic. But less bad does not equate to good, as Wynder and others would eventually learn.

Like much tobacco propaganda, the reality was quite different from the impressions that were carefully nurtured about filter-tip cigarettes. The story of what has become known as the Cigarette Papers or the Cigarette Century has been widely told in broad brush strokes. One of the little-known parts of that story is how the tobacco industry tried to have it both ways. At the same time they were assuring the public that their product was safe, many in the tobacco industry in England and the United States used the cover of trade secrets to carry out expensive, clandestine efforts to design a less harmful cigarette. They did this, in part, with the full cooperation of officials and scientists working for the U.S. and British governments.

A memo of March 1, 1957, discussed research underway at the British American Tobacco group in Southampton, England. Here's how the strangely coded language appears in some internal documents from this secret project.

> As a result of several statistical surveys, the idea has risen that there is a causal relationship between Zephyr and tobacco smoking, particularly cigarette smoking. Various hypotheses have been propounded, from time to time, as explanations of this conception. The two which seem most important and present are:
>
> (i) tobacco smoke contains a substance or substances which may cause Zephyr
>
> (ii) substances which can cause Zephyr are inhaled through the atmosphere, e.g., in the floor of sort.[1]

Elsewhere in this amazing memo some of the suspected causes of Zephyr/lung cancer are identified, including Borstal and 3,4,9,10-

DBP, code for benzo(a)pyrene, then and now well known to cause cancer in lab mice.

The language used here could have been adapted from George Orwell's novel *1984,* where words convey the opposite of their plain meaning. Let me explain. For lung cancer, the term Zephyr is applied. Zephyr comes from Greek and Latin words meaning a warm west breeze, something that those with lung cancer seldom fully experienced. For suspected carcinogens, the word Borstal is used. In British slang, a borstal is a reform school. How strange a name to give to agents that could never reform anyone, but rather were deforming the ability of the body to breathe. The research plan included studying how different burning temperatures for tobacco affected the formation of various cancer-causing compounds. Of course, using secret words to disguise their efforts seems childish in retrospect, but they proved, like much of the tobacco story, to have deadly consequences. As part of this program, research at Southampton looked into whether or not the amount of various carcinogens formed could be lowered by different shapes or designs of the cigarette. At one point, they tried to rebuild the cigarette itself, creating a coaxial design in which the tobacco core was completely blanketed by thicker filter-like materials. They also developed a wide variety and length of filters. Each variant was tested for the amount of carcinogens released. None worked completely.

To carry out the objective of crafting a safer cigarette, the industry tapped respected researchers at private institutions, including the University of London, Wynder's American Health Foundation in New York, the American Medical Association and the American Cancer Society, as well as the U.S. government. Not only did all of these groups work with the tobacco industry to engineer a better smoke, they also came up with methods by which the chemical engineering and epidemiological safety of the product could be evaluated.

Much to the shock of the tobacco firms in England, the Royal College of Physicians came out with its report declaring that smoking damaged human health in 1962. Richard Doll, the man who championed British research on the dangers of tobacco, told me that the report was delayed close to five years because of the tremendous influence of the industry. The tobacco manufacturers in Britain had long assumed that because the government depended so heavily on revenues from tobacco to fund the health system, among other national services, they would be immune from direct government control. When it finally appeared, the Royal College report sent them reeling. Right afterward, a major powwow was held by the British American tobacco research and development leaders. At this conference in July 1962, Sir Charles Ellis, a leader of the British American Tobacco Company, explained the challenge was to come up with a basic change in the nature of cigarettes:

> The board recognizes this problem must be tackled from two sides, and the first being at [sic] medical research on the origin of lung cancer and bio assay on the biological effects of smoke, and the second being the composition of smoke and the possibilities of modifying it[2]

Sir Charles went on to promise that if any new toxicology research was found that was relevant to improving the health of its product, the board would share this information with other tobacco companies, rather than seeking any commercial competitive advantage. Whether this information, if it were ever obtained, would be shared with a public or public health authorities was not even considered.[3]

Nowhere are the clever and complex strategies to manipulate the public mind more clear than in the prolonged, failed and costly campaign to produce a safe cigarette. Once again, the fingerprints of Clarence Cook Little, the former chief of the American Cancer Society, who became director of the Tobacco Industry Research Council, are evident. Little came up with a policy modeled on a well-known axiom that continually resurfaces in the war on cancer: If you can't beat them, join them. He promised that if research did prove a direct relationship between cancer and smoking, "The next job tackled will be to determine how to eliminate the danger from tobacco."[4]

Little was not alone in this view. He had some powerful allies, including many of those whose work had shown the damaging effects of tobacco. A darling of some in the cancer establishment, Wynder had enormous cachet, which he did not hesitate to use. In private meetings and extensive interviews with the media, he championed efforts to make smoking safer. U.S. and British government scientists and those at leading private universities worked directly with the tobacco industry for more than a decade at a cost of millions with the goal of producing a less dangerous cigarette.

By 1957 the notion that tobacco smoking could be considered a healthful habit—as many modern ads promised—was beginning to come undone in many quarters. Doctors may have smoked Camels, along with other cigarettes, at that point, but growing numbers of them and others were beginning to realize the absurdity of their dependence. That year, in two separate stories, the popular *Reader's Digest* broke what looked like a fatal blow to the tobacco industry. The magazine revealed a set of supposed secrets of the tobacco companies. They were holding out on what was alleged to be a safer smoke. As with many of its promised advances, the industry was blowing more smoke than it was clearing.

The first *Reader's Digest* story revealed laboratory tests proving that the amount of nicotine and tars inhaled from filter-tip cigarettes was no less, and was sometimes far greater, than that taken in from plain smokes. In some cases, switching from a regular size plain cigarette to a king-size filter actually increased the average amounts of tars and nicotine inhaled. In fact, the filtered King and Hit Parade cigarettes contained 30 percent more nicotine and tar than unfiltered Camels.

In 1953, when filters were just beginning, the American Medical Association, reflecting its close connection to tobacco research at the time, tested three new brands and found that one actually did remove 55 percent of all tars and nicotine. This turned out to remove much of the taste as well. Sales of this newly designed, filtered Kent cigarette tanked. The industry learned from this experience that smokers wanted the taste that came with tars and nicotine.

So what was in these too efficient Kent filters? That was disclosed in the second story. The Atomic Energy Commission had recently declassified a report about a remarkable aerosol filter that removed radioactive particles from the air in nuclear power plants. This extraordinary material was crocidolite—a bluish kind of asbestos. The company making Kent cigarettes, P.J. Lorillard, decided to use this new material in its brand-new cigarettes in 1952. Nearly 12 billion cigarettes—

about 585 million packs of these asbestos-filtered cigarettes—were sold in the United States until 1956. Ads assured smokers that these filters provided health protection. Laboratory tests using smoking machines to simulate human exposures finally published in 1995 prove that this was not at all the case. A typical smoker would have inhaled considerable amounts of asbestos, known now to induce lung cancer and mesothelioma—a tumor of the lining of the internal organs that basically causes people to suffocate to death.[5]

There is a price to be paid for being an early adopter. For Kent, its filters proved to be too efficient. Smokers complained that the cigarettes just didn't have that tobacco taste. So the first battle of the filter tips began: how to design a filter that looks like it's doing the right thing, but doesn't really remove too much of whatever it is that makes a cigarette a cigarette and keeps people craving more?[6]

At first reading, these articles from *Reader's Digest* in 1957 looked like a heavy hit on the industry. In fact, they proved to be a setup. The first article in this series ends with a tantalizing report. There is a trade secret that nobody wants you to know. Most tobacco companies had begun using what once was considered scrap pieces of tobacco stems formerly used for landfills and blending this into their cigarettes, along with the fine leaves of tobacco.

Using tobacco scraps in cigarettes proved to be useful on several fronts. It was cheaper to use something that had previously been thrown out. But it also turned out that smoking machines, used to measure amounts of tar and nicotine by the Federal Trade Commission, found that cigarettes which used what was later called reconstituted tobacco looked healthier. The industry was basically recycling what had once been tobacco garbage into cigarettes, and producing smokes that looked better and had less tars. It seemed a financial and public health triumph: less costly and less potent cigarettes could be crafted. Like much of the tobacco story, it proved too good to be true. But that realization would take years to unfold. In the meantime, the economy and its addicted citizens would continue to grow more dependent on what would prove to be an unhealthy compulsion.

In testimony before Congress, the chairman of R. J. Reynolds, John Whitaker, admitted that most of their cigarettes at the time used what had once been tossed away. When asked whether filter tips had more scrap tobacco than plain cigarettes, Whitaker refused to answer. "You are getting into trade secrets now."[7]

What was this trade secret? Lois Mattox Miller and James Monahan, the *Reader's Digest* reporters who wrote this article, would go on to receive an award from the American Cancer Society for the work, which included this big disclosure: "It is entirely possible to manufacture filter tips much more efficient than any now on the market." Their big scoop played right into the hands of the tobacco industry. They revealed that improved filters already existed, would cost no more to produce and would remove 40 percent or more of all of the tars. How did they learn this?

The article closes with a confident quote from the man widely known as one of America's most prominent antitobacco scientists at the time, Wynder, then with Sloan-Kettering, one of the top cancer facilities in the nation. "Such a filter, placed on a regular-size cigarette which normally yields 30 mg of tar in its smoke would reduce the smokers tar to 18 mg. A reduction to that level, as shown by both animal experiments and human statistics, would be a significant reduction in cancer risk."[8]

As a young medical student in 1950, Wynder had prodded the chain-smoking Evarts Graham into studying how smoking affects lung cancer. Wynder was the opposite of the studious, introverted scientist. He not only produced major scientific studies on tobacco in humans and the cancer-causing impact of tars in animals, he also made sure that anyone he met knew of his considerable accomplishments. His message, replete with an authoritative-sounding German accent, was not only heard at scientific meetings but was played out on television and radio, and in the Manhattan social scene in which he was so much at home. A charming man with a broad intellectual reach and glamorous social ties, Wynder didn't mind being told he resembled the handsome movie star Ricardo Montalban.

Wynder was a larger-than-life character and a man with a vision—a big vision. Former colleague Frank J. Rauscher quipped, "He's a hard S.O.B.—he's good, he thinks he's good, and he'll tell you he's good. But there is nothing phony about Ernie Wynder and his flag-waving on smoking and health."[9] A scientific showboat, he once boasted he had the best of two worlds—a German education and American opportunities.

This did not endear him to his fellow scientists. Some of them ridiculed Wynder as being as "much a scientist as a bomber pilot is."[10]

How this bomber pilot scientist and the chain-smoking director of the NCI, Ken Endicott, became staunch allies in the effort to make smoking safer is a remarkable story. Not a smoker himself, Wynder believed that cigarettes could and should be made safer. He knew that the Nazis had failed to ban smoking and acknowledged some early German work in his own publications. As someone who was comfortable in the fast-moving worlds of the Hamptons and Hollywood, Wynder talked and people listened. And if they didn't listen, he'd just keep coming back until he gained a footing. This is a man who didn't hear no. Wynder came to genuinely believe that smokers would not give up their lousy habit, no matter how much evidence he or others produced on the dangers of smoking. The death of his own mentor, the heavy-smoking surgeon Evarts Graham, from lung cancer, doubtlessly colored his views on the subject.

Despite the envy and resentment of some of his peers, Wynder knew how to get people to pay attention. He also had an aura of great authenticity coming from his slightly patrician demeanor and commanding presence. The package was very impressive. When Wynder began to argue for a safer cigarette, he found an amenable collaborator in the chain-smoking Endicott. Many of the scientists at the NCI were not convinced. The idea that the government should spend millions of dollars to come up with a less toxic cigarette struck many as absurd. The notion that the brilliant work at the NCI would serve an industry that at the time was the source of considerable skepticism did not sit well with the working scientists of the government.

During the tumultuous 1960s, the tobacco-addicted Endicott was at the helm of the U.S. NCI. During this time, public concerns over cancer rose, as did the rate of the disease. With the full support of the American Cancer Society board, and members of Congress eager to appear engaged in the issue, funding for the NCI's research program grew dramatically. From 1960 to 1969, the institute budget reached $1.8 billion. According to those who worked with him at the time, Endicott was a four-pack-a-day man. He seldom needed cigarette lighters or matches because he used the end of one cigarette to light another. His staff joked that the only time Endicott ever was known to stop smoking was when he testified before Congress. He even had some creative ways of using the bathroom without losing his light. Not surprising for someone so hooked on nicotine, Endicott believed that you could never get people to stop smoking. He knew that he could never give up the fiercely addictive habit. But Endicott was intrigued by Wynder's idea. In 1967, three years after the surgeon general's report had linked smoking with lung cancer, Endicott appointed the NCI Lung Cancer Task Force to create a government research program to deal with the smoking issue. This was not your usual critical group. On the contrary, a subcommittee of the task force was set up with an explicit goal—make cigarettes safer. The Less Hazardous Cigarette Working Group, formed in March 1968, began what turned into a decade-long search to figure out how to make a cigarette that would be less dangerous than those used at the time.

At the time the tobacco industry was basically trying to have it both ways. On the one hand, it continued to carry out a vigorous campaign to sell the public on the idea that the science just wasn't there to make a connection between smoking and lung cancer. It did this by funding reputable scientists at major institutions who continued to study and study and study various aspects of tobacco chemistry. The industry also funded many major figures in epidemiology like Tom Mancuso, a former government scientist whose work explored all of the other things that contribute to cancer besides tobacco, such as working with ionizing radiation or mining. They fixed their hopes on those instances where men got lung cancer or lung and heart disease but did not smoke, like those working in uranium or coal mines, or other dangerous workplaces. For these men, where they worked could be shown to be far more important for their health than smoking. So what about the rest of us?

At the same time, well behind closed doors, the tobacco industry carried out a major research program of its own aimed at engineering what was variously termed a better cigarette, a less toxic cigarette, or a safer cigarette. The ability of the industry to simultaneously maintain these two different tacks is a testimony to its ingenuity and to the power of its purse to dictate the nature of research.

According to one supposed inside report, the tobacco companies already knew how to make cigarettes safer—by using filters. Their reluctance to make these safer cigarettes could be handily overcome by concerted public pressure and demands for filtered smokes. This is precisely what happened after the *Reader's Digest* series appeared, as did similar articles in England. Remember that at the time, literate men smoked. They figured out the intended message brilliantly. If you want to smoke and smoke safely, get yourself a good filtered cigarette. Demand for filtered cigarettes soared.

The tobacco industry responded to initial reports of hazards by just saying no. Trust us, we've got and bought other experts who can explain why these reports are just wrong. In 1969, a congressional hearing featured testimony from highly credentialed medical specialists, couched in the soothing, certain language of scientific discourse that assured the Congress the surgeon general's 1964 report had been wrong. Dr. Sheldon C. Sommers appeared with his impressive qualifications: pathologist and director of laboratories, Lenox Hill Hospital, New York, clinical professor of pathology, Columbia University College of Physicians and Surgeons, clinical professor of pathology, University of Southern California School of Medicine, teacher, Cornell Medical School, Tufts-New England Medical Center and New York Medical College, and most recently, research director, Council for Tobacco Research, USA. He advised, "To claim there is now sufficient scientific evidence to establish that cigarette smoking causes disease is, in my opinion, unjustified."

If the British effort appeared to be independent of its government, that was not the case in America, where the public-private partnership on the safe cigarette appears to have been given full rein. Most of the NCI's official staff had little interest or enthusiasm for such applied work. Then as now the NCI prided itself on having the best and the brightest scientific minds working on some of the most important basic scientific issues. The idea that they would turn to doing the work that the tobacco companies ought to be doing for themselves was especially unappealing to most of the agency scientists at the time.

The scientists who were appointed to the working group on tobacco were charged with three objectives: engineer a less hazardous cigarette, identify people at increased risk of tobacco-related diseases, and develop drugs to help people quit. The committee focused chiefly on the first priority—making a safer cigarette. The committee defined such a cigarette as one that would yield fewer tumors when tars from smoke were painted on the skin of mice than did the typical cigarette then in common use. Privately, many scientists muttered that the term "safe cigarette" was an oxymoron. There was no way that such an entity could ever exist.

Gio Gori became the leader of the Less Hazardous Cigarette Working Group in 1969. You knew right away when you met him that he was different from most of those in the government. Others have described him in less flattering terms. "His medical training had been at a backwater school, he had no scholarly publications to speak of, and he brought no depth of knowledge on the nuances of cancer."[11] It's never been clear to me how Gori ended up at the helm of a major U.S. government program to design a safer cigarette. But it's clear at this point what his leanings were all along. He now works directly for the tobacco industry and edits the journal *Regulatory Toxicology and Pharmacology,* which receives some funding directly from tobacco giant R.J. Reynolds.

What Gori lacked in scientific pedigrees at the time, he more than made up for in schmoozing ability. Frank Rauscher succeeded Endicott as chief of the NCI in 1971 and apparently found Gori an impressive manager of a program that someone needed to carry out.

A number of people felt that Gori was a lesser version of Wynder. The man was certainly smooth, maybe too smooth, with an aura of superiority that was never justified and the suits to go with it. But where Wynder was a serious intellectual with real achievements and a flare for showmanship, Gori had much more of the latter and little of the former; many of his colleagues found him to be an overdressed bureaucrat with an exaggerated sense of his own importance. He also developed unusually close relationships with the industry and contractors working under his supervision. In a short period of time, Gori's name appeared as coauthor of some twenty articles. Before that, he had not written a single publication on his own.[12]

The floor began to drop out from under Gori when he boasted of progress in efforts to make a safer cigarette. In September 1971, he told the Second London Conference on Smoking that "based on epidemiological studies, filter cigarettes delivering less tar and nicotine show a remarkably decreased risk of disease; the studies give unequivocal proof in man that reduced tar and nicotine provide a first model of a less hazardous cigarette." This was a gross overstatement. Even though filters removed some toxic ingredients, there was little indication that this radically reduced the burden of lung cancer or any other diseases.[13] Smoke is a mixture of several hundred different chemicals. If anything, what evidence there was about the impact of filters pointed in the opposite direction. Those who smoked filtered cigarettes tended to inhale more deeply in order to get as much nicotine as their brains needed for a buzz. To compensate for the fact that filters made it harder to get the right amount of nicotine, filtered smokers tended to inhale harder and longer. Also, the numbers of cigarettes smoked by each person rose, along with the growth of filtered cigarettes. People were upping their use of cigarettes to get the same buzz from filtered cigarettes that they previously got from those without filters.

Don Shopland, the hardworking and respected official with the Office of Smoking and Health in the Department of Health, Education and Welfare (HEW), offered this sardonic interpretation of Gori's message right after it appeared: "I may not like these weak smokes and may get a hernia drawing on them to extract any pleasure whenever, but wow! They're practically risk-free." In fact, it was later proven that Shopland had been dead on. In 1954 the average smoker consumed twenty-two cigarettes a day. By 1978, when most cigarettes were filtered, the average smoker consumed thirty cigarettes each day.[14]

In 1976, when I first met him, Gori began publicly promising that a safer cigarette was around the corner. He appeared to be more of a salesman than a researcher. This infuriated a number of his colleagues at the NCI. I worked closely with Marvin Schneiderman at the time. A world-class statistician with a Borscht Belt sense of humor, Schneiderman had been an associate director of the NCI and a member of the smoking and health working group. He candidly told me he thought it was beyond chutzpah for a nickel of government money to be spent trying to make smoking less harmful. Politics regarding scientific research, as with much else, is quintessentially the art of the possible. It was more expedient to look like we were doing something than to admit there was little we could do at the time. The Yiddish expression he used was "kenn garnisht helfen," literally, there is nothing to be done here. Umberto Saffiotti, another distinguished NCI researcher, shared Schneiderman's disdain for Gori and the entire effort to produce a safe cigarette and made his views known throughout the NCI, though in less colorful terms.

An exposé on the safe cigarette program in 1979 by reporters Frank Tursi, Susan E. White and Steve McQuilkin, of the *Winston-Salem Journal,* portrayed the inner workings of this committee and its demise.[15]

> Meeting two or three times a year starting in 1969, the committee authorized hundreds of experiments that were aimed at trying to understand how cigarette design might reduce the risks of smoking. Scientists working under contract with the committee tried to extract from tobacco the precursors of what were considered cancer-causing chemicals. The committee also tested various parameters in tobacco growing and cigarette production: nicotine content, fertilizer application, artificial substitutes, paper porosity, pesticides and such additives as sugar and cocoa. They tried homogenizing tobacco, removing nitrates and adding nitrates. From all that came more than 100 experimental cigarettes that were tested on mice against two control cigarettes, a standard experimental cigarette and one developed at the University of Kentucky.

The mouse experiments proved nothing the industry didn't already know. They showed, for instance, that cigarettes with very porous paper or blends made only from tobacco stems produced fewer tumors than the standard blends. Low-nicotine blends and those made from puffed tobacco also produced significantly fewer tumors on the backs of mice.

In fact, about the only tangible result of the committee's work was that it served as a backdrop for a theory eagerly advanced by Gori that a "socially tolerable" smoking limit was possible. As early as 1971, he cited epidemiological studies that he claimed showed low-tar cigarettes were less harmful than regular cigarettes. He even went as far as to suggest a legal limit: 20 milligrams of tar and 1 milligram of nicotine.

All along the strategy of those pushing to make smoking acceptable had been to cobble together what looked like hard-core proof that cigarettes could become less dangerous and to get this out in reputable scientific publications. *Science* magazine generally publishes the most important work in science. In December of 1976, this highly regarded journal featured Gori's argument for a safe cigarette, putting forth the notion of "tolerable levels" of smoking. A similar analysis appeared in the *Journal of the American Medical Association* in September 1978. On August 10, 1978, the *Washington Post* headlined its front-page story on smoking, "Some cigarettes now 'tolerable,' doctor says."

The public response to these carefully positioned pronouncements was a public relations dream. Sales of American Tobacco's Carlton brand—the lowest-tar smokes identified in these papers—doubled within a week.

The reaction among doctors and antismoking groups was immediate. They said that all cigarettes, regardless of tar and nicotine levels, posed some risk. They accused Gori of putting the antismoking movement back years and demanded that Joseph A. Califano Jr., the secretary of HEW, discipline or fire Gori.

Califano, a Harvard-educated lawyer and skilled litigator, was not known for possessing diplomatic skills or lacking toughness or courage. As a powerful aide to President Lyndon Johnson, he had helped shepherd through a balky Congress many of Johnson's Great Society programs. As a recent former heavy smoker, Califano was no friend to tobacco. He had proposed, on January 11, 1978 (the fourteenth anniversary of the first surgeon general's report on smoking), the most strident antitobacco program to ever come out of a cabinet-level office. Calling cigarettes "public enemy number 1," Califano wanted every school in America to teach kids about the consequences of smoking. He wanted a higher federal excise tax on cigarettes and called on the Civil Aeronautics Board to ban smoking on all commercial flights.

The breadth of Califano's proposal left the tobacco barons and their defenders seething. Sen. Jesse Helms chided Califano for "demonstrating callous disregard for economic realities, particularly for the economy of North Carolina." Governor Jim Hunt urged Califano to visit so he could learn what tobacco meant to North Carolinians. Representative Charlie Rose, a Democrat who represented tobacco farmers in the eastern part of the state, elevated the whole discussion when he promised, "We're going to have to educate Mr. Califano with a two-by-four, not a trip."

At one point, Bertolt Brecht, the famed German playwright who spoofed Stalinist policies, argued that if the government didn't like what the people wanted, then perhaps the government should elect another people. At this point in American history, Califano's efforts to get the government to tackle tobacco made it clear that he was one person who could not continue to be part of that government. President Carter later that year visited the state of North Carolina—then the top tobacco-growing state in the nation—as part of his campaign for reelection in 1979. He joked to the crowd that he had planned to bring along that infamous former smoker, Joe Califano. The president explained that he changed his mind when he realized that North Carolina not only produces more tobacco than any other state but makes "more bricks than anyone in the nation as well."[16]

The same day that the president was mocking Califano's antismoking efforts during his visit to North Carolina, the AMA Education and Research Fund released a lavishly printed report on a study that took four years and considerable money to complete, with more than eight hundred researchers and untold numbers of lobbyists. This much ballyhooed AMA report consisted of nothing but a potpourri of mostly unrelated studies that reached a rather lame conclusion. "The bulk of the research . . . supports the contention that cigarette smoking played an important role in the development of chronic obstructive pulmonary diseases and constitutes a great danger to individuals with pre-existing diseases of the coronary arteries."[17]

Wait a minute. It's 1979, nearly fifteen years *after* the surgeon general's report on lung cancer and smoking. The AMA finally decides to acknowledge that smoking is bad for the lungs. This costly AMA report completely ignored the issue of lung cancer on the flimsy excuse that the NCI was already studying the problem. Within days after this report was issued by the AMA, Gori and one of his main contractors, Cornelius J. Lynch, published yet another article in the organization's main journal, *Journal of the American Medical Association,* "Toward Less Hazardous Cigarettes: Current Advances." They claimed that, thanks to the work their program had been carrying out, a major breakthrough was at hand. Modern filters, some of which they named by brand, could be smoked to yield a "tolerable risk."

When these news reports about Gori's declarations came out appearing to endorse smoking, Califano was still head of Health, Education and Welfare. He was furious. He asked Arthur Upton, the director of the NCI, what the hell was going on. Upton called Schneiderman from his vacation home in Salisbury Cove, Maine, and asked if he had any idea what Gori was talking about. Yes, he had seen the calculations, Schneiderman said, and he didn't agree with them at all. It would not be the first time Gori went his own way, but it would be one of the last when he would do so and remain in the federal government.

What makes this story so remarkable is that the tobacco industry knew all along that filters did not make enough of a difference. While the *Reader's Digest* articles appeared hard-hitting, there was in fact a cordial relationship between the writer and the tobacco industry. Miller sent advance copies of her work to the tobacco industry and to the American Cancer Society. That was just the way things were done then. These articles paved the way for the tobacco industry to say it was going to take seriously the need to come up with a safer cigarette.

Not everyone in the tobacco industry was convinced of the value of pursuing a "safe cigarette." Peter Sheehy of the British American Tobacco Company argued on December 29, 1986, that the principal objective of the industry research program should be to make smoking acceptable. If people came to accept their right to assume a risk to their health, a risk that was demonstrably rather low, then much of the debate about smoking would be moot. At the time Sheehy wrote

> Understandably, the causation issue in relation to several diseases is important and we have to take note of all relevant publications that can throw light on this issue. We sponsor research work on mechanism of disease, including psychological or genetic predisposition, as well as probing the simple conclusions of what is probably rather poor epidemiology . . . another important issue affecting acceptability is passive smoking. Our current initiatives are to challenge the whole area of "low risk epidemiology."

Here the message is clear. If the industry can succeed in debunking and dismissing risks that seem relatively small, such as passive smoking, then the public will ignore the evidence of harm. Sheehy was a smart man. He pointed out the hypocrisy underlying the whole idea of a safe cigarette. "In attempting to develop a 'safe cigarette' you are, by invitation, in danger of being interpreted as excepting [sic] that the current product is 'unsafe' and this is not a position that I think we should take."[18]

For Gori and the Tobacco Working Group, the end was near. On August 28, 1978, HEW said that Gori was taking a leave of absence to get a master's degree in public health at Johns Hopkins University. He would never return to the NCI, but he did start a career as a tobacco industry adviser. The working group he had once headed up, after spending more than $32 million, was dissolved in 1979 after HEW refused to continue paying for it.[19]

In the end, the tobacco story is not just about tobacco. Rather, it is a lesson in how public access to information about any suspect hazard can be skewed, bent and twisted to suit other interests. Here we learn that the same tactics that delayed public action against tobacco also played a role in laying down the foundations of what is considered proof in epidemiological research. For years, the industry maintained, with some credibility, that proof of harm about the dangers of tobacco hadn't been established. This was technically correct but morally wrong. The explicit duplicity of the tobacco industry is no longer surprising, having been well established in courts of law, and finally in the most important court of all—that of public opinion. What is not at all appreciated is that there have been many other consequences of the tobacco struggle. As part of its program to see science as nothing but a tool for public relations, tobacco industry lobbyists succeeded in using the absence of human studies as proof that there was no harm. What does this mean for other modern hazards? Wherever proof of human harm is lacking—as it typically is—the conditions that may give rise to this harm can continue without interruption. "Show me the proof" becomes the mantra of those who would have us do nothing. What is seldom appreciated is that sufficient numbers of ill or dead people with clearly defined exposures constitute the only proof that counts under these conditions.

The story of how the basic principles of epidemiology were played out by public relations gurus provides a revealing glimpse into the egos and values of science and advertising. It is not clear which of these major forces of public life was more compromised in the process. Scientists love to argue. Appealing to their inherent drive to dispute and split hairs, the debate about whether we had enough proof of the dangers of tobacco ultimately laid the foundations for how proof in public health research is ever determined. So, what does it matter if a given compound causes cancer in animals? The brilliance of the public relations specialist was dazzling and damning. You just can't trust those sneaky little rats and mice.

Today we know that only three hundred genes differ between the noble human and the lowly rodent.[20] But part of the last gasp of smoking advertisements was to make sure every man, woman and child in America and Europe who ever bothered to look at their increasingly colorful displays understood that people should never be compared with rats. Fair enough. But most people never imagined that the refusal to heed what the studies in animals told us forced us to rely on an even more complicated and less predictable experimental subject— the human being.

Shortly before he died, Sir Bradford Hill despaired over the way public health studies were being manipulated in the debate over smoking. Hill saw that the disputes over human evidence on the dangers of smoking or other hazards were being shaped and prolonged by well-paid, industry-supported scientific experts. To put an end to this cultivated and protracted public confusion, he laid down some principles for deciding whether a given exposure can be deemed to cause any specific health condition. Hill knew that of the many variables that affect our health, not all can be measured.

Hill also appreciated that, unlike precisely controlled laboratory studies on our rodent relatives, the real world is not simple.[21] But saying that things are complicated does not mean they can't be understood at all. Figuring out whether observed differences in public health are merely the luck of the draw or a result of some underlying environmental conditions requires that we do the best we can to make sense of messy, large-scale data of real lives. In public health research, as in much of life, the perfect is the enemy of the good.

Before any two groups can be said to differ, statisticians start out assuming that they are, in fact, the same. Then they use a variety of tools, developed over the past century, for deciding whether or not the difference that appears between them is likely to be real or not. In making choices in our lives every day, we use similar principles of sorting through information, although few of us are aware of the informal calculations we make to do so.

When we look at patterns in human populations of any age-group, we rely on two basic principles to see whether what we observed is important. First, where the differences between any two groups are big, they are less likely to be random. Big differences delight epidemiologists, precisely because they are so unusual in public health. When they occur, they tend to signal something important is going on. Second, your chances of finding a difference are greater the more times or things you observe. This is called the law of large numbers. The two principles are related: the larger the sample that you get to look at, the smaller a real observed difference can be found. Roughly speaking, statistical testing asks whether the differences that are seen are not just some fluke. If it is pretty unlikely that the difference would have arisen by chance alone, then the difference is called "statistically significant."

For large populations, like those of industrial countries with hundreds of millions of citizens with common exposures, what is significant in statistics may be a small number but a very big deal. For instance, a 10 percent increase in deaths from current patterns of air pollution for persons living in the most polluted areas of industrial nations today translates in the United States into about 60,000 extra deaths a year. An extra death is one that we would expect not to happen if people had lived under other conditions. This is a statistical concept that is hard to wrap one's mind around. My friend Kathy Kashrovian's previously healthy father suddenly died from a heart attack during the worst air pollution in Tehran's history around Christmas of 2000. His death, we later learned, was one of hundreds more that week. All the statistical modeling in the world does not change the loss entailed in such personal tragedies. Yet it does allow us to grasp that deaths and births are part of larger patterns.

Every child is born alone. Seen through the prism of statistical analysis, the unique events of our lives can be understood to follow patterns that provide clues about the world in which we live. Normally for every male born there are slightly fewer females. Evolutionary theorists speculate on why this is the case. Greater numbers of males are born but fewer survive to old age. In modern human populations studied so far, the proportion of males born relative to females in a healthy human population is .515. This can also be presented as the sex ratio of males to females of 1.06. For the past three decades in many industrial countries, this number is declining by very small amounts. Today there are nearly 4 million births in a single year in the United States. Since 1970, the proportion of baby boys born has fallen by just one out of every thousand births. This small number in such large populations over the past thirty years means that 135,000 fewer baby boys have been born as a result of this perplexing trend.

For many patterns in public health, like the puzzling drop in boys being born, significant effects are not always easily explained. For other associations, like the increased rates of lung cancer or heart disease in smokers, the links are more obvious. But, even for these, we use various statistical tools to make sense of the information. How do we know whether the differences in lung cancer or heart disease between smokers and nonsmokers are incidental? We calculate the odds that you would see a difference between groups at least as big as the one you actually obtain. This probability is called the p-value. P can range from 0 to 1. A value of 1 means there is a 100 percent chance that the findings are inadvertent; .5 indicates there is a fifty-fifty chance the findings are accidental; and .05 signifies that there is only a 5 percent chance, or one in twenty, that the findings are just random. According to scientific convention in public health research, a value of .05 is usually accepted as statistically significant. But this is merely a convention. In physics the upper limit for statistical significance is $p = .1$, meaning there is a 10 percent chance the results could have happened randomly. Public health scientists accept a p of .05, not because this is a magic number but because this is the one experts have agreed to. What if the p is .06 or .09? That's where judgment becomes important.

Public health importance should not be confused with statistical significance. You can have one without the other.

Where small numbers of persons or very rare events are involved, using the p-value alone may be plain wrong. In these situations, epidemiologists sometimes rely on "confidence intervals" that are calculated to show the values that are likely to have occurred. Basically a confidence interval lets us know the range within which a given result is likely to fall about 95 percent of the time. In other words, there's just a .05, or one in twenty, chance the result falls outside that range.

Another thing public health scientists look for when deciding whether they have found a true connection between the environment and health is any evidence that with more exposure you get a stronger effect. This is referred to as a relationship between dose, or the amount of exposure, and the risk of the response, or the health end-point under study. For instance, we know that those who smoke two packs of cigarettes a day tend to get sicker sooner and die younger than those who smoke one pack a day. But the real world throws us lots of curve balls. It turns out, for instance, that those who smoked four packs a day in the 1950s actually lived longer than those who smoked a bit less. The reasons are pretty clear. Anyone with lungs and a heart capable of sustaining the continual bombardment of so much smoking probably had some resistance to these toxins. Even smoking, one of the nastiest and most important environmental hazards we know of, doesn't kill everyone.

The advertising industry may not have understood confidence intervals, but it did know how to shake public confidence in science. It did this by regularly distorting scientific analysis of the dangers of tobacco in ads that peppered television, radio and print. Numbers from public health research were no match for the robust energy of the Marlboro man's spirited dash across the screen. What would you rather watch or hear about: The call of the West with pounding hoof beats or a recitation of cancer risks? You've probably never heard of Wayne McClaren. McClaren was the horseback-riding, hard-smoking Marlboro cowboy in 1976—the year Gori was championing a safer smoke in major scientific publications. Before dying of lung cancer in 1992, McClaren became an antismoking activist. This poster, showing only what was left of him—his boots and hat—was prepared as a public service announcement against smoking, but didn't get the air or print coverage accorded earlier rides.

Appalled by the undercutting of public health research in tobacco advertisements on television, radio and print, Hill, then one of the world's leading statisticians in 1967 issued a warning. At the time, respected statisticians like R. J. Fisher and Nathan Mantel worked directly for the tobacco industry. I never could tell whether Mantel truly

Figure 7-1 *An anti-tobacco ad from 1992 showing the remnants of the Marlboro Man, who died of lung cancer.*

believed what he was doing or not. He was the sort of fellow who relished a good intellectual fight. He kept coming up with more ways to skin a statistical cat, to parse through studies and find some weakness that had not been taken into account. This paid well and drove many to despair.

Dismayed by such technical caterwauling, Hill weighed in with a clear warning. He urged that statistical evidence should never be considered the sole litmus test for public health analysis.

I wonder whether the pendulum has not swung too far—not only with the attentive pupils but even with the statisticians themselves. . . . There are innumerable situations in which [statistical tests] are totally unnecessary—because the difference is grotesquely obvious, because it is negligible, or because, whether it be formally significant or not, it is too small to be of any practical importance. What is worse, the glitter of the table diverts attention from the inadequacies of the fare. Only a tithe, and an unknown tithe, of the factory personnel volunteer for some procedure or interview, 20% of patients treated in some particular way are lost to sight, 30% of a randomly-drawn sample are never contacted. The sample may, indeed, be akin to that of the man who, according to Swift, had a mind to sell his house and carried a piece of brick in his pocket, which he showed as a pattern to encourage purchasers. The writer, the editor and the reader are unmoved. The magic formulas are there.

Of course I exaggerate. Yet too often I suspect we waste a deal of time, we grasp the shadow and lose the substance, we weaken our capacity to interpret data and to take reasonable decisions whatever the value of P. And far too often we deduce "no difference" from "no significant difference." Like fire, the chi square test [of significance] is an excellent servant and a bad master.

Just imagine a physician telling a patient with a monstrous headache to wait five years for a study to be completed before getting treated for his malady. If medicine were practiced like that, it would have long ago gone out of business. Decisions about what to do for any illness must be based on judgment and inference while the work proceeds to resolve questions at hand. When groups of people are affected, the need to draw on such judgment remains great.

Hill explains that on fair evidence we might take action on an apparent occupational hazard.

For example, we might change from a probably carcinogenic oil to a non-carcinogenic oil in a limited environment, and without too much injustice if we are wrong. But we should need very strong evidence before we made people burn a fuel in their homes that they do not like or stop smoking the cigarettes and eating the fats and sugar that they do like. In asking for very strong evidence, I would however, repeat emphatically that this does not imply crossing every t, and swords with every critic, before we act.

All scientific work is incomplete—whether it be observational or experimental. All scientific work is liable to be upset or modified by advancing knowledge. That does not confer upon us a freedom to ignore the knowledge we already have, or to postpone the action that it appears to demand at a given time.

"Who knows," asked Robert Browning, "but the world may end tonight"? True, but on available evidence most of us make ready to commute on the 8:30 next day.

Hill closes his comments on this topic by invoking Sherlock Holmes's advice to Dr. Watson: "When you have eliminated the impossible, whatever remains, however improbable, must be the truth."[22]

If statistical issues were the only challenges to conducting epidemiology, they would be daunting enough. As Hill knew at the time, the real difficulties of the field have been complicated by a stream of disinformation fueled by short-term economic interests of those who stand to profit from keeping matters unresolved. There is no way to know whether or not the delays in controlling cigarette smoke in the United States and Britain had anything to do with the nefarious origins of the Nazi studies on the problem in the 1930s. Certainly the German studies remained obscure. A 1931 study by the insurance analyst Frederick Hoffman found increased health risks for smokers. Raymond Pearl's report on the higher death rates of smokers in America was presented in 1938 in *Science* magazine and included statistically robust calculations.[23] By 1950, several different reports appeared in a single issue of *JAMA,* all showing increased deaths from lung cancer for smokers. The larger question that must be asked is, How did it happen that democratic countries took such a long time to act against smoking?

Like much in the war on cancer, the failure to control tobacco and other causes of the disease was not incidental but deliberate. As Allen Brandt has shown in his definitive historical review of the subject, the reach of Madison Avenue into the realm of the scientific debate on tobacco remained sure and insidious—often invisible even to those on the frontier of research.[24]

The controversy over smoking did not end with British and U.S. official reports on the dangers of smoking in the 1960s, because the private sector and the government, under both Republicans and Democrats, combined forces to keep cigarettes safe from control until the 1990s.

In the late 1970s, I worked briefly in the administration of President Jimmy Carter. Today President Carter is known for his global interests. Back then he was struggling politically with many different

agendas. Tobacco industry influence permeated the Carter administration like a smelly cigar. It certainly appeared that the requirement
for a political appointment to a health post in the Carter administration was being a chain-smoker. Douglas Costle headed up the Environmental Protection Agency. David Rall directed the National
Institute of Environmental Health Sciences. Joseph Califano ran the
Department of Health, Education and Welfare. Each one of these men
smoked more than a pack a day at the time of their appointment to
the government.

By the time they left office, each had quit smoking. But the controversies about smoking had not quit them. The battle over passive
smoking had become the next front. EPA scientists had shown that indoor levels of fine particles from cigarette smoking could far exceed
those for which they had already set standards outdoors. Costle was
ordered not to measure the amounts of fine particles released by cigarette smoke in the indoor environment—an order that still stands today. Although he headed the National Toxicology Program, Rall
couldn't study the ability of sidestream or passive tobacco smoke to
cause cancer in animals in experimental studies. Gori, the former
head of the government's safe cigarette program, eventually emerged
as one of the tobacco industry leaders fighting against any government
effort to declare passive smoke a health hazard.

In his book *Propaganda,* written in 1928, Edward Bernays argued
that democracy depended on the successful control of public opinion.

The conscious and intelligent manipulation of the organized habits and
opinions of the masses is an important element of Democratic society.
Those who manipulate this unseen mechanism of society constitute an
invisible government which is the true ruling power of our country. . . . We are governed, our minds are molded, our taste formed,
our ideas suggested, largely by men we have never heard of. This is a
logical result of the way in which our Democratic society is organized. . . . In almost every act of our daily lives, whether in this year of
politics or business, in our social conduct or our ethical thinking, we
are dominated by the relatively small number of persons . . . who understand the mental processes and social patterns of the masses. It is
they who pull the wires which control the public mind.[25]

Bernays was a very clever man, as he never hesitated to let others know. In the United States, the burgeoning tobacco industry became one of his best clients after World War II ended. Bernays developed a major tactic for shaping public beliefs. Advertising expenses for tobacco in the decade after the surgeon general's report came out in 1964 was close to $3 billion, much of which supported the high-minded scientific declarations of professors and medical researchers from some of the nation's leading universities.

In truth, the great bulk of the "science" of tobacco that appeared in print was actually well-disguised public relations work. Public doubt about the dangers of tobacco had long been carefully nurtured under the aegis of various medical experts. In some real sense the modern field of epidemiology grew up in response to the tactics of the tobacco industry. The basis for determining what constitutes evidence of human harm within the canons of public health research was crafted in large part under terms set out by the tobacco industry. A well-engineered campaign of unsuspecting experts fomented doubt about what sort of evidence is needed to establish proof of human harm regarding any public exposure. Based on what now appears to have been at best misguided and at worst delusional thinking, the American and British governments spent millions trying to help the tobacco industry come up with what we now know is an impossibility—a safe cigarette. How so much money was spent on such a bad idea for such a long time offers a moral tale of considerable relevance to other realms today.

It's not clear what was more foolish: the idea that a healthy cigarette could be designed using filters and newly configured types of tobacco waste or the notion that the government ought to pay to come up with such an entity to benefit what was already a heavily subsidized, multi-billion-dollar, multinational industry. The decision to attempt to engineer a safe smoke can be viewed as the triumph of wishful thinking over reality. This has got to be seen as one of the more perverse results of tobacco's grip on public thinking. Where did this notion that cigarettes can be made safer come from? It was fueled in part by scientists who were themselves heavy smokers, and by a naive sense that a safe cigarette would be far easier to create than a major program to discourage most men and growing numbers of women from smoking.

The long battle to gain public acceptance of the dangers of tobacco lasted as long as it did because of two things. Tobacco is a highly addictive product and was made more so by the ways the industry physically manipulated the cigarette. The addiction to tobacco for smokers was chemical, but for the rest of society it was just simple economics. Tobacco advertising and sponsorships invaded every part of modern life. The workings of the invisible wires of the mass media about which the public relations guru Bernays waxed did not end on Madison Avenue but ran straight through the academic world. Goaded by a bevy of expertly expressed technical doubts, public opinion came to wonder whether proof could ever be developed that smoking is harmful. Each time a report appeared showing that smokers faced increased risks of lung cancer or other diseases, experts were tapped to explain how the given study had not quite got it right. Crucial evidence was missing. The groups looked into were not really the right ones. Some basic flaw kept the results from being deemed definitive proof. The costly, decade-long effort to craft a healthier cigarette can be seen as further proof of how scientists are just like anybody else. They want to believe in what they know probably isn't true, if it makes their world a bit easier to live with.

As the systematic study of how anything affects health, epidemiology became an important public force in large part because of its role in exposing the way smoking affects health. The triumph of the field in clarifying the impact of smoking also becomes part of its limitation. Why did it take so long for the world to come round about the dangers of tobacco? Did those scientists who kept repeating studies until more than fifty different papers showed the same results truly believe it was necessary to find that not only American and British, but Dutch and German and French and Canadian lungs all behaved the same way?

The presumption that your health can be controlled by your personal behavior was not widely shared in the 1950s. Disease was thought of as bad luck or a sign of moral failure. Within the Judeo-Christian tradition, if you got sick, it was assumed that somewhere along the line you had done something wrong. All this would change with the development of modern medicine. Antibiotics and life-saving surgery and life-sparing anesthesia made it possible to prolong life.

Questions about what people could do to promote their health became more important once some of these modern medical interventions became widely available. Austin Bradford Hill, Richard Doll, Thomas Chalmers, Archie Cochrane and other clinical pioneers created a system that had never existed before. Information could be gathered to determine what medical practices actually worked and which didn't. The radical impact of their work is hard to appreciate.

Internal documents have revealed that in response to overwhelming information surfacing on the dangers of tobacco, the industry changed its tune. Perhaps older cigarettes had been dangerous, but new ones would be tastier and healthier. If they could argue that they had made a safer cigarette, one that was less harmful, their market would grow. Filter cigarettes were to be the salvation of the tobacco industry. They intended to show that smokers of filters lived better, longer and with fewer health problems. For this to be established, it would be necessary to rely on studies of humans, the same studies that they had generally dismissed as lacking in rigor whenever they pointed to the hazards of tobacco.

Ultimately, despite all the public relations gurus in the world, public health research eventually led to the undoing of the tobacco industry. The road to demonstrating the harm of tobacco was not short or simple and is littered with some surprising political and scientific bodies. That struggle has become global. Tobacco remains a growth industry in China, India and Latin America. In these rapidly growing nations, some leaders are persuaded that short-term economic gain from expanding sales will offset the longer-term public health damage. Some calculating cynics once quipped to me in China that having people die quickly from lung cancer is better than having them live longer and requiring more money for their care.

Bernays, the man who pioneered public relations, didn't smoke. He died at 103 in March 1995. He lived long enough to see his work used by the Nazis and the tobacco industry. "When the profession of public relations was first outlined," Bernays wrote, "it was envisioned as other professions functioned: That is, as it was applied to a science, in this case social science, and in which the primary motivation was the public interest and not pecuniary motivation. . . . No reputable public relations organization would today accept a cigarette account, since their cancer-causing fact has been proven."[26]

Judge Gladys Kessler of the Federal District Court of the District of Columbia recently detailed the seamy network on which tobacco disinformation has rested for nearly half a century. Her opinion should be required reading for anyone concerned with the ability of democratic societies to rely on expert advice.

For their concerted, coordinated efforts to sow doubt about the dangers of smoking, she found several major tobacco firms guilty of racketeering "with zeal, with deception, and with a single-minded focus on their financial success, without regards for the human tragedy and social costs that success exacted." Kessler's lengthy, thoughtful decision shows that the manipulation of science on the dangers of tobacco was not limited to one nation or a few academic centers. Instead, the tobacco firms built an international network that fomented phony scientific debates, engaging some of the world's top scientists along the way.

Just south of the northern Italian city of Ravenna runs a small stream called the Rubicon. During the heyday of the Roman Republic, it marked the northern boundary of Italy and the city of Rome. Under ancient Roman law, any invading general who crossed the Rubicon could never turn back and had to be prepared to face draconian consequences. We may have crossed the Rubicon in public health research on cancer-causing materials today. It is not possible for any major public health issue involving millions of dollars and lives to be resolved without a major press of public relations and scores of cutting-edge scientific studies being funded by those who run the invisible government. The dreams of an open society where the marketplace of ideas governs looks faint and ephemeral. What information is permitted to get to the marketplace, who decides when to release findings about public health hazards, all these things are not determined by scientific inquiry but by the social and economic realities that constrain them.

A 1977 Herblock Cartoon, copyright by The Herb Block Foundation.

8

The Good War

The most exciting phrase to hear in science, the one that heralds
new discoveries, is not "Eureka!" but "That's funny . . ."
—— ISAAC ASIMOV

ON ITS WESTERN FRONT, World War I was largely fought along a line
that snaked from the North Sea to France's mountainous border with
Switzerland. Men sat for hours in sandbag-fortified trenches four to
eight feet deep, waiting for the occasional volley of gunfire to ring out
across the no-man's-land separating enemies. At first the war was
fought with quaintly old-fashioned etiquette. In the early days of the
conflict on the Austrian front, according to one Viennese veteran, a flag
went up daily to signal when it was time to break hostilities for lunch.

That civility ended weeks after the war started, when the French
lobbed tear gas grenades full of irritating xylylbromide into German
trenches. In April 1915, the Germans sent thick, yellowish green mists
of chlorine gas wafting over the French dugouts. Later that year, the
British killed 2,000 of their own men with the same gas when the
wind changed direction. Some of the troops on both sides got gas
masks. Hot and cumbersome, the masks didn't work well.

Chlorine gas arrived in thick, yellowish-green clouds that left soldiers unable to see or move. But these mists could be detected from a
distance and thus, with luck and a fair wind, avoided. Mustard gas
proved far more insidious. This gas was odorless and could be taken
deeply into the lungs. Those who gulped in the most gas were the
worst for it. Parts of the lung would die. The throat would tighten,

lungs would spasm. Pneumonia followed. Eventually a man would suffocate. In the phrase of the day, those who survived these attacks
would have lungs that were "completely knackered."[1]

Those who were not killed in a few hours were left with telltale
blisters and scars on the outside of the body and on whatever surfaces
of the lung the gas reached. Many were permanently disabled. But
their survival provided some fundamental lessons about how the human body responds to chemicals. A strange and important clue about
what might help treat cancer came out of some of the awful experiences of this first war to use poison gas.

In 1919, with Europe awash in the residues of the war to end all wars,
an astute U.S. Army captain named E. B. Krumbharr noticed a pattern
among the men who had been gassed and made it back to his medical
post at a French army base. The white blood cell counts of these gassed
survivors were amazingly low. At first he thought there had been some
mistake, but when he repeated the test, he saw the same thing. These
men had the lowest counts of white blood cells he had ever seen.
Krumbharr reported on thirty-four such survivors. Many of them made
it back to the military base on stretchers, pale, nauseated, and unable to
stand or sit. The Captain suggested that if anything would ever help
these men, it would be some treatment to replenish their white cells.[2]

RED AND WHITE blood cells are made in the bone marrow, a spongy,
soft cavity that runs down the center of long bones such as the femur
and provides the vital core of our hips, sternum, ribs and vertebrae. At
birth, all human marrow is bright red—packed with red blood cells;
but with age, more and more of it turns yellow. About five pounds of
bone marrow is all most of us need to keep the right balance of red
and white blood cells. When the marrow is damaged and loses its ability to maintain that combination, we can bruise or bleed easily, come
down with infections, or worse.

Some nineteenth-century physicians published case reports on patients with way too many white blood cells. Writing in German, then
the common language of science, they named the disease *Weisses Blut,*
meaning "white blood." The term *leukemia,* which simply translates this
phrase into Greek—*leukos* (white) *heima* (blood)—was first proposed

by the Scottish physician John Hughes Bennett in 1845 to describe a form of cancer in which the bone marrow makes too many white cells and not enough red ones. People with leukemia tend to bruise easily and look pale because they lack the red cells which contain hemoglobin, which transports oxygen. Leukemia patients are literally anemic, or without sufficient amounts of red cells.

White blood cells are measured in terms of how many of them can fit into a perfect liquid cube that is a millimeter, one thousandth of a meter long, on all sides. Today these cells are counted by machine, but in Captain Krumbhaar's day it was done by eye. He would have used a microscope to peer at a drop of blood sandwiched between two glass slides. The upper slide was etched with a finely lined grid, each square precisely 10 microns on a side. He carefully counted the number of white blood cells that appeared within a number of grid squares and used the average value to calculate the total number in a thousand-micron square. From this number he would have calculated the number of white cells per cubic millimeter. This is called the white cell count. A cubic millimeter of blood from a healthy person usually contains between 4,000 and 11,000 white cells. Patients with leukemia can have white cell counts as high as several hundred thousand. Survivors of poison gas attacks, on the other hand, might have close to zero.

LONG AFTER NATIONS had carried out uncontrolled experiments on soldiers at war, when mustard gas was finally studied in a laboratory it would prove to be a Janus-faced compound. On the one hand, those who worked to produce poison gas during World Wars I and II under what were believed to be safe conditions would eventually be found not to have been safe at all. Lung damage and cancer would show up years later, both in men who had made the gas and in those who lived through attacks of it. On the other hand, the devastatingly low white blood cell counts first seen in survivors of mustard gas at the end of World War I provided the basis for the golden age of chemotherapy. While mustard gas attacked the lung, it also suppressed the bone marrow's ability to make white blood cells.

People with leukemia have too many white blood cells. Those who made it through poison gas attacks had far too few. It is an irony of medical history that the natural experiment of war and poison gas provided the foundation for modern chemotherapy. Physicians working for the U.S. and British armies were among the first to put the two observations together, but it would be more than a decade before they could talk about it.

Their work to create drugs to treat cancer remained top secret military research for close to a decade as part of a covert program conducted at a number of leading medical centers in America. In the 1930s and 1940s, the Yale anatomist Thomas Dougherty asked a question that was both fundamental and simple: Could poison gas be altered so that it killed cancer without killing the patient? Two young assistant professors at Yale, Lewis Goodman and Alfred Gilman (who would later be known to generations of medical students as the authors of a widely used textbook of pharmacology), conducted the first studies on nitrogen mustard in a single mouse. Barely able to walk, this unlucky rodent had a tumor twice the size of its body. What happened to it would send shock waves through the research community. Two decades later, when Gilman looked back on this work from the 1930s, he noted that this particular mouse had proved especially lucky for this research. After just two treatments, its massive tumor began to shrink. The tumor returned about a month later:

> We then treated the animal again and a regression appeared again, although it was not as complete as the first time. In any case, the tumor did decrease and finally began to grow again, at which time further treatment brought about no inhibition of growth. This animal lived eighty-four days following implantation of the tumor, which was a very remarkable prolongation of survival time . . .
>
> Many animal experiments followed in which we varied the doses, number of administrations of the same dose etc., in order to attempt to find a proper method of treatment during the course of the lymphoma growth . . . Among these results was the interesting fact that the very first mouse treated turned out to give the best result. We never achieved an eighty day prolongation of life in any other animal. The best we did was some forty day prolongation which, of course, we now know is highly effective. However, in most of the murine leukemias, particularly those which metastasized rapidly, we frequently obtained no effect at all.

I had often thought that if we had by accident chosen one of these leukemias, in which there was absolutely no therapeutic effect, we might possibly have dropped the entire project.[3]

At the start of 1942, working under a secret army contract, Goodman and Gilman applied this same remedy to a forty-eight-year-old silversmith whose cancer, a lymphosarcoma, had spread throughout his body. Tumor masses extended to his armpits, belly, face, and upper chest. He could not chew or swallow.[4] Using a derivative of nitrogen mustard, they added a month to his life. For a while the tumors completely disappeared. This was a phenomenal and unprecedented success.

Although this patient later died, the fact that he had lived at all was enough to start a revolution in thinking about bone marrow cancer. Around the country, secret military research began on sixty-seven patients with advanced and terminal forms of cancer. Because it involved poison gas, extraordinary precautions had to be taken with the agents. No one on the medical staff who administered the first drugs had any idea what they were using; things were labeled with codes, not names.[5] While many of these patients died quickly, some survived. Physicians began to believe that chemical agents would eventually prove successful.

A landmark article by Goodman and Gilman in the September 21, 1946, issue of the *Journal of the American Medical Association* reported that intravenous nitrogen mustard kept a few terminal cancer sufferers alive. The treatment worked best with patients who had Hodgkin's disease. Many cancer patients died soon after the infusions began. Their veins shut down before the drugs could course through their bodies.[6] This ground-breaking paper described a fifty-two-year-old man with chronic myelocytic leukemia and a severely swollen belly. His white cell count had soared to 293,000 per cubic millimeter, and radiation had stopped keeping his disease in check. Nine months after a series of intravenous treatments with an adapted version of nitrogen mustard, he was reported to look, feel and act disease free. His blood counts were normal.

Years later, people would point out that the disease he had been cured of—chronic myelocytic leukemia—is not always fatal. Sometimes white counts spontaneously stabilize, though they never return to normal. But at the time, this man had been close to death. His survival was seen as proof of what lay ahead. Two years later, Sidney Farber stunned the medical world with a report in the *New England Journal of Medicine* in 1948. At the time, children with leukemia were expected to die quietly and were known to die more quickly when given folic acid. Farber reported that giving them agents that blocked the formation of folic acid basically extinguished their cancers.

These discoveries rocked the cancer world. They showed that Mary Lasker had been right about one thing: medicine could come up with drugs to treat cancer. Within ten years, death rates from leukemia in children had begun to fall. Where the disease had been a death sentence for most, it became treatable in many instances. People were brought back from near death to relatively normal lives. Gradually, however, the human price of this miracle treatment became known. Doctors were not surprised to see hair loss, pale skin, wrenching nausea, and deadened feet and hands—the hallmarks of cancer-treating drugs. The side effects of chemotherapy were accepted, then as now, by those who lived through it. Life is worth any price.

After years in which cancer was a death sentence, chemotherapy offered real hope. Buoyed by these phenomenal results, the American Cancer Society began funding its own program in 1958 to look for chemicals that would attack cancer. The ACS program was larger than the government effort.

As a young boy in 1920, my husband's Uncle Phil left his dirt-floor home in the small farming village near Bialystok, a region that was either Russian or Polish depending on the tumult of history. Phil and his older brother Ralph had learned to avoid the bodies on the streets of those who had been felled by bullets or poison gas warfare against the Russian army. The Morgenstern boys and their mother Fannie spent two very long, stomach-churning weeks in third class on a packed steamer. None of them would ever set foot on an ocean vessel again. Neither they nor their mother, nor most of the thousands they sailed with, spoke a word of English. Their father had left for America just before Phil was born. By 1920, he had saved up enough money working odd jobs in the Flatbush section of Brooklyn to bring them all over. At Ellis Island, they passed inspection for obvious defects, three among hundreds that day. Phil was seven and Ralph was sixteen.

Seventeen years later, Phil graduated from George Washington University Medical School and Ralph, a Columbia University law graduate, was working as an attorney in New York City. From 1943 to 1947, Phil served as an army physician, stationed at Redstone arsenal near Huntsville, Alabama. This rustic area of sharecropper cabins and magnolia-scented gardens housed the nation's producers of phosgene, chlorine, and mustard gas. The military base at Redstone arsenal was the largest poison gas facility in the United States, set up after President Franklin D. Roosevelt declared a national state of unlimited emergency on May 27, 1941. After the Japanese attacked the U.S. military base at Pearl Harbor, Hawaii, arms production became a 24/7 affair. By 1941 the government budget for chemical gas production reached $34 million. By the war's end it would be more than $1 billion. The facility in Alabama produced millions of gas grenades, bombs, shells, and tons of chlorine and mustard gas, much of which was shipped to Britain.

Morgenstern provided medical care to the arsenal workers in Building 117. Most were young men and women from the surrounding farms, some still teenagers. Their job, for which they received no particular training, was to produce poison gas, place it into steel vats set in caves that were dug into the local mountains, and take care that nothing and nobody slipped up while doing so. These were good-paying military jobs for hard-working men and women who otherwise had few opportunities in the rural, Jim Crow South. At first only local white men and women were hired for the plants, along with a few black men. Segregated facilities were maintained, including separate day care and school programs for the workers' children. When production needs grew, black women were allowed into the workforce.

A half century later, Phil recalled the patients he would see at the arsenal. "We had a hospital on the grounds to take care of all casualties—black or white—that would arise from routine army work, including that at the nearby arsenal. Most of the patients I saw in the dispensary came from the chemical plant."

"But they must have been using gas masks even back then?" I asked. "Everyone knew this was poisonous material to be around, right?"

"Of course they were given masks," he replied. "That didn't mean they used them. Many workers found the masks hard and cumbersome. No one had much experience with these things. We couldn't measure the gases because we didn't have tools for doing so. But we could certainly see their traces. These men and women had raw, red eyes, violent coughing spells that would last two to three minutes or even longer. Some of them could barely walk.

"It was clear to me that this had to be related to what they were doing on the job. Some of these guys would get better every weekend and be sick as a dog again by the end of the week. I would recommend to the supervisor that they be given rest and sent to another part of the plant, where they would avoid these exposures. After a few days, their symptoms would clear, and they would return to work and the coughing would begin all over again.

"Basically, more spills would occur and by the end of the week, they usually were back in the clinic with acute respiratory symptoms. Every weekend they would feel better, and the cycle would go on again."

Diagnostic equipment was limited then and kept in areas where soldiers might need it, but was not available to the men and women working in chemical plants in Alabama. Some workers were sent to Memphis for chest x-rays. Their films did not reveal any real shadow such as you get with tuberculosis or pneumonia. The radiologists believed that since they could not see anything on x-ray, no real problem existed. Phil Morgenstern looked at things differently. Many of the workers in the chemical weapons facility eventually started coughing most of the time. The plant supervisors assured Morgenstern that these men could not possibly have been exposed to any of these agents. If there had been even the slightest exposures, he was told, they would have been obviously and immediately sick. Morgenstern wasn't so sure. He asked whether gas escaping at levels too low to be detected could leave these people with impaired breathing.

All these so-called specialists saw were slight markings on the bronchi [the lung]. They believed these men were faking how sick they felt. So they said this was not a service-connected disability. They refused to provide any support to them. . . . I appealed the decision. I didn't agree. I thought the injuries clearly were due to their work and could result in more serious health problems down the road. I had had extensive experience as chief of medicine in a 2,000-bed hospital. The government finally sent a senior man to check my claim that these lung disorders had arisen from repeated exposures. After many months, some of the workers finally got relief pay. As far as I know nobody ever followed them up to see what happened to them.

Morgenstern had seen the reports that white cell counts dropped to the floor in soldiers who came out of poison gas attacks. He knew the lungs of those who worked producing such gases had been compromised and wondered what this would portend in the long run. As the war effort grew and even after it ended, the nation remained on alert, fixated on immediate threats—real or imagined. Morgenstern told me years later that he puzzled about the lingering residues of something that could cause deep disturbance in the lung at levels that could not be measured then. These were not questions others were willing to pose. The symptoms Morgenstern was seeing would later become familiar to cancer doctors. As an apt Chinese proverb advises, if you don't want to know, don't ask.

THE LASKERS AND their powerful friends may have had somewhat different expectations of what the national effort against cancer could deliver. When Mary Lasker created the famed Lasker Awards for Cancer Research in 1946 as a birthday present for her husband, Albert, she fully expected a cure for the disease within a decade. Those who ran the ACS were skeptical but understood that belief in an imminent breakthrough guaranteed steady financial and political support. Their language made no bones about what lay ahead. The slogan "Beating cancer!" rallied volunteers and fund-raisers throughout the nation. A country that had just beaten Hitler, Mussolini and Hirohito could certainly conquer cancer. The new chemotherapy agents provided valuable fodder in the public campaign to generate more funds for research on the disease.

When the ACS began to flex its national muscle in the 1950s and 1960s, medicine was coming into its own as a field that could incorporate stunningly sophisticated techniques. Advances in surgery for cancer and other problems had been breathtaking. Better anesthesia and improved technologies for saving patients both during and after surgery markedly reduced the risks of operations. Transfusions of blood and saline solutions reduced death from shock. With the new antibiotics, previously fatal postsurgical infections fell rapidly. By 1950 about half of all women with localized breast cancer survived five years. The number of patients living for five years after other forms of surgery also rose substantially. The percentage of medical doctors who were surgeons doubled between 1935 and 1965.

As a cure, however, surgery had its limits. People whose solid tumors could be removed were often left with permanent scars and disabilities. Surgery offered no help to those with cancer of the blood-forming organs, like leukemia. Nor did it provide much hope to the growing number of men who were contracting lung cancer or to those with stomach cancer. Against this backdrop, and with the intense personal lobbying of Mary Lasker and the ACS, Congress launched one of history's most intensive medical dragnets, giving the National Cancer Institute $25 million in 1958 to find chemicals to cure or treat cancer. The NCI began examining thousands of potential treatments, more than half of which came from natural products.

Enthusiasm for the cancer effort crossed political and economic boundaries. Americans and their British counterparts in the cancer world rallied to a great cause—conquering a universal enemy. Elmer Bobst, a major force in the ACS, the chief of Warner-Lambert Pharmaceutical and a staunch ally of Richard Nixon, explained his abiding commitment to increased funding for cancer research.

> However partisan I have been, I believe the most rewarding of my political activities were nonpartisan once performed, as a result of my work in the American Cancer Society, on behalf of American health care generally. My involvement with the cancer society brought me into contact with a small group of active people—someone called them "benevolent plotters"[7]—who set out during and after World War II to revolutionize American medical education, research, and health care . . . All of them [Albert and Mary Lasker, Michael DeBakey, Alton Ochsner, and others] did not agree all of the time, but all were united in one purpose: to stimulate federal support of medical research and education.

Boosters of the search for drugs to treat cancer were nothing if not supremely confident. The business executives who led the ACS had a simple plan: they would mimic the enviable track record of American engineering in space and war. These early leaders of the cancer war were not scientists. Most of them regarded scientists as bothersome nuisances with no grasp of how the business world got things done. They had little patience for the incremental pace of normal science. They wanted a breakthrough, a D-day–style offensive, an all-out invasion against the enemy. Dazzled by the success of the space program, they were sure something comparable could be achieved against cancer. Bobst helped persuade Nixon to make it his administration's top public health priority.

Yet despite the enthusiasm over chemotherapy, despite even the 1964 surgeon general's report on smoking and cancer, by the time Congress began to debate the National Cancer Act in 1970, funding for the National Institutes of Health was dropping fast. The Laskers and their influential associates knew they had to drum up public support. Pulling a lesson from the master of public relations, Edward Bernays, they did not employ media experts to make the case for increased cancer research funding. Instead, they tapped scientists and physicians who had a knack for the quick retort and snappy phrase, turning medical leaders into public voices of authority. Michael DeBakey and Manuel Farber, the medical rock stars of their time whose stunning feats saving lives with surgery and new drug regimens were widely reported, proved invaluable spokesmen. One person watching them work over the Congress put it this way: "DeBakey is unique; he has the aura of the surgeon, he's articulate, enthusiastic. Most doctors are not enthusiastic, not used to the verbal give-and-take. The Rusks, Farbers, DeBakeys have evangelistic pizzazz. Put a tambourine in their hands and they go to work."[8]

When he signed the National Cancer Act officially launching the war on cancer in 1971, President Nixon was riding a public groundswell that it had taken the ACS over twenty-five years to build. He was also trying to head off his political rival, Sen. Edward Kennedy, who was viewed as a champion of medical research. The president would lead a war everybody could love, a popular war that would divert the public from the one America was then losing in Southeast Asia. Nixon had been well briefed by scientists and leaders of the American Cancer Society. Cancer, he declared, was no longer a death sentence. Chemotherapy was taking people from their death beds back to their dinner tables around the world. The miraculous successes against leukemia were a forecast of the future. The war on cancer would be fueled by the belief that a cure for cancer was within reach, simply a matter of a little more time and a lot more money.

Nixon promised the country that the official national cancer enterprise would be more than a patchwork cobbled together from disjointed pieces of medical research. This was to be a big, popular war that trained its gun sights on the disease itself. It did not target any of the things known or suspected of causing cancer at the time. Neither tobacco smoking nor chemical exposure to the likes of asbestos or synthetic fabric dyes was put on the table as a matter of serious concern. With that can-do confidence peculiar to Americans, the president echoed the view that the same ingenuity that put a man on the moon and crafted the atomic bomb would wipe cancer from the earth within the decade.

One of the only researchers to go on record as opposing the notion of a war on cancer was Philip Randolph Lee, a physician who would become assistant secretary for health under President Carter. "When they made the Cancer Institute director, for example, independent of the director of the National Institutes of Health, I was one of the few people against the war on cancer," he noted years later. "Gaylord Nelson was the only person in the Senate to vote against it. When I went to see Sen. Allen Cranston to say that he should oppose it, he said he had 6,000 letters from constituents in California telling him he should support it."[9]

What was Lee concerned about? "I just didn't see the logic. I thought we were promising people things we could never deliver."

We will never know whether any of the ardent supporters of the cancer war realized how shortsighted the enterprise was. Clearly some outside observers understood the truth. A Herbert Block cartoon from 1980 features a group of enthusiastic, well-dressed men peering over the shoulder of a beleaguered, bald, bespectacled scientist sitting at a microscope. "Could you hurry and find a cure for cancer?" they ask. Their shirts are labeled Cigarettes, Asbestos, Radiation, Chemicals—the industries that stood to gain the most from a cure.

Early planners of the national strategy against cancer brought in battle plans and even personnel straight from NASA. For analysts who had managed to put our man on the moon before the Russians, cancer seemed like an easy second shot. These rocket engineers were a little different from the scientists at the Centers for Disease Control and the

National Institutes of Health. For Clark Heath, in 1971 a young epidemiology intelligence officer in Georgia and now executive vice president of the American Cancer Society, the drumbeat of advocacy proved very convincing. "One day some of the seasoned managers from NASA marched into the CDC and set up easels which held several large overlapping diagrams with lots of charts and boxes with arrows. They had the entire war on cancer mapped out on one huge drawing. All the known technologies about cancer and viruses were on the bottom left, and all the things we needed to fill in, so to speak, were at the top right. To get from one to the other, it was all just supposed to be a matter of filling in the blanks to comply with their engineering requirements."

The U.S. government devoted more money to the war on cancer than it had ever committed to a medical problem. Ronald B. Herberman, now the director of the University of Pittsburgh Cancer Institute, where I work, was a young investigator at the National Cancer Institute in the early 1970s. "I remember going to a big meeting at the NCI, which was pretty heady stuff for someone at my level. I will never forget that they brought in all these charts, with a series of concentric circles. These became widely referred to as the bull's-eye charts. Right there in the center was our target—the cure for cancer."

Many of the bright young scientists with whom he worked were incredulous. "We knew how complicated it all was. We understood that cancer was not going to be a simple problem, no matter what these engineering diagrams showed."

The idea of a single remedy for cancer struck most serious cancer researchers as preposterous. Cancer was a disease of inherent complexity—really more than two hundred different diseases that had to have different causes and would require different treatments. There were a lot of smirks and rolled eyeballs in the scientific community. To Lee, "it all seemed like just a political fuss. It was pretty clear that declaring this as an official war was just a smokescreen for President Nixon. Remember, things are going pretty badly in Southeast Asia at the time. The White House was surrounded with protests. Beating cancer is something that everybody wanted to do. Some of us realized it was mostly a political show."

Herberman shared some of Lee's skepticism, which few then dared to express. He also realized that it would be necessary to think outside the boxes and circles being proposed. It made little sense to try to kill a single enemy with a gigantic bomb or cannon, when the enemy was not one target but many. "You really can't organize basic research the same way you go about putting a man on the moon. In that case we got the right rocket engine, found the right fuel, built the right rocket, set the right trajectory, all that sort of logistic stuff. Found the right people with the right stuff to get there. We actually did it!" The new war had more complex and elusive goals. "Fighting cancer involves far more than taking stock knowledge off the shelf of the laboratories and slapping it into some design problems. We are taking a deeper look into what can be done to strengthen the body's defenses to cope with the disease and not merely coming up with ways of killing the many different types of ailments that have arisen from unregulated cell growth. Of course we have to pay more attention to prevention, to how the things people do every day affect their chances of getting cancer. Exactly what that means and how to go about it is not at all clear."

By the 1950s, researchers had a number of clues about how to treat cancer and it was well known that people in certain jobs had more cancer. But there was no clear idea of the cellular mechanisms involved. The absence of technical information about the underlying processes that give rise to cancer—whether those tied with tobacco or industrial processes or radiation—was carefully cultivated into a simple argument: Cancer is so complex. We need more scientific research before we can act to control cancer, so let's wait. And wait. And wait.

In the 1960s and 1970s there were other young doctors like Herberman, people with research aspirations who joined the growing band of clinicians in the new specialty of treating cancer patients. Excited and inspired by the spectacular leukemia cures, they devoted themselves to searching for novel ways to cure other forms of cancer.

My cousin Mark Tuckfelt was part of this golden age. As a boy in the small Monongahela valley town of Belle Vernon, Pennsylvania, Tuckfelt survived a few close calls with homemade rocket firings to graduate at the top of his high school class. His entry into medical school vindicated all the teachers who had looked the other way when he "borrowed" chemicals with which to experiment throughout his high school years. I knew all about his various adventures and close calls. By the standards of childhood, Mark was way older than I—almost three years. He lived with his family in a big new brick house across the river and up the road from our smaller, less prosperous town of Donora.

After the usual adrenaline-powered, low-sleep, high-stakes internship in medicine, Tuckfelt enlisted in the search for new ways to treat cancer. In 1972 he signed on for a two-year residency in hematology, the study of blood disorders. The newly expanding National Cancer Institute provided his highly coveted annual salary of $14,000.

"I worked with smart people at Sinai. It was exciting. I worked with clinicians trying different combinations of cancer-killing drugs to control leukemias and lymphomas. I worked with laboratory scientists studying the mechanisms by which cancer growth genes were turned on and off. We thought that bit by bit, in the laboratory and on the clinical services, we would find the answers to the cancer question."

Science historian Thomas Kuhn believed that scientific routines are not much different from run-of-the-mill puzzle solving. True breakthroughs in science are achieved by radically changing the shape and rules of the puzzle and the nature of what is even thought of as real and possible. Until the sixteenth century, in Western Europe the church tightly controlled views of how the world worked and the place of people in it. By the end of that century, some clerics advanced the radical idea that "the Bible was written to tell us how to go to Heaven, not to tell us how the Heavens go."

In the century after Christ, Claudius Ptolemy had advanced the idea that the sun circled the motionless earth. This view agreed neatly with the church fathers' sense that humans stood at the center of the divine world. In the seventeenth century, Galileo Galilei observed that the tides changed because the earth turns round every day as it circles the sun—an opinion that the Polish astronomer Nicolai Copernicus had offered earlier that same century. The acceptance of the Copernican idea that the earth revolves around the sun provided a fundamentally different sense of how the world operates from that time forward, creating what Kuhn called a paradigm shift—a major change in how we look at the world.

Like Herberman, Tuckfelt understood that medicine is also subject to paradigm shifts. Doctors still treat patients one at a time, but the things they believe about what causes disease and how best to approach it can shift radically. Tuckfelt felt his field needed a revolution in thinking about cancer that would not come just from putting more money into attacking the disease, but from taking a different look at controlling and identifying what gives rise to cancer.

"The drugs we used sometimes stopped previously untreatable cancers, but they also had new and previously unimagined, sometimes disabling, and deeply disturbing side effects. They worked by killing the things that grow the fastest. When we were lucky, we destroyed the tumors before we killed the patients. Cancer survivors had to be tough. Many of them were young. Those who came through chemotherapy tended to look like aged newborns—bald and hairless. Eventually their hair, nails would grow back, and when we got it right, their cancers did not.

"It seemed to me that there were two separate ways to attack cancer. On the one hand, we could try drugs to wipe out the disease. At the same time, I had a sense that there were some really big pieces that remained completely undiscovered. I was especially interested in learning what determines how the body tells the difference between our own cells and foreign ones and what prevents us from recognizing cancer cells as foreign and destroying them. If we could figure out the things that make immune systems do their job of keeping things in balance, we would have a completely new set of tools to fight cancer, not just the traditional poisons that came out of toxic war gases. We needed to turn around the way we looked at things. We needed to understand what fueled cancer and what would lead to its natural destruction within the body. We basically needed an entirely new paradigm—not one based on cutting, burning or poisoning."

Tuckfelt remembers his impatience with the incremental, puzzle-solving approach of normal science. "If you keep chipping away at a big hard rock, eventually it will break. It will take a long time but the important pieces will come into view, one at a time. But if you know exactly how to cut the same big rock, you can give it one tap at the right place and the whole thing falls apart! I thought that I knew what one of those places was. It involved a different way of looking at how the immune system works and a better sense of what we can do to strengthen our ability to fight. If I was right about it, it would have changed the world. It's still a possibility."

Tapped for the sort of job that many physicians run from, for the past two decades Tuckfelt has served as chief of medicine and vice president for medical affairs at Orange Regional Medical Center in Goshen, New York. He is in charge of the problems in medicine that nobody wants to admit exist—mistakes in medication or surgery that can kill or sicken people. The practice of oncology has changed but not as much as it needs to, since the old killing paradigm still has a stronger grip on cancer medicine than it should.

Tuckfelt looks back wistfully. "Of all those chipping away at the big rock, a small number were really searching for the plane of cleavage in order to open the rock and find the diamonds. Progress in that kind of research is very unpredictable. The diligent chipper will always have little pieces to show. Searchers for a plane of cleavage may have nothing to show for years.

"One of the things that I thought was really out of line back then was the idea we could actually carry out a war on cancer—go ahead and wipe out the disease. Some people looked on this all as but another chance to apply science big, just as we did in making the atom bomb and putting a man on the moon. But the analogy was way off. In both those cases, we already understood the fundamental science—Einsteinian and Newtonian physics respectively—and merely needed to develop a technology to use them. In the case of cancer, we lacked the basic science. We seldom asked what had allowed cancer to occur. We rarely thought that we might cause other cancers by curing some. Development of new fundamental science is erratic and unpredictable. You have to fund the cleavers, the guys who can figure out just where to tap the rock to make the whole thing split apart. A lot of this work leads to dead ends."

"But what about what was causing cancer? Weren't you thinking about that at the time?" I asked.

"Frankly, Dev," he admitted, "that wasn't at all on our radar. Our job was to get rid of the illness. Not many of us paid much attention to what things around us made a difference in who got the disease. If we thought about anything, of course, we obsessed about the magnificent genes and the brilliant people discovering the ways that these affected cancer. Those were the guys who would get Nobel Prizes for their work."

The personal toll for Tuckfelt was wrenching. "Patients expect you to be like a god to be able to zap whatever is wrong. This is pretty hard to deal with when you're not yet thirty and just starting to think about your own mortality. Suddenly people look to you to save their lives.

When you fail, and we failed a lot, you never escape the haunting looks of despair from your patients and their families."

Some three decades later, Tuckfelt has no doubt that a new paradigm must be devised. He argues that we must move beyond refining our ability to find and treat the disease. "We have to find out why some people who may be drenched in cancer-causing agents, like my ninety-six-year-old grandfather who drank half a pint of whiskey a day, chewed tobacco, and worked around heavy machinery, oils and grease much of his life, don't die of cancer, and why others who may have worked with a toxic chemical like vinyl chloride for just one unlucky month and may have been vegetarians end up dying from the disease."

AFTER YEARS TRACKING his own research on cancers of the blood, Herberman wanted to know what made the blood cells of those with cancer so peculiar. What turned normal white cells into exuberant abnormal ones, leaving the marrow of leukemia patients bleached and pale? The immune system is one of the most complicated parts of the human body, working like a complex cellular policeman. It regulates what can get through the blood and heart, and what gets gobbled up and eliminated through various waste streams. Herberman did not become a rocket scientist despite his boyhood enthusiasm for building model airplanes. Today Herberman is recognized as one of the world's experts on how natural immunity works and what can happen to upset it. As a young researcher, he asked why some patients appeared to recover completely from cancer after brief bouts of treatment with chemotherapy, while others perished. Could there be something different about those who survived—something that could be used to help others overcome the illness? He started out by looking at blood from those who had gone into remission, meaning that they had no sign of disease at all. He found something odd. Those who survived appeared to have established their own unique ways of combating cancer. Looking at those who had come through acute leukemia, he came up with a brilliant strategy.

Identical twins occur when one plump fertilized egg literally splits apart and grows into two embryos. At birth, they share more genes than any two nontwins possibly can. Herberman found identical twins in whom one had survived leukemia and one hadn't developed any disease. The existence of such pairs of twins, one with and one without cancer, made it clear that environmental forces play a role in whether or not the disease arises. Herberman compared the ways that the cancerous white cells of those twins who had survived leukemia reacted against the normal white cells of their healthy twins. He also contrasted how normal white cells from the cancer-surviving twins matched up with their own cancerous white cells. He thought he would find big differences. After all, the leukemic twins had been making their own abnormal cells for some time and had ample opportunity to mount an immune reaction against them. Of course, the healthy twins had no contact with leukemia cells. What he found was not at all what he had expected. In both the formerly diseased but surviving twin and in the healthy counterpart, there were normal blood cells whose sole job appeared to be to fight off anything abnormal. These fighting cells—termed natural killer cells—proved vital. Having more of them might mean that cancer patients will survive longer and better. And these killer cells are made, not born. While they come from the bone marrow, like most of what gets into the blood, these natural killer cells circulate through other parts of the immune system—those soft places doctors feel for signs of a bulge or a lump, our tonsils, lymph nodes, and spleen. Once they exist, these cells go into standby mode. They are spontaneously "on," like the body's own Pac-man defense system, ready to detect and attack abnormal cells whenever and wherever they find them. Once activated, much like stealth missiles, the natural killer cells lock on to the intruders, go into full battle mode, and destroy abnormal cells. They wait until they're called into action by tiny signs, or chemical signals and imbalances, that indicate something has gone wrong. Chemical messengers pulse through the body, organizing responses to keep us in balance, at least in most of us most of the time.

Today a lot of the action in cancer research asks whether or not natural killer cells can play a role in preventing tumors from spreading. By the end of the 1970s, Herberman's lab at the National Cancer Institute showed that natural killers are on high alert for cells that are in the wrong place at the wrong time. They can immediately identify potentially metastatic cells that have broken out of their original location and entered the blood, and quickly eliminate most if not all of them. This provided a paradigm shift as well, showing that under the right circumstances, natural killer cells can keep cancer from spreading. An entire field is growing up looking for other kinds of markers in the blood that can predict who will get cancer and how they will do. Several hundred different markers or signs of activity in the blood are being tested and retested in groups with cancer and those without to see whether we can find the right combination of these signals to tell us who will or will not get the disease.

Some cancer researchers understand that this new information means we have the ability to do something that has never been possible before. We can examine hundreds of markers in the blood—whether from genes or hormones or proteins or byproducts of digestion—in order to predict who is likely to get cancer and come up with ways to prevent that from happening. We can literally change the dice by taking what is known about how these complex factors combine to affect the chance that cancer will happen and block the process at many different steps. Using these same tools, we can also affect the chances that cancer will come back.

H. Leon Bradlow does not look like a revolutionary, but he has worked outside the box of cancer therapy for most of his life. A spry eighty-something, a trim version of Santa Claus, he has authored more than three hundred publications, many of them devoted to showing that things we eat or are exposed to early in life affect the chances we will get cancer, or that it will return once it's been treated. He works with hormones, small molecules that are the road cops of the body. He has shown that the way the body processes hormones can predict the chance that cancers associated with them will occur. The world is finally catching up with him. Recent studies in England have shown that the higher the levels of certain metabolites of hormones found in women's blood forty years ago, the greater their chances of developing breast cancer later on. Other work at the U.S. National Cancer Institute and in Scotland is showing remarkably similar results.

But there is no reason to think that this work on markers in our blood means that our fate is fixed. Bradlow and his colleagues have also shown that these metabolites can be changed. Women who eat diets high in cruciferae—vegetables such as broccoli, cauliflower, brussels sprouts and cabbage—produce less of the bad hormones and more of the good. Their immune systems appear to work better to fend off cancer. This may be one important reason why women and men in Asian nations tend to have four to five times less prostate and breast cancer than those in industrial countries, where fewer vegetables are regularly eaten.

Researchers are already putting these findings to use. In Scotland, trials are under way using diindolylmethane—a compound derived from cabbage-like vegetables called cruciferae—to make precancerous conditions of the cervix go away. In Baltimore, Paul Talalay has come up with a specially designed form of broccoli-based sprouts rich in sulforaphane that looks like it can convert bad cells back to good ones. In New York, teams are finding that young patients with a precancerous condition of the larynx called laryngeal papilloma, which normally requires repeated operations to remove small growths on the voice box, don't need surgery after drinking enriched cabbage juice. At Pittsburgh, Chivendra Singh and Sanja Srivastava have extinguished cancer in cell cultures grown from cancerous ovaries and the prostate by giving them concentrated chemicals taken from these same vegetables. Others are showing that extracts made from red wine and chocolate—two basic food groups for many of us—also look promising in their ability to turn on good properties and turn off bad ones.

This work is opening the way to fundamentally new ways of thinking about cancer, like those Tuckfelt and Herberman began to work on more than thirty years ago. Rather than waiting in high-tech ambulances ready to roll out when incidents occur, researchers are developing ways to set up traffic signals to lower the chance that car wrecks of cancer will happen. Food is one logical focus, as more and more information develops on the ability of good nutrition to fight off various insults of modern life—whether they result from drinking too much alcohol or the bad luck of growing up in high-traffic areas. Vitamin D, whether from food or inexpensive supplements, also plays a part. Getting the right amounts of sunlight every day so that the liver can make the needed amounts of this essential vitamin may turn out to be vital to boosting the body's ability to fend off pollution. Finding ways to reduce attacks on our bodies from tiny amounts of combined pollutants in the air and water from our homes and workplaces is obviously important. Coming up with ways to make more natural killer cells is another part of this promising new equation—to mobilize the body's own defenses and change the environment in which cancer cells grow.

It turns out that sixty years ago when expressing fears about the eventual effects of making poison gas on his patients' health, Morgenstern was right all along.[10] He suspected that chronic coughs in these men could be a sign of something worse to come. He also worried that lung irritation might be setting up conditions for more grave and long-term health problems to arise.[11] As best I can find out, nobody in the American defense research system has actually looked at what happened to those former sharecroppers Morgenstern took care of who worked in the factories making poison gas in the 1940s in Alabama.

But others have examined the records of those who did similar work halfway around the world.[12] In 1996 a team of researchers looked into the health of men who had worked at the Poison Gas Resource Center located on Okuno Island, right at the harbor of Hiroshima, one of Japan's centers of military production during World War II. The allies had been hesitant to bomb the concentration camps and gas factories of the Nazis, perhaps because they were so close to major economic centers of great interest to Allied firms. They also heeded pleas from the Rockefellers and Roosevelts to leave the ancient temples of Kyoto untouched. They showed no such reluctance with Hiroshima. With numerous industrial facilities going full bore, that city became, with Nagasaki, a target for the devastating blast of the world's first atomic bombs in 1945.

The ability of the Japanese to send missiles with poisonous materials from submarines onto the Pacific coast of America was rightly feared. They had shown no hesitation in using such gases on their enemies in various Asian theaters, and had also used gas on a large scale to kill prisoners of war. As in Germany, the Japanese use of deadly gas was not the rogue action of deranged individuals but part of an organized and officially sanctioned set of activities. Even though the Versailles peace treaty specifically banned its use, Emperor Hirohito personally signed imperial orders *(rinsanmei)* allowing poison gas attacks against China on more than 375 separate occasions.[13] The Japanese would commit other atrocities, including conducting autopsies on living patients. They would subject their enemies to experiments just as barbaric as those for which the Nazis became much better known. Out of these horrors, some six decades later, information has

emerged regarding the long-term consequences of some of these activities that remain relevant to the world of cancer treatment today.

By 1958 Gilbert Beebe, a researcher with the U.S. National Cancer
Institute, had put together enough information on men who had survived poison gas attacks from World War I to report that forty years
later, they had much greater chances of developing lung cancer.[14] By the
1960s, several other reports confirmed that poison gas had done more
than permanently impair or kill thousands in combat. A series of reports on American and Japanese production workers found that they all
had more lung and other cancers as a result of where they had worked.

In 1996 Japanese researchers took a long look at the health records
of nearly 2,000 men who had labored in poison gas factories between
1927 and 1945.[15] They compared their chances of getting cancer to
those of the general population. These poison gas workers had five
times more lung cancer than those without such exposure.

Other evidence of the lingering impact of poison gas on the lungs
comes more recently from Iranian soldiers exposed to a single heavy
dose of mustard gas during the Iran-Iraq War in 1986. Less than ten
years after this event, the lungs of those who survived were smaller,
sicker, and less resilient than those of persons who eluded such
attack.[16] Reportedly rates of cancer in children and bystanders to recent wars in the Middle East have also skyrocketed, though precise
numbers are not surprisingly hard to come by.[17] The types of cancers
and sorts of lung damage found in poison gas survivors in Iran and Iraq
are similar to those Morgenstern reported seeing in some of his production workers in Alabama in the 1940s. Just like so many other
workplace hazards, cancers tied to poison gas exposure on the battlefield can take decades to become evident.

The crippling agents that proved so frightening in all twentieth-
century wars left deadly residues in those who originally produced
them. The continuing studies of poison gas workers and long-term
survivors of chemotherapy confirm what we've seen elsewhere: things
that don't necessarily kill you right away can determine your fate later
on. This is a lesson that echoes throughout efforts to control and reduce the burden of cancer to this day. The chemicals that were first
used to fight wars and later provided the foundation for fighting cancer half a century ago leave long and lethal legacies.

In the winter of 1950, Harvey (the taller) and Ron Herberman played in Cropsey Avenue Park, in the Bensonhurst section of Brooklyn, New York, where they sometimes flew and repaired model airplanes. They also spent part of their childhood in southern California, close to heavily sprayed fields. They both became physicians and did laboratory research using chemicals now known to cause cancer. As middle-aged adults, both developed the same form of leukemia, although they have no family history of this rare disease.

9

Cancer Doctoring

I don't want to achieve immortality through my work;
I want to achieve immortality through not dying.
—WOODY ALLEN

SINCE THE SEVENTEENTH CENTURY, some have understood that those who earn their living with their hands face bold, palpable and unsurprising hazards. The shortened lives and wizened bodies of miners, blacksmiths, printers, and leather tanners were the stuff of literature and occasional dark paintings or grim woodcuts. Dickens, Shaw and Marx wrote of the vile working conditions of girls and boys in the mines, foundries and metal works. But what about people who sit behind desks, dispense medical advice, or work their cell phones, computers and GPS device sometimes all at once? Could their health be endangered by how they earn their living? It turns out that some prestigious professions that no one thinks of as unsafe are in fact riddled with risk.

We're not used to thinking that doctors or scientists face greater odds of developing cancer and other diseases tied to the way they spend their daily lives. But in fact more white-collar professionals, including those we turn to for help in treating disease, are dying of ailments previously thought to be much more common in blue-collar workers.

As you get older, you tend to read the obituary pages, maybe not every day, but often enough that you start to see patterns. Forty years ago cancer was rarely listed as a cause of death in major newspapers. That's changed completely. Two students working with me in Pittsburgh, Matt Zurenski and Jen Powers, conducted an informal study. They looked at obituaries of MDs and PhDs in the *New York Times* and the *Pittsburgh Post Gazette* over the past five years. For these men and women of science and medicine, of those whose obituary listed any cause of death, more than half died from cancer.

Of course, this is not a scientific study. It may be that people who die of cancer have family members who just want to tell the world about it. It could be that people who work as scientists and physicians exercise more and drink less and thus don't die of heart disease. But why do so many of them die of cancer? Could something about the things people do to become or to work as scientists and doctors affect their risk of cancer?

In centuries past, patterns of infection proved to have broad, underlying social and economic causes. As open sewers stopped flowing into local rivers, refrigeration provided more fresh fruits and vegetables, and workplaces stopped burning wastes in cities, deaths from once fatal infections like tuberculosis and typhoid dropped in industrialized nations. Chronic illnesses also can have broad social roots in both developed and developing countries. Just having the bad luck to grow up in a place downstream from heavy industry can add to this cancer burden. Lately the Chinese government has admitted that cancers in people living along some of its heavily polluted rivers and streams in Huangmengying in the Huai River basin follow the steady flow of heavy metals, leather tanning, paper and pulp mills, and other uncontrolled pollutants that render half of its waters undrinkable.[1]

Modern nations and modern professions face risks that may be more complex in origins but no less drastic in implications. Some of the doctors I know who have cancer asked me not to write about their struggles. They aren't ashamed, but they want to keep their private challenges out of their patients' worlds. One surgeon explains, "I think of my cancer diagnosis like a rare object kept in a box on a shelf in a museum I rarely go to. Occasionally, late at night, I will take the box down, open the various tests and residuals of my cancer treatment, look at them, and think how lucky I am to have this in a box. Then I close the box, put it back, and never think about it again, until another late at night time, when I remember that I too have been there."

He has an extraordinary capacity to listen to patients rail at him. "I love it when they tell me I just can't understand what it's like to have cancer, to have young children you're not sure you will see get married. I can look at them with more empathy than they could ever imagine, and say, 'I know how hard this is for you. And I know you will get through this.' But what I don't ever tell them is that the reason I know what they are going through is because I am there too."

My friend is entitled to his privacy, but he does understand that he is not alone. He and a number of researchers and physicians with cancer—many of them women with breast cancer or cancers of the blood—grasp that the reasons for their disease may come directly from their training and the things they have done in their medical work. A few of them have allowed me to tell how they contracted the same disease they are working to combat. This is a price that no one ever expects to pay.

As physicians today operate with video game–like equipment, they have acquired the ability to see and do things formerly only imagined. The intricate ebb and flow of signals that keep cells under control can be turned upside down by the loss of a single molecule. Rather than waiting for the full expression of diseases like cancer, scientists can now measure early changes in cells and blood that signal whether disease is present or likely to arise. This has created an intense concentration on finding microscopic indications early enough so that we can come up with treatments to keep people healthy. Scientists are eagerly looking for fine disturbances of the body's billions of cells and molecules called biomarkers.

What happens when doctors or scientists learn that they have biomarkers of cancer? Often they pick up the first signs themselves. Simple blood tests or slow to heal or persistent bruises can signal something awry. Surprise is usually not an option. Most doctors today are trained in taking finer and finer looks into smaller and smaller parts of the human system. Skilled at ferreting out disease and getting rid of it, immersed in a can-do culture where technology and pills are the first round of defense, physicians know full well that tiny perturbations can have ominous portents. They also understand that in many cases cancer treatment remains a crapshoot. Maybe they have fewer delusions that can be sustained when the prognosis is grim. The fact is no matter how capable we have become at describing and finding disease, no matter how heroic we can be at treating some forms of cancer today, we still struggle to understand why any individual comes down with cancer.

In one basic way, doctors and scientists are just like the rest of us. They go through the same system where one suspicious test leads to another. When confronting a cancer diagnosis, doctors and scientists, including oncologists, take the approach that many patients do: they deny it.

After recovering from the shock of the pronouncement, doctors with cancer turn the microscopes on themselves and ask the familiar question: Why? Searching for what they hope isn't there, they probe deep into every tissue and cell, into spaces and places that were once only imagined. We have an entire arsenal of weapons for detecting and combating cancer. Those who spend their lives tending this arsenal tend to be pretty prosaic when they have to use these tools on themselves. They know exactly what they're facing.

SOMETIMES THE CAUSES of cancer are not hard to decipher. Li Dou is a doctor of modern Chinese medicine who teaches at the Maryland Acupuncture College along with her good friend, the charismatic Yiping Hu. Trained at Shanghai Medical School, one of China's best colleges of traditional medicine, they each took the usual courses in Chinese *materia medica*. While Chinese medicine goes back more than 5,000 years to the Yellow Emperor's classic treatise, modern Chinese medical training includes up-to-date chemistry. Just like medical apprentices in other traditions, Chinese medical students work with acids and metals and solvents to extract and prepare compounds. Phenols, formaldehyde and benzene are workhorse chemicals for many procedures, whether in Western or Chinese medicine. The two women worked with these chemicals at metal benches affixed to sinks. There were no hoods to take away fumes. The smells around the bench were something most people grew accustomed to if they wanted to make it through the program.

About ten years ago, each moved to America. Like many who go into medical fields, both have treated many cancer patients, often easing the discomfort and nausea that frequently go along with modern treatment. Neither ever imagined that one of them might end up with the illness.

Li grew up in Taiyuan, China, the capitol of Shanxi province and a city with 2,500 years of history. Framed in the east by the Taihang mountains and in the west by the Lulian mountains, the city of Taiyuan rests at the bottom of a large geological basin where the air can be still for days. In the early seventh century, the city was the center of a peasant uprising that overthrew the Sui regime. At the start of the twentieth century, Taiyuan was central to the nationalistic Boxer Rebellion, in which all foreign missionaries were killed on orders of the provincial governor. Today it is famous as one of the dirtiest cities in China, perhaps the world. Inversions of cold air from the surrounding hills trap the fumes from the city's huge steel mills, coke ovens and foundries. City residents regularly breathe air full of fine particulates, aromatic hydrocarbons, diesel exhaust and other industrial residues. Even with infra-red probes, the ground of the city itself sometimes cannot be seen from space, because it is swathed in opaque fumes and dusts.

In the 1960s, when Li Dou was a girl, like every young child on this planet, she took in many more breaths every day than adults around her. Each minute, the typical adult inhales about sixteen breaths while awake and about half as many when asleep. Children take in more pollutants than adults because their lungs are smaller and work faster. Infants at rest can take about forty-four breaths a minute, toddlers about half that. With each breath, microscopic solids and invisible gases enter the body. Particles that can be fifty times smaller than a strand of hair—some 10 microns in diameter—get trapped by those annoying hairs in the nose. Those that are smaller—like the ones made by engine exhausts and coal-fired power plants—can slip through the nose or the mouth and end up coating the throat lining or lung surfaces.

Modern life has given us smaller particles than humans ever encountered before, which can pass through cell walls. Those who grow up playing hard in polluted air, like Li Dou, who was a basketball star in China, develop lungs that are especially strong and also stippled with soot. Air goes in and out all the time. The smallest of particles enter and stay. Lungs that grow up dirty tend to blacken over time. Even though any single breath may take in less than a part per billion of a given pollutant—like one drop of water in an Olympic swimming pool—over a lifetime, things add up.

Healthy lungs look like spongy, pink-tinged treetops and are made up of more than 50,000 sections. If your lungs were removed from your chest and splayed out flat, they would cover about 50 square meters—the size of a tennis court. The airways of the lungs are 2,000 miles long—the distance from Miami to New York and back. Each year, the average adult breathes about 7 million liters of air, more or less depending on their activity level. Whether we think about it or not, an active person takes in seven to fourteen liters of air each minute, some 10,000 to 20,000 liters every day. For those whose lungs are still growing, whose hearts are already damaged, or whose airways are a bit narrowed by other disease, regular breathing of dirty air can be especially tough.

Smokers' lungs start out just as complicated as those of nonsmokers. The upper part of the respiratory tract is lined with tiny, invisible hairs (cilia) that work like an escalator to get rid of things that don't belong in the lungs. But after years of tobacco smoke, the marvelous system of invisible fine hairs loses the ability to get rid of dirt, viruses and bacteria. Mucus, that slippery stuff we cough up when we have a cold, is our first line of defense against things we breathe in but don't need. This thin layer of liquid protects the lungs from damaging particles or gases, but it gets dried out by hot fumes and particles. Lungs damaged by regular smoke inhalation lose their resilience. In healthy lungs, an invisible escalator of cilia regularly shuttles about 100 cubic centimeters of viruses and bacteria trapped in this mucus out of the lungs every day. This shuttle gets slowed and can even shut down totally when a person has bronchitis or pneumonia or is a heavy smoker. That is why sick people and smokers cough so hard and often. Their lungs are in a constant state of auto rebellion.

Rats cannot cough. The Brazilian physician researcher Paulo Saldiva showed just how vulnerable rodents can be to the air around them. He took two groups of caged rats and fed them food and water for a year. One group was placed in the church steeple of the Ordem Terceira do Carmo in the center of São Paulo, one of Brazil's most polluted cities. The other was set up in the countryside. By the end of the year, the lungs of the city rats looked like those of heavy cigarette smokers. The fluffy linings that sweep out pollutants—the cilia—had become shrunken, shriveled and clotted. Those from the countryside had rich, plush fuzz that scooped up bacteria and viruses, along with pollutants, and escorted them out of the lungs.

The people of Taiyuan live in air that is much worse than that of São Paulo. The dark, smoky haze that covers the region can even prevent satellites from imaging the ground. Conditions are so dire that in 2000 the United Nations and Asian Development Bank got together to do something about it. A massive economic experiment is under way to change local industry. Millions of dollars are being spent in a major effort to clean up the skies and get industry fully engaged in the process.

The entire world has a stake in the effort to turn this prosperous, dirty town into a productive, green city where people will not be afraid to send their children out to play. China, India and other developing countries are full of Taiyuans, places where tremendous economic growth has created massive degradation of air and water.

You can move from where you grew up, but your body takes with it all the things that ever went into it that were too small or too well entrenched to come out. Li now breathes the much cleaner air of suburban Maryland, where her husband is a scientist at the National Institutes of Health. She remembers as a child in Taiyuan that whenever she wore white clothing, it would become covered with black dirt within a few hours. Sometimes the sun did not shine for days.

A relatively recent report based on information provided to the stalwart volunteers of the ACS has shown that the chance of getting lung cancer, a disease that can take decades to develop, is greatest for those who have lived in places with the dirtiest air in the United States.[2] Of a half million volunteers sampled in recent years, those who lived in cities with the highest levels of air pollution had about 30 percent more lung cancer. Similar work has been done in Sweden, England, and more than twenty other nations. For Li, breathing the air of Taiyuan for the first thirty years of her life was like living with someone who smoked two packs of cigarettes every day.

Yiping Hu will never forget seeing her friend Li Dou two months before she was diagnosed with cancer. At that time, Hu told me, she had this awful feeling that something was not right. "She looked like cancer."

I asked, "What does cancer look like?"

"In Chinese medicine," Hu explained, "we watch the face and skin carefully. We don't just ask what is the disease, we look at color; we measure the strength of pulses at various arteries. With cancer, the color of the face is gray or yellow. The blood looks sick at the surface of the skin. I could tell that something was wrong, but I had no idea what. I wondered whether I was wrong, because Li is such a strong and healthy woman. So I didn't say a word. Then, when I heard from other professors at the school that Li had lung cancer that had spread to her bones, I felt sick myself. This was such a young woman, a good doctor. I knew she didn't smoke. I wished I had been wrong."

Today, almost three years after she was told she had three months to live, Li is very much alive. Surgery removed her biggest tumors. Chemotherapy shrank them further and took her hair. Every day she performs Qi Gong exercises to mobilize her spirit and energies. Hu and Doi have brought all of their training in acupuncture and Chinese herbs to bear on a disease that usually kills people quickly. The special medicines Hu prepares for Li seem to make the body stronger. The tumor keeps shrinking. So far, so good. Li's Western-trained surgeon and oncologist don't quite know what to make of this, but nobody argues with success.

How can we explain what happened? Li led a clean life. She ate well. Nobody in her family ever had cancer. Yet she ended up with lung cancer in her early forties. Li is now wondering what happened.

"The idea that my home or my work could have anything to do with my sickness, this is not something anyone ever told me. But now that I understand what was in that air, it starts to make sense. When I think about some of the chemicals we used in medical training, I begin to see what could have brought this cancer on." As time passes, Li Dou is beginning to think she may survive her lung cancer. She is glad to be far away from the ground of her city that still sometimes can't be seen from space. Staying out of polluted air has become a life-or-death matter for her, and so is bolstering her body's ability to get rid of old poisons. She hopes she can restore the health she once had, but she knows she can never go home.

My DEAR FRIEND Deborah Axelrod is one of New York's top breast cancer surgeons. For years, we've been plowing through inconsistent studies on environment and cancer, frustrated by the limited efforts to understand the many connections. We know that benzene is one of more than two hundred different chemicals studied that causes mammary tumors in male rodents. We know that many solvents and some plastics can distort the way the body produces hormones. Life is a complex mixture of good and bad things we can't control. We rarely can get the right amount of information together on any group of people with shared exposures to learn whether their chances of getting cancer have been affected by the mixtures with which they live and work. Our scientific knowledge has not brought us very far toward understanding why any given individual gets breast cancer.

Deborah is a very careful surgeon who trained with one of the world's top breast surgeons, Michael Osborne at Cornell Medical School. In a meticulous, beautifully designed study, she showed that breast cells from women with breast cancer had abnormal levels of some hormones compared with those found in women who had their breasts reduced.

During slow moments in her grueling work schedule—early morning when most of the world is asleep or late at night when she wishes she was—she writes articles and kicks around research projects. She called me one day in 2002 with a troubling report.

"I think I've got a smoking gun."

I had no idea what she was talking about.

"Listen. I've got this amazing case. I've just seen this woman, seems pretty intelligent. She has had some really rotten luck. Pretty classic breast cancer. She tells me that she worked in the Soviet Union as a professor of chemistry. In Russia, they had lots of hoods. They were careful. She says she gets to this country and she's working with benzene, ethylene oxide and methyl chloride. No hoods to trap and remove the fumes! Can you believe that?!"

I was incredulous. In its three decades of existence, the U.S. Occupational Safety and Health Administration has set standards on about two dozen chemicals known to cause cancer in humans. Benzene and ethylene oxide are among them. Anything containing these chemicals has to be labeled for limited and controlled use. A detailed hazardous material safety data sheet has to be available on any worksite in the event an exposure occurs. The air where they are used is supposed to be monitored regularly. Workers are to use heavy-duty respirators, gloves and powerful negative-pressure hoods that keep fumes from their lungs.

I asked her, "Are you sure?"

She replied, "I really think you need to talk to her."

Yelena B. is a Russian-Jewish-American original who fled the Soviet Union in December 1990, just when the country was collapsing. When she got to America, she was delighted to get a top job in her field in New Jersey, even though she barely spoke English. In Russia, with a doctorate in organic chemistry, she was a senior researcher. Today she is wrestling with breast cancer and unemployed.

She does not regret leaving Russia. "I came here from Russia, with my two old parents and my son, and my English was just terrible. So lucky, I was I thought so very lucky, to get a good job in a good company.

"To America I came in 1990. The Jewish Federation gave me food. Red Cross paid my rent. I spent $600 on the grave for my father, who died the first year. In Russia, I was chemist and professor teaching organic chemistry. Here I was a nudnick, a nobody. My language was so poor. I never went to school for English. I learned on my own to read and do technical writing. My writing skills were so poor."

Within months of working at the specialty chemicals firm, which she is not now allowed to name, she began to wonder why practices were so different from those in Russia. "We had hoods, where I worked in Moscow University making chemicals for food processing. I worked with lactic acid and citric acid and other food-related agents, making esters for the food industry," she remembers. "Nobody would ever have asked us to work with the things I handled in New Jersey without giving us protection like chemical hoods, respirators, strong gloves and plenty of good air. As poor as Russia was, we chemists received half a gallon of milk a day for detoxification."

"On the first month in New Jersey, I told them, you must buy hood and they put me on probation right away. At first, they bought canopy type restaurant hoods which did not provide us with any protection and did not fall under the definition of chemical hoods. I told them it was not good enough. 'You must purchase the right one.' They finally purchased a secondhand hood which was made out of asbestos fiber. The company made us install this hood, including fifty holes drilled right through what turned out to be asbestos-containing slabs. We were working with asbestos dust and debris around us for decades. Every time the hoods were turned on, the air turbulence dispersed the asbestos dust. When I protested and contacted OSHA, they put me on suspension (administrative leave) right away."

Yelena is no slouch scientifically. She holds patents in the United States from her work at this firm that supplies chemicals used in personal care products to some of the leading cosmetic companies in the country. Globally, the market for cosmetics is huge—more than $100 billion a year. With prosperity comes an interest and a capacity to spend money on ways to enhance appearance. You know how some shampoos leave the hair shiny and smooth? That's because they contain things that bond to the surface of the hair shaft, leaving it silky and under control. But the ways these magical beauty treatments are crafted can be quite unhealthy.

Whether male or female, chemists tend to be a bit macho about their work. My esteemed colleague H. Leon Bradlow boasts of regularly needing new trousers as a young chemist because his pant legs would be full of tiny holes etched by drops of acid from his daily work in the lab. One sure-fire way to clean glass tubes was to drop them into a boiling bath of sulfuric acid. The glass came out spotless every time.

The French gave us the wonderful word "pipette," a tiny glass tube used to pick up liquids and transfer them from one place to another in the laboratory. Squeezing a tight-fitting air-filled rubber bulb at the end of the glass tube is a reliable way of pulling the liquid up, but the rubber bulbs tend to lose their grip. When I was learning how to construct chemicals as a postdoctoral fellow at Johns Hopkins in the early 1980s, it was not at all uncommon to take direct action. One's lips, properly applied with a slight constant suction, could pull liquid into the glass.

We all learned the hard way not to suck too fast. Budding chemists like me might swallow whatever was being transported. Bradlow still boasts of being a champ at mouth pipetting, though he hasn't had to do it for years. Among the basic workhorses of chemistry that some of us may well have mouth-pipetted are compounds that today are understood to slice through bone and genes, like benzene. Today many of these materials are handled by robots. Robotic sensors don't need visual cues and they don't need to mouth pipette.

Ads tell us that modern cosmetics can put a shiny coating on dull, lifeless hair and other parts of our bodies where an extra glow may be nice. What they don't say is what it takes to make these ingredients.

Most people can't stand the smell of ammonia for long. But ammonia, an atom of nitrogen surrounded by three atoms of hydrogen, is an essential chemical for many uses. Nitrogen, properly tooled, has some terrific properties, for refrigeration and for taming snarled, frizzy hair. Chemists have learned how to put a positive charge on the nitrogen in ammonia, using extremely reactive compounds that allow polymers to be bound up by the negative charge of the keratin protein that makes up the hair shaft. The physical result of putting this charged nitrogen on the hair is that things look soft and shiny. But this compound is not simple to create.

To make some of the chemicals that are widely used in personal care products, toxic materials like ethylene oxide and propylene oxide are placed in oversize pressure reactors and pushed to react with oils or with amines, while they cook for a day. Long, high-weight chains of carbon and hydrogen, which are attracted to fat, are added to other compounds that are attracted to water. Making oil mix with water yields some very impressive cosmetics.

Yelena remembers being told at work to be careful with the pressure reactor, which sat on a metal table. Photos she has of what she says was the lab where she worked show that in the place above the reactor where a hood should have been installed to remove the noxious fumes, there was nothing. The table was next to a sink. "We were supposed to seal the metal pressure reactor and fill the lines which go from the holding tanks with ethylene oxide, or propylene oxide, to the reactor by opening the valves to let the gas flow into the line. When we were finished, the manager told us when liquid ethylene oxide came out that hadn't been used in the reaction—and it always did— we should just dump it into the sink."

She reports that reactors used with ethylene oxide, propylene oxide, methyl chloride and methyl bromide routinely leaked large quantities of highly toxic substances and gasses into the lab environment. The laboratory lacked early detection alarm systems and/or continuous personal monitoring to find leaks in their earliest stage.

Yelena had never worked with ethylene oxide before. A single parent with a family to support, she did what she was told. Eventually, she learned the state of California, the U.S. Occupational Safety and Health Administration and the World Health Organization all consider ethylene oxide carcinogenic in humans. Unlike some things that have only been shown to induce cancer in animals, ethylene oxide is considered proven to do so in humans. This means that, in contrast to the nearly 3,000 other high-volume chemicals in use today, ethylene oxide exposures have been documented to induce cancer in people who have worked with it to sterilize medical equipment. The amount of ethylene oxide that was needed for the reaction Yelena was producing was very small—one part per million. This also happens to be the maximum level under U.S. law that a person may be exposed to in eight hours.[3]

Ethylene oxide comes stored in heavy metal tanks as a liquid under lots of pressure with skull and crossbones labels on them. On the top of the tank there's a valve that is supposed to be shut tight. When it's opened, ethylene oxide comes shooting out into a closed chamber. When the reaction is complete, there is always something left.

What was left over also added to Yelena's workplace hazards. Ethylene oxide can strip out the eye-burning chemicals from children's shampoos by adding some of its oxygen to the stew that makes soap bubbly. At the point where the reaction happens, ethylene oxide takes on a twin and becomes di-ethylene oxide—literally two paired ethylene oxides. Another name for this compound is 1,4-dioxane. This is a chemical known to the state of California, the U.S. government, and the World Health Organization to cause cancer in male and female rats and mice. The European Union bans it from personal care products. The United States does not have the authority to do so, does not monitor levels in baby shampoos and bubble baths, and does not recommend limits for this substance in these products.

David Steinman, leader of the Green Patriot movement and author of a new book *Safe Trip to Eden,* is also the father of young children. As an environmental journalist he knew enough to ask about complicated long-named chemicals that could be in children's bubble baths. He paid for private tests on baby shampoos and soaps in 2006 and found what Yelena had feared. The European Union recalled some children's bath products because they were found to contain 1,4-dioxane. European manufacturers had known for some time why 1,4-dioxane forms and also how to get rid of it cheaply. Until now, American manufacturers haven't needed to do so.

People like my son and daughter-in-law have assumed that products they buy to bathe my young grandchildren in would be especially safe. Like me, they were shocked to find that this was not the case.

When Steinman's book came out in February 2007 reporting that tests showed 1,4-dioxane in U.S. bubble baths, I wrote a *Newsweek* column about it in which I referred to FDA recommendations that levels of 1,4-dioxane not exceed ten parts per million. Even though the levels of any single hazard may be very small at any one time, the combined impacts of such contaminants over a lifetime should not be dismissed as unimportant.

In response, the government swung nimbly into action. The agency responsible for advising the public of environmental hazards, the Agency for Toxic Substances and Disease Registry (ATSDR), pulled the compound's toxicological profile from its website after the Food and Drug Administration (FDA) asked it to issue a correction: there was no official FDA-advised level for this cancer-causing ingredient in shampoos and bubble baths.

A revised posting on the toxicological profile of 1,4-dioxane appeared in April 2007. The explanation of the FDA reads like a spoof dreamed up by a geeky Monty Python.

> In February, 2007, ATSDR, an agency charged with evaluating toxic hazards, was informed by the Director of the Food and Drug Administration's (FDA) Office of Cosmetics and Colors, that an error was present in the Public Health Statement of the toxicological profile. *The FDA pointed out to ATSDR that FDA had not recommended a limit for 1,4-dioxane in cosmetic products.* (Italics added.)

This notice neglects to mention that the FDA has, in fact, set standards for 1,4-dioxane, but those are only for residues of this chemical that could get into food additives and adhesives. It also does not explain that European children are not exposed to this risk at all, and that EU regulatory agencies have recalled products from their shelves that have been found to have levels of 1,4-dioxane that are legally allowed in America and Canada today.

Instead, the FDA is now telling consumers to read labels and decide for themselves. But again, look carefully at what's on the official FDA website as of June 2007:

1,4-dioxane may be a contaminant in cosmetics, detergents, and shampoos that contain the following ingredients (which may be listed on the product label): "PEG," "polyethylene," "polyethylene glycol," "polyoxyethylene," "-eth-," or "-oxynol-." *Most manufacturers remove 1,4-dioxane from these ingredients to concentrations recommended by the FDA as safe. Thus, most products on the market today contain 1,4-dioxane in very small amounts or not at all.* However, some cosmetics, detergents, and shampoos may contain 1,4-dioxane at levels higher than recommended by the FDA. Because products contaminated at concentrations higher than the FDA-recommended levels are not possible to determine without testing, families should avoid using products containing the ingredients listed above unless the manufacturer can guarantee that 1,4-dioxane is below the FDA-recommended level. *(Italics added.)*[4]

In short, the FDA asked the ATSDR to remove its warning on 1,4 dioxane in February 2007 because that warning said the FDA had set a recommended standard in shampoos for 1,4-dioxane and there is no such standard. Two months later, the revised warning tells families that some shampoos can contain levels of 1,4-dioxane at *concentrations higher than the FDA-recommended levels*—according to recommendations that don't, apparently, exist—and that these levels cannot be determined without testing. Families are therefore urged to ask manufacturers to prove that their products are free of 1,4-dioxane.

In a nation where more than half of all parents work outside the home and barely have enough time to change their children's and their own underwear, the agency charged with keeping our personal care products safe is asking parents to write to each manufacturer of the more than two hundred different personal care products that can be applied to babies' bottoms. What would these letters say? "Dear Sirs: I note that your product lists the following ingredients in type that I can barely read: PEG, polyethylene, polyethylene glycol, polyoxyethylene, -eth-, -or, -oxynol. Please let me know whether your product has been tested for any of these ingredients and does not contain any 1,4-dioxane at levels above what FDA recommends, even though there is no standard, no monitoring, and no consequence of your not doing so."

A MEMO FROM Yelena's boss, which I have a copy of, instructed her to let the leftover liquid from the pressure reactor of ethylene oxide run right into the sink. She became a whistleblower right away. Yelena claims that in 1999 her company did in-house testing for ethylene oxide for the first time, monitoring the air near the pressure reactor where she worked. Yelena claims that her boss showed her a report saying they had found 10,000 parts per million when leaks happened, which was about once a month. The company's own monitoring noted that Yelena was personally exposed to 17 parts per million when leaks were not taking place.

Rats that breathe in similar concentrations of ethylene oxide for two years lose the ability to make normal blood.[5] Male rodents develop mammary tumors. This is pretty odd. In the past two decades, a number of technical papers have appeared showing that ethylene oxide is not just a problem for lab rats.[6] Women who work with it regularly have more breast cancer.[7] Now Yelena worries about herself and the two women who worked with her. The office adjoining their lab space was full of the smells from spills and splashes. Both of them are under thirty-five, and both of them, like Yelena, have had breast tumors. One was a low-grade carcinoma; the other was benign.

Cancer patients tend either to refuse to pay attention to their disease or else be hypervigilant. Yelena falls into the latter category and is now watching her body change. It's hard to know what's real and what's imagined. She sees signs that her bone marrow is taxed. "I have mouth bleeding every single day. My blood pressure is very, very low; sometimes 90 over 60. Nosebleeds are happening a lot, and my body hurts all over."

Yelena is not an objective observer. She thinks the company caused her cancer. This is a subject of more than academic interest. "This was my mistake that I worked for these cheap bastards. Now, they pay almost $200,000 a week to defend this company. They have an in-house lawyer and they hired a law firm. They just built a new building. The air supply for their administration area is completely separate from the one for my old lab. Do you think that's an accident?"

"This is the United States of America. This is not China. This is not Cuba. This is not Russia. What is going on here?"

When this book was in final editing, Yelena's legal claims were no longer something she could talk about with me. Although we met more than two years ago, and she gave me documents about her experiences that are redacted on this book's website, she is not allowed to discuss what happened ever again. She is not allowed to speak with the press. If any government agency wants to know how a small chemicals firm making products for beauty endangered the lives of its workers, they have to subpoena her to hear what it was like working directly to produce chemicals that are put onto our bodies every day. Only if she is forced to talk will she be able to do so. Otherwise her lips are sealed. That's the law.

Rᴜᴛʜ Sᴘᴇᴄᴛᴏʀ ᴡᴀꜱ an anesthesiologist at Long Island Jewish Hospital. As the mother of three young children, she was no stranger to fatigue. Figuring she might be low in iron, she sent a sample of her blood to the hospital laboratory. Some types of iron deficiencies can be remedied fairly easily with supplements.

The results were worse than she could have imagined. Spector's blood count was all out of whack. She first thought it was a great big lab error, the sort of thing that occurs more often than anyone likes to admit. The white cell counts were so high she thought they couldn't be hers.

But they were.

Ruth Spector did not get to the twentieth reunion of her class at the University of Pennsylvania in 2002. She had more important matters to deal with. She faced a tough choice: she could die within a year from an aggressive form of acute myeloblastic leukemia, or she could try a dangerous procedure that might kill her but also might extend her life. There was no way to sugar coat this one. She opted for the high risk–high payoff option and made a date for a bone marrow transplant at Memorial Sloan-Kettering Cancer Center in New York City. She turned out to be one of the lucky ones.

The disease Spector developed was a cancer of the blood that occurs when the bone marrow goes into overdrive and starts spewing out thousands of not-ready-for-prime time white blood cells that crowd out healthy ones. The prospects are not great. Overwhelmed by cells that are supposed to fight off infection but do not, the immune system stops working well. Whenever a virus or bacterium enters the body of a healthy person, it gets zapped by a complement of cellular policemen.

Leukemia upsets the immune system's ability to know when to turn itself on or off. Some of the first signs of leukemia emerge when ordinary cuts and bruises and other simple health problems put extraordinary burdens on the body. Infections are not fought off. Bruises develop easily. Joints ache from things that a well body would not even feel.

In some ways, physicians have it worse than others dealing with a serious disease. Hope remains the most powerful drug in the world. Doctors are trained to look at numbers, but sometimes the numbers suck the hope out. Spector told a local newspaper, "As doctors, we were able to look things up easily, and we were not intimidated to call up other physicians and medical centers. But our professions were a disadvantage too, because I think we understood the prognosis more clearly than other people might have."

We can't be sure what causes most individual cases of AML. But we can learn something about what unleashes the disease by looking at groups where the illness occurs more often. Refinery workers, shoemakers, machinists and others regularly working with solvents have a much greater chance of developing AML than those without such exposures. Anesthesiologists have chemical exposures of their own. Some of the agents used to induce anesthesia in years past, like trichloroethylene, are now understood to break the back of genetic material in ways that can lead to cancer. Other anesthetic gases have been tied with damage to the nervous system.

Benzene, that remarkable chemical that has fostered so much of the modern revolution in chemistry, can slip into the bloodstream and into the bone marrow. Since 1928, scientists have understood that benzene can cause AML in those who work with it regularly. Why doctors should be among them may not be that difficult to understand. People like Ruth Spector, who are middle-age physicians today, learned to make chemicals the old-fashioned way, just as Li Dou and Yiping Hu did in China—they built them. In addition, physicians work with compounds like alcohols, anesthetics and chemotherapeutic drugs that are also recognized to cause cancer.

RON HERBERMAN is hardly your typical physician. Like Spector and Yelena, he's spent lots of time in the laboratory. His claim to fame in the research community is that he figured out that what some people thought was a basic lab error turned out to be phenomenally important. He isolated white blood cells from cancer survivors and people without disease. Most scientists had assumed that cells from healthy people would have little if any ability to recognize and kill cancer cells because they had not been exposed to any tumors. Not Herberman. In a detailed series of studies, he showed that cells taken from healthy people did a better job of attacking cancer cells than those from cancer patients. He also showed that normal mice and rats had a remarkable ability to clear cancer cells after being injected with them into the bloodstream, because the mice and rats carried Natural Killer (NK) cells. Herberman proved that these NK cells had a distinctive size and appearance, and were jam-packed with granules of potent cancer-killing substances. These NK cells got their name honestly, by zeroing in where tumors had taken hold and releasing the equivalent of small, focused bombs that destroyed much, but not all, of the cancer.

Nearly two decades ago, Herberman's older brother, Harvey, also a physician, came down with leukemia—another cancer of the immune system. Every day our bodies make more than 10 billion new blood cells. Lymphoma occurs when a rare and powerful cell that is the grandmother of all cells—stem cells—can't stop making abnormal lymph tissue. Eventually Harvey's cancer was transformed to chronic lymphocytic leukemia (CLL). Lymphoma cells and CLL cells both come from lymphocytes, which normally are responsible for making protective antibodies against a wide variety of foreign materials and infectious diseases. The carcinogens that may lay behind both these cancers are unclear, but there are hints that radiation and exposure to some chemicals, especially pesticides and solvents, are involved.

About five years ago, Ron Herberman joined his brother as a member of the club neither of them wanted to belong to—he also developed CLL. When two brothers from the same family with no known history of disease develop the same disorder of the blood and immune system, we have to ask, Is this just a coincidence? Is it some statistical fluke? Or did something happen to each of these men earlier in life to put them at risk of cancer? In thinking about their shared disease, Harvey was not surprised. He remembered the planes spraying clouds of pesticides overhead when the two brothers were growing up in Southern California. But Harvey hadn't thought about all the other things that could have seeped into their bones or damaged their immune systems. What about those model airplanes they loved to build and rebuild, and fly and repair and repair again for as long as they could still be put back together? What about the glues, the fuels and grease for the engines? What about all that time spent riding around in cars that belched considerably more fumes and vapors than they do today? What about those hours doing the repairs and projects that busy young homeowners do on weekends, working over dank, smelly tiles and paint and solder and plaster in small spaces with lousy air flow?

Could their early lives as budding scientists with no precautions or work on home improvement projects with all sorts of smelly goo have anything to do with why they both contracted the same chronic problem with their immune system? Nobody can say. Any good scientist will tell you that we can rarely know what causes cancer in any one person. But more and more doctors and scientists with cancer are beginning to think that the reasons for their disease are not so mysterious.

A newspaper story that is more than fifty years old provides another clue. It shows the earnest, young Herberman peering into a glass beaker that contains a fish. Herberman is described as on his way to New York University Medical School as a Phi Beta Kappa graduate at the grand age of nineteen. In the photo he is completing a summer project at the Museum of Natural History. Unearthing this photo, Herberman remembered, "I anesthetized the fish with urethane, and as you know, urethane is a carcinogen in mice. I handled the urethane and the solutions without gloves or any other precautions."

What caused these two physicians to come down with the same cancer? We can't be sure. But both came to believe that something in their life experiences could have kicked in. Brothers have similar genes, but they are not identical. Something happened to the genes of the brothers Herberman to give them both the same disease. Harvey, who also had Hodgkin's disease, died of a bacterial infection early in 2007.

MANY ONCOLOGISTS wonder why they are seeing so many more cases of cancer in young persons and why so many more colleagues and scientists are affected. When the disease hits home, as it increasingly does for many of those who care for others with cancer, the answers may also be uncomfortably close.

In one respect physicians and researchers who become cancer patients are revolutionizing the way people are thinking about the disease. After they have finished the body scans and contrast studies with the latest dyes and studies, many of these doctors come to appreciate that they have lots of company. More often than physicians even could expect, they gain expertise waiting for cancer treatments.

We can't be sure why so many more physicians and researchers are coming down with cancer.[8] Ask the researchers themselves and they have strong opinions. One colleague, who told me she would kill me if I used her name, thinks she knows why she got cancer. When multiple myeloma shut down her bone marrow, she survived a life-threatening transplant. Back at work coming up with new drugs for other cancer patients today, she's convinced of what happened. "I used to mix my own chemotherapy cocktails for patients two decades ago. You know that stuff is really foul. I did this with no hoods. No masks. No nothing, just sloshing around with all those nice, nasty killer compounds that we would prepare to inject into our patients.

"Do I know what caused my cancer? Of course I can't tell you which one did it. But all those toxic compounds I played with for all those years didn't help my bones."

We can't be sure what's causing what looks like an increase in cancer in researchers and physicians, but we've got some engrossing clues, and they come from more than newspaper obituaries. Finding patterns of illness, whether for physicians, scientists, or heavy smokers, is never a simple matter. First of all you have to be able to look. Second, you have to have a group that cares enough about the answers to make it possible to publish the results. Finally, you have to be prepared to deal with a phalanx of skeptics that will arise as certain as the sun, whenever a report suggests that a highly profitable business or technology may be the source of illness.

My father, Harry Davis, as a toddler in the family-owned dry-cleaning and tailor shop in 1923, the year before he survived a massive explosion of benzene.

10

Deconstructing Cancer Statistics

In G-d we trust. All others must provide data.

In 1981 I was fired by Ronald Reagan. It was nothing personal. Thousands of us were let go at the same time. The president canceled all federal positions that had been offered just prior to his election in November 1980. Among the many posts eliminated was my three-year term as a National Institutes of Health senior postdoctoral training fellow in epidemiology. I hadn't even reported for work and I was already out of a job.

Housing contracts, marriages and weddings were put on hold as folks suddenly found themselves cast adrift by what became known as the Reagan rescissions. The real estate market in D.C. took a tumble. I was one of the lucky ones. My husband was still employed and we had enough income to make our mortgage payments.

Abe Lilienfeld, a professor at Johns Hopkins University and the dean of American epidemiology, bailed me out using funds he had previously secured from the National Cancer Institute. With his support, I became a senior fellow in epidemiology at one of the world's top programs in public health. He had hired me, he said, because my experience working in the federal government in the 1970s gave me a perspective that few researchers have. I was to focus on the really big picture. What do patterns of cancer look like after you've allowed for the fact that more of us are living longer?

Even though most of us don't think about it, we are all part of health statistics every day. If we're lucky, we belong to the population that is healthy, in which case we form part of the denominator against which cancer is gauged. If we're unlucky, we become part of the numerator—the top part of the fraction, the segment of the population that has the disease.

That same year my hardy, guy's guy dad became one of the cancer statistics I would be working on. At first his doctor, who was also his tennis partner, did not want to give him the news. Dad faced what was then a death sentence—a diagnosis of multiple myeloma, a cancer of the bone marrow. His response was simple: if nothing changed in his life, maybe the disease would just go away. Maybe, just maybe, he could stay healthy and somehow jump back to the denominator.

Lilienfeld was fond of paraphrasing the famed nineteenth-century English statistician William Farr: "Look at the patterns of disease, the grand scenario. Everything else is conjecture." The first chief of statistics for England's Office of the Registrar General, Farr made his life's work the finding of patterns in lives and deaths. England was one of the first nations to collect such information and Farr led efforts to make sure the standards and systems were applied as uniformly as possible in those days before typewriters. A stylish if somewhat florid writer, he did not hesitate to describe his subject in Dickensian terms:

> Every population throws off insensibly an atmosphere of organic matter . . . this atmosphere hangs over cities like a light cloud, slowly spreading, driven about, falling dispersed by the winds, washed down by showers . . . to connect by a subtle, sickly medium, the people agglomerated in narrow streets and courts, down which the wind does not blow, and upon which the sun seldom shines.
>
> This disease mist, arising from the breath of . . . millions of people, from open sewers and cesspools, graves and slaughter-houses, is continually kept up and undergoing changes; in one season it was pervaded by Cholera, in another by Influenza; at one time it bears Smallpox, Measles, Scarlatina and Whooping Cough among your chil-

dren; at another it carries fever on its wings. Like an angel of death it has hovered for centuries over London.[1]

Farr spent his life collecting and analyzing what we now call vital statistics, numbers about the most critical aspects of what it means to be alive: birth, illness and death. Lilienfeld, a demanding, aloof mentor who was not at all disposed toward literary elaboration, insisted that the only way to understand the true rates of a disease was to build those rates by hand, slowly, again and again.

Everybody knows there are more cases of cancer because there are more people, and especially because there are more older people. Though only 15 percent of the population is over age sixty-five, they develop 80 percent of all cancers. Because cancer is a disease of aging, the only way to figure out whether there is a real increase in the disease is to adjust for the aging of the population. In a sense, each person dies alone, but the circumstances under which these signal events occur tell us a lot about the nature of the world in which we live. To make sense of the patterns of births and deaths and disease, epidemiologists have devised different ways of assembling information. We start out with simple things. What's the average age of the entire population? In relatively young and fast growing Mexico, half of all people are under twenty-five, and about a third are under age fifteen. In America the average person, if we could find her, would be 35.3. We have more older people—more of whom are female—and a lot more of us hoping to get even older. In fact, the fastest growing part of the population is now between sixty-five and seventy-four—the age range the baby boom generation is now entering.

In any population, if most people are under age thirty, then the rate of cancer for the entire population will chiefly tell you what's happening to the young. But not all cancer happens just because some of us manage to get older. To take into account the fact that more of us are living longer, epidemiologists come up with techniques that adjust rates for the numbers of people who are alive at any given time in any given age-group called an age-specific rate. For every 100,000 people between the ages of sixty-five and seventy-four in 1980, we can ask how many of them will develop cancer of the bone marrow. We then ask the same question for 1990 and in that manner come up with a sense of whether or not the rate of the disease is changing independent of the fact that more of us are reaching greater age.

Our challenge was to find a way to determine whether cancer was really increasing or just appeared to be doing so because there were more people. For me this meant going to the offices of the National Center for Health Statistics, then located in beautiful suburban Hyattsville, Maryland, amid shopping centers and concrete office towers. At the NCHS I pored over black books so huge they had to be placed on tables to be opened. They contained the counts of the entire population in any year.

After Hyattsville I would go to Bethesda, to the National Cancer Institute on the growing campus of the National Institutes of Health, to beg and plead for the release of reports on cancer cases then being provided to the national Surveillance Epidemiology and End-Results from about 10 percent of the country. SEER had begun listing all cancer cases that occurred since January 1, 1973, in Connecticut, Iowa, New Mexico, Utah and Hawaii, as well as in Detroit and San Francisco-Oakland. The next year Atlanta and the thirteen counties of the Seattle-Puget Sound area were added. In 1978, ten predominantly black rural counties in Georgia were added. Not until 1980 did the system include any of the First Americans, when Native Americans from Arizona were added. Each SEER site had its own registrar who oversaw the recording of every case of cancer in the region and made a painstaking effort to get slides, x-rays and other medical tests to confirm each diagnosis.

With all this information assembled, I returned to my cramped office in Baltimore to put it all together and calculate the rates of specific cancers for specific age-groups. When I had finished, Lilienfeld made me do it all over again from scratch. That way, he told me, I would never forget what an age-specific rate really was. After months of toiling alone at Lilienfeld's direction, I felt in my lower back what others knew only abstractly: statistics are built from the ground up, or at least they were in the days before much of this work became automated.

When the customary complications of treatment for my father's cancer began putting him in and out of the intensive care unit at the University of Pittsburgh, I found myself spending weekends driving the Pennsylvania Turnpike, sometimes with but more often without Richard and our two young children. I would return from these exhausting treks to the exciting but grueling work at Hopkins. Moving mountains of cancer data gave me a sense that somehow the work I was doing could help my dad and others like him. When reports surfaced that lung cancer deaths had increased in the 1930s and 1940s, it strengthened the resolve of public health scientists to find out what lay behind these patterns.

Lilienfeld insisted that evaluating cancer was not simple. Once I had allowed for the effects of aging, I then had to make sure doctors were not simply counting more cases because they were getting better at finding them. We had to find out whether deaths that had in the past been attributed to tuberculosis had really been lung cancer in disguise. I needed some way to look at what was happening to cancer over time that took into account not only that more people were living longer but that we had better methods of finding the disease. And after I had allowed for those two factors, there was yet a third complication: I had to learn about styles and fads of diagnosis. Physicians are like everybody else, Lilienfeld advised; they are influenced by fashions.

Meanwhile, the various crises that attend cancer treatment were unfolding in my family's life.

ONE DAY LILIENFELD called me into his office.

"I have to tell you something I don't usually discuss with anybody." He motioned me inside and gestured for me to close the door. I sat down on a hard wooden chair and waited. His intense blue eyes signaled this was a serious conversation. "You have to talk to your father."

I was bewildered. He and I had never spoken about my frequent trips to my dad's sickbed. Lilienfeld was not exactly the warm and cuddly type. Still, he had heard my dad was very ill, and he knew I was driving back and forth to Pittsburgh a lot. I bit my lips nervously, unsure how much he had been told, what he meant or how I should respond to this sudden display of concern.

I decided that perhaps he was not aware how grave the situation was. "But Professor Lilienfeld," I said, "my dad can't hear. He's in a coma. They've got more lines in and out of his body than I've ever seen. He doesn't move at all."

"I know that," he replied. "That's why I'm telling you, you must talk to him." This was getting weird. My reserved mentor tells me that I should talk to my comatose father.

"Look. Let me tell you something that I know about personally. He can hear you!" I began to think that Lilienfeld had lost it. He went on. "Two years ago I had a heart attack, right in the middle of a medical school lecture I was giving. The students began to work on me with CPR. My heart stopped three times, but each time, they brought me back.

I ended up in the intensive care unit here at Hopkins. My wife, Lorraine, came in. I heard the physicians tell her, 'We think he's brain dead.'"

As I listened to his story, I began to shiver. Even now, as I write about it some twenty-five years later, the chills come on. I blurted, "What did you do? That must have felt dreadful."

"You can't imagine. I was trapped in my own body. I could hear *everything* they were saying. I could hear the noises on the unit, the ticking of the monitors, the squeaking of the wheels of machines being moved around. I tried as hard as I could to open my eyes, to move a hand, a foot. I couldn't move a thing. That's how I came to realize that hearing is the last sense to go. So you go and talk to your dad. Be careful what people say around him. He needs to hear you. He needs to know that you are there."

At the time I was in the midst of deciding on a major project for my postdoctoral research. I was thinking about conducting a detailed study of workplace and other environmental causes of multiple myeloma. This would have involved finding and talking with people who had the disease or with their surviving family members, developing precise measures of their workplace and other experiences. I wasn't at all sure I could do this while my own father lay so ill with myeloma himself.

When I was wrestling with this question, Colin Soskolne, a South African then doing graduate work at the University of Pennsylvania, had come up with a system for understanding workplace conditions quite similar to those my dad and many other steelworkers, machinists and foundry workers had faced. For almost one hundred years the South African Medical Bureau for Occupational Diseases had kept meticulous health records on all those who ever worked in mining. Soskolne had earned his fellowship at Penn by designing the software that made this century's worth of records accessible on a computer, winning a national award for revolutionizing the country's occupational health research.

Now Soskolne was under contract to Exxon to create a similar database for the workers at its refinery in Baton Rouge. He amassed records on some 10,000 people, tracing specific workers until they died or retired. For how long had they smoked? How much alcohol did they drink? Where and how long had they worked with specific chemicals or acid mists? What did they eat? Did their parents have cancer? How tall and fat were they? Where did they live? What salary did they earn?

The answer to each of these questions can be turned into a number. The average place and type of job can reveal the most frequent type of exposures to various chemicals. The average number of cigarettes smoked per year or per day can be counted or reasonably well estimated. Even weight and height can be combined to yield an index of fatness or thinness. How to analyze all this was not at all obvious. Simple mathematics was not up to the task.

From the thousands of past and present employees of the Exxon Baton Rouge plant, every one of those with cancer of the upper respiratory tract was matched with at least three comparable workers who never became ill. This method is called a nested case-control study.

At the time that the lives and deaths of these refinery workers were being turned into numbers, public health research was itself in the midst of a quiet revolution about how to analyze such information. Working with researchers in North Carolina, Soskolne was one of the first persons ever to apply a new computerized technique in statistics that calculates the odds that a given group with a specific history of working, eating, playing and living would develop or die of a given disease. Using a technique that goes by the daunting name of "conditional logistic regression," he could lay the odds that those who had worked with inorganic acid aerosols developed upper respiratory cancer (URC) simply because of their work.

The logistic regression models allowed him to assess the relative role of any given set of activities in determining the chance that a group of fellows would all come down with URC. The technique was then fairly new and Soskolne was one of a few dozen people in the world who understood how it worked. His dissertation in 1982 was awarded the Society for Epidemiologic Research's annual prize for best thesis of the year.

Using this advanced model, Soskolne reported that men working with strong inorganic acid mists had about four times the chance of developing URC compared to those without such experiences. Those with the highest exposures had nearly seven times more cancer. Smoking, being chubby or lean, or drinking alcohol did not alter that relationship.

Exxon was not pleased. We will never know whether they were surprised.

After his results had been disclosed to the firm, Soskolne reports that "scientists from the company began to ask questions of my work that seemed a bit odd. But I took them all very seriously. Each time one of them raised an issue, I would redo, revise, and recalculate the work. This involved several more trips to Baton Rouge at Exxon's expense to extract more data. All these queries had the intent of trying to make the big risks we had found go away. Under this pressure, of course, I basically redid everything. I had to go back and check information that had been provided by the company's own doctors in the first place. This didn't really seem unreasonable at the time. After all, in science we have to be certain of any information that we use. In fact, I relished the opportunity to be sure that what I had discovered was indeed correct."

Things went on for a few years in this vein. Every time Soskolne reaffirmed his basic analysis, new issues would be raised that had somehow escaped the company's attention. Ironically, with many of the adjustments that Exxon sought, the risks did change. They grew, sometimes reaching as high as thirteenfold, meaning that subgroups of the men with URC who were exposed to acid mists had more than thirteen times as high a chance of developing laryngeal cancer compared with those who did not work with acid mists. For those with the highest exposures the risks truly were higher, while risks for those with lower exposures were somewhat less. This neat finding strengthened the case that inorganic acid mists were causing cancer, because it showed that with more exposure the response was even greater.

Though I knew nothing about Exxon's reaction, I certainly knew of the stir Soskolne's work was creating within epidemiology. I was eager to try the new methods for myself, but I didn't get the chance. After my dad made it out of intensive care, Lilienfeld, who rarely gave orders, told me I was not the right person to focus on interviewing people with a specific illness or building the sorts of models that Soskolne was working with.

"Look," he said. "There are lots of people here who can do case-control studies. There are very few who have the really broad perspective. You've worked at the capital. You know how our work gets used and abused. Focus on the big issues. Find out the answer to a really important question: Is there an increase in cancer today independent of that tied with smoking?"

This question also involved multiple myeloma, but at a more abstract level. Why were more and more men and some women developing myeloma and other cancers at younger and younger ages? Lilienfeld plunged me into arcane reports of several centuries earlier, and shaped my postdoctoral studies to seek the temporal and spatial patterns of cancer that emerge out of the world in which we live and work.

He also imbued me with a wonder for medical history, showing me original drawings and diaries of early medical experts who understood that cancer has an array of social roots. In the seventeenth century, the Italian physician Bernardino Ramazzini asked whether nuns had more breast cancer because their menstrual cycles ran without interruption, unlike most Italian women of the day, who typically went through up to a dozen pregnancies. In the eighteenth century, the British clinician Percival Pott found that chimney sweeps had higher rates of scrotal cancer than other men; those who bathed the least had the highest rates. French sweeps, who bathed more often, and Germans, who wore leather trousers, had far fewer cases than the British.

Lilienfeld is the person who first made me aware of the relative neglect of the long history of understanding the social causes of cancer. If earlier observers had figured out some of the environmental causes of cancer, why then have modern cancer efforts focused so heavily on finding and treating the disease in each person? Doctors treat individuals, not the world into which they were born. Those confronted with cancer, like my dad, want a cure right now. When you've got the disease, how you got it is a secondary concern.

Many things entered my dad's body that could have led to his cancer. As a toddler, he survived a benzene explosion in his family home-based dry cleaning plant. (See the photo that opens this chapter.) "Yes, benzene," Aunt Bertha told me. "It was for sure benzene they had been using, because it cleaned up the clothes real good. They had fought the fire by themselves, because they didn't want nobody to know what they'd done." Benzene basically slices through membranes and gets into the part of the bone that makes blood—our marrow. Once there, it can keep iron from getting into red blood cells, rendering them sickly white. Anything that gets into the body of a young child can be especially dangerous. Children are not just little adults. They breathe faster and can absorb relatively greater amounts of poisonous materials around them.

When he was five, my dad sold newspapers every day. Papers back then were printed with lead-laden inks and smelly solvents that came off on your hands. Today we know that these solvents not only cut through grime and grease; they can slip right through the skin and get into the bloodstream. Dad started out, while still in high school, working as a chemist in the steel plant, where he was exposed to more solvents. This was not a job usually given to young people, but my dad was bright and interested in chemistry. I can remember him washing his slimy hands and mine in gasoline, when I was a young tomboy. Back then, gasoline could be nearly 10 percent benzene. Did my grandparents' basement explode because they were cleaning clothes with gasoline? I'm sure it would have done the trick of removing dirt and grit.

During World War II, Dad worked as a welder on ships being built for the navy. Welds were often checked with x-rays, using crude machines that delivered heavy, scattered rays of ionizing radiation. Later, as company commander of the 110th Infantry, Company D, of the Pennsylvania National Guard, Dad proudly went to army camp every summer for more than thirty years. Each year, sometimes twice or more, he had x-rays to make sure he stayed in the best of health.

MARVIN SCHNEIDERMAN, a senior researcher at NCI, once defined an epidemiologist as someone who can find something wrong with something that someone else does, who also calls herself an epidemiologist. Schneiderman also joked that epidemiologists eat their young. It is not a field for the faint of heart. In the early 1980s, it was wrestling with a major controversy.

Joseph Califano, the activist secretary of Health Education and Welfare under President Jimmy Carter, had not only taken on Big Tobacco, as we saw in Chapter 7, he also went after the chemical industry. In testimony before Congress in 1978, Califano stated that up to 20 percent of cancer in the future would be due to workplace exposure. This shocking number sent the public relations industry into full battle mode. Others have written about this conflict. In making this prediction, Califano was trying to prevent more cancer. But the debate soon was turned upside down.

Richard Doll and his collaborator at the time, a brilliant young epidemiologist named Richard Peto, were asked by the U.S. Office of Technology Assessment to weigh in on workplace cancer. To do so, they asked an entirely different question—what percentage of past cancer deaths could be attributed to various known causes? They answered this question by examining patterns of cancer death in white persons under sixty-five from 1950 to 1977. They concluded that smoking was the single most important preventable cause of cancer deaths up to that time. Excepting what was caused by smoking and perhaps diet, as well as pesticides in some small populations, they assured the Congress there was no increase in cancer.

I was confused. Lilienfeld had taught me that the incidence of cancer—the rate of *new* cases of the disease—was the most important indicator of factors affecting the disease. Incidence also was the best predictor of future demands for medical care. Yet Doll and Peto had not looked at incidence at all, nor had they included the growing rates of cancer in blacks. Instead, they had considered the other end of the process: causes of death and only in whites. We both knew that at the time, four out of every five cancer deaths occurred in people over sixty-five. Why, then, did Doll and Peto restrict their work to deaths

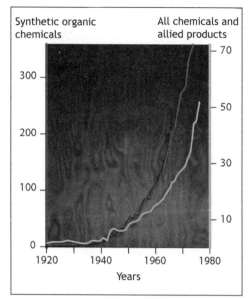

Figure 10-1 Exponential growth in
U.S. industrial chemical production.[2]
Source: Davis and Magee,
Science, 1979.

that occurred in whites under sixty-five? Finally, we knew that four decades or more could pass between a person's first exposure to a hazard, like asbestos or a solvent, and the onset of cancer. The use of synthetic organic chemicals had grown explosively in the 1960s and 1970s. Blacks then and now worked in dirtier jobs. By ignoring everyone who had cancer over sixty-five and was still alive, and by excluding blacks altogether, Doll and Peto systematically underestimated the effects of recent chemicals.

It's hard to grasp how different the modern world is from that of my grandparents. My grandfather grew up driving a horse and buggy and ended up flying across the country. Synthetic chemicals have made jet planes and many modern developments possible. Production and use of these materials during the last century grew exponentially in the developed world, and continues at this pace in the developing world today.

These tables and graphs show that at the time and in the manner that Doll and Peto did their report, the full brunt of cancer from synthetic organic materials could not possibly have been detected. By looking at deaths in whites up to 1977, their report told us the cancer burden that arose from exposures dating from the 1930s and 1940s.

Their methods were incapable of addressing the health consequences of more recent materials.

Schneiderman was convinced there was a lot more to the story. At his urging I began to look for a different way to analyze cancer statistics. My colleague Joel Schwartz, a talented statistician on his way to becoming one of the nation's top epidemiologists (now at Harvard), and I asked whether the incidence of multiple myeloma and brain cancer in men between the ages of forty-five and eighty-four had changed over time. We found that both of these usually fatal cancers had grown by more than a third in less than two decades. This work was published in the *Lancet* and became the basis for an entire volume of the *Annals of the New York Academy of Sciences*.

The *Lancet* article made headlines. The conclusions pronounced by Doll and Peto, whose authority was beyond question, had become the accepted wisdom as soon as they were published. Yet we had contradicted them. I had no intention of challenging these eminent, respected men. Lilienfeld urged me to just stick with the facts.

One evening, after a symposium in Lyon, France, I was thrilled to find myself having drinks with none other than Sir Richard Doll. His entry in *Who's Who* listed conversation as one of his hobbies, and sure enough, he was a captivating, engaging and scintillating man to talk with. Doll assured me that he was taking the time to speak with me because he wanted to help. I was honored by the attention. I could do good work in the future, he explained, but my work contained a fundamental mistake, a colossal error, which I'd made even worse by repeating it in other publications. I had reported increases in specific types of cancer in men ages forty-five to eighty-four. All of these increases, Doll assured me, came from one simple cause: medical record keepers were doing a better job of reporting cancer, and doctors were doing a better job of finding it. If I would look at the data more closely I would realize my mistake. There was a cause of death, he explained, called senility, that was listed on a death certificate when doctors had no idea what really killed someone. There was another cause of death called "cancer of unspecified site." This meant that a person had obviously died of cancer, but the doctors couldn't identify the original primary site.

Thus the increases in so many kinds of cancer I had reported were nothing but improvements in finding disease. The way we could show

Table 10-1 Risk Factors Associated with Workplace Exposures to High-Volume Carcinogens[3]

Chemical	Sites of Primary Cancers[a,b,c]	Other Chronic Health Effects[d,e]	Occupations at Risk[b]	Latency Period for Cancer (years)[b,f]	Risk Ratios for Cancer[b,f]	1981 NIOSH Estimated Number of Workers Exposed[9]	
						full + part time	full time
Acrylonitrile	Colon, lung	Eye and nose irritant, gastrointestinal effects, jaundice, mild anemia	Chemical workers and plastics workers	20+	4-6	374,345	55,706
Arsenic	Skin, lung, liver, lymphatic system	Gastrointestinal disturbances, hyperpigmentation, peripheral neuropathy, hemolytic anemia, dermatitis, bronchitis, nasal system ulceration	Miners, smelters, insecticide makers and sprayers, chemical workers, oil refiners, vintners	10+	3-8	255,277 432,017 (arsenic oxides)	5,926 596
Asbestos	Lung, pleural and peritoneal mesothelioma, gastrointestinal tract	Asbestosis (pulmonary fibrosis, pleural plaques, and pleural calcification), anorexia, weight loss	Miners, millers, textile, insulation and shipyard workers	4-40	1.5-12	1,280,202	449,960

Beryllium	Lung	Dermatitis, bronchitis, respiratory effects	Beryllium workers, defense and aerospace industry, nuclear industry	15+	1.5–2	855,189	632
Benzene	Bone marrow (leukemia)	Central nervous system and gastrointestinal effects, blood abnormalities (anemia, leukopenia, and thrombocytopenia)	Explosives, benzene and rubber cement workers, distillers, dye users, printers, shoemakers	6–14	2–3	1,495,706	147,604
Cadmium	Prostate, respiratory tract, renal	Renal disease, respiratory effects	Electrical workers, painters, battery plant and alloy workers		2.5	1,376,871	38,433
Carbon Tetra-chloride	Liver	Cirrhosis and liver disease, kidney and gastro-intestinal effects, dermatitis, jaundice	Drycleaning, machinists			1,380,232	64,023
Chromium	Nasal cavity and sinuses, lung, larynx	Dermatitis, skin ulceration, nasal system ulceration, bronchitis, bronchopneumonia, inflammation of the larynx and liver	Producers, processors, and users of Cr: acetylene and aniline workers; bleachers; glass, pottery, and linoleum workers; battery makers	5–15	3–4	1,451,631 (oxides)	59,946

continues

Table 10-1 (continued)

Chemical	Sites of Primary Cancers[a,b,c]	Other Chronic Health Effects[d,e]	Occupations at Risk[b]	Latency Period for Cancer (years)[b,f]	Risk Ratios for Cancer[b,f]	1981 NIOSH Estimated Number of Workers Exposed[g]	
						full + part time	full time
Ethylene Oxide	Leukemia, gastric cancer (suggested)	Mutagenic, respiratory irritant	Hospital workers, laboratory workers, fumigators			144,152	107,455
Nickel	Nasal cacity and sinuses, lung	Dermatitis	Nickel smelters, mixers and roasters, electrolysis workers	3–30	5–10 (Lung) 100+ (nasal, sinuses)	1,369,278 (oxides)	51,840
Vinyl Chloride	Angiosarcoma-lung, brain, haematolymphopoietic	Reproductive and central nervous system, Reynaud's syndrome, acroosteolysis	Plastics industry	20+		239,375	29,838

[a] Devra Lee Davis and David P. Rall, "Risk Assessment for Disease Prevention" in Lorenz K. Y. Ng and Devra Lee Davis, eds., *Strategies for Public Health*, Van Nostrand Reinhold, New York, 1981.

[b] Philip Cole and Marvin Goldman in Joseph Fraumeni, ed., *Persons at High Risk of Cancer*, New York: Academic Press, 1975.

[c] Occupational Diseases, Washington, D.C.: U.S. Department of Health, Education, and Welfare, 1977.

[d] I. Casarett and John Doull, eds., *Toxicology*, New York: Macmillan, 1975.

[e] George Waldbott, *Health Effects of Environmental Pollutants*, Saint Louis: C. V. Mosby, 1978.

[f] International Agency for Research on Cancer, Chemicals & Industrial Processes Associated with Cancer in Humans, supplement to vols. 1–20, Lyon, 1979.

[g] National Institute for Occupational Safety and Health, Interim Estimate, 1981.

Source: Davis, *Environment*, 1981.

this, Doll said, was to look at the number of deaths attributed to senility and cancer of unspecified site over time. We would find that these poorly diagnosed causes of death were dropping, while those tied with specific cancers were growing. QED.

I was flabbergasted and flattered. I had spent an evening with the great Sir Richard. He had told me what to do, and his argument was persuasive. I knew there had been great advances in computerized x-rays and other methods of finding cancer. It made a great deal of sense that our findings were rooted in one giant artifact—something that gets created by error as a result of a mistake in thinking and therefore is not a real phenomenon.

I spent the next four years looking into all of the ideas Doll had suggested. By this time Lilienfeld also was engaged in the question. He had served as a reviewer for Oxford University Press when it published Doll's and Peto's book *The Causes of Cancer,* which quickly became a bible of cancer epidemiology. As our work progressed, Lilienfeld confided that he regretted not having challenged some of Doll's and Peto's extreme conclusions more vigorously earlier on. He especially questioned their view that workplace cancers account for less than 5 percent of all cancer. With his support I expanded my research on patterns of cancer incidence to other countries.

In science, if you come up with a result that a giant in the field has told you is wrong, it's a good idea to pay a lot of attention to getting it right. With help from Lilienfeld and Allen Gittelsohn, a biostatistics professor at Hopkins, I examined everything Doll told me to look at. Was there a major decline in deaths from senility and from cancers of unspecified site? Not at all. When we looked at the data from 1968 to 1978, we saw exactly the opposite. Deaths from these unspecified causes of cancer and deaths from senility had not dropped at all in older whites. They had gone up and so had those from the specific types of cancer we'd reported earlier. There was a slight drop in these poorly diagnosed causes of death in blacks, but Doll and Peto had never analyzed them at all.

I began drafting a manuscript with Lilienfeld and Gittelsohn in which we laid out our analyses. We showed that there were continuing and unexplained increases in specific sites of cancer that could not be

TABLE 1
Percent Change[a] in Age-Specific Site-Specific Cancers
and Potentially Miscoded Causes of Death from 1968 to 1978

TABLE 1

	Ten-year Age Groups				
Cause of Death	35-44	45-54	55-64	65-74	75-84
Nonspecified Cancers					
WF[b]	-12%	-10%	10%	7%	14%
WM[c]	-9	18	15	17	30
NF[d]	-14	-10	3	9	42
NM[e]	-17	30	15	23	40
Pneumonia					
WF	-48	-43	-36	-33	-24
WM	-47	-46	-39	-33	-13
NF	-60	-49	-49	-47	-26
NM	-53	-44	-37	-36	-25
Senility					
WF	10	13	8	26	22
WM	-4	-5	-5	16	18
NF	-41	-35	-28	-24	-29
NM	-29	-24	-28	-31	-20
Brain Cancer					
WF	-21	-5	6	43	143
WM	-16	-8	4	34	115
NF	-17	21	5	133	255
NM	-37	-14	20	91	54
Multiple Myeloma					
WF	17	18	15	18	38
WM	-5	5	16	18	40
NF	1	-4	28	27	82
NM	21	-24	6	31	67
Lung Cancer					
WF	23	72	105	104	66
WM	-19	16	12	24	55
NF	22	83	92	72	52
NM	-12	15	40	48	60

[a]Percent change $= \dfrac{(1977+1978) - (1968+1969)}{(1968+1969)}$

[b]White Females
[c]White Males
[d]Nonwhite Females
[e]Nonwhite Males

Figure 10-2 Table from Abraham D. Lilienfeld's last publication on cancer patterns.[4]
Source: Davis and Lilienfield, *Toxicology and Industrial Health*, 1986.

due to better reporting and that these occurred most in those who had lived long enough to experience cancers tied with industrial experiences. We suggested that these patterns could indicate past problems with the rapidly industrializing workforce. We even predicted future rates might drop as a result of reductions in toxic chemicals in the workplace then being proposed.

Scientific papers can take years to complete. By the time ours was finished, I had left Johns Hopkins and was working at the National Academy of Sciences in Washington, D.C. Lilienfeld, one of the eminent advisers to the academy, sometimes took the train from Baltimore to Washington to meet with various committees. One day as he was leaving the academy's headquarters, he spied the FedEx package I had left earlier in the day for pickup. Seeing his name on the front, he took the envelope and said to the security guard with some amusement, "She finds me, no matter where I am."

Lilienfeld suffered his last heart attack as he was getting off the train in Baltimore. His son David told me that the final, opened galleys for our accepted manuscript were in his briefcase.

As my father lay in a coma for three weeks, few people thought he would come out. I would spend hours sitting with him, sometimes holding his hand. I'd close my eyes and softly describe the slowly rising sun as it turned warm red, and tell him to see it flow throughout his skull into his body. The sun would soften to buttery yellow that melted, pouring over his limbs. I'd tell him of floating on a quiet green pond, surrounded by shimmering light over summer green grass. Slowly, gently, a sweet blue sky opened overhead. Softly, slowly, I would talk about drifting into that sky; bathed by beautiful, light, holy clouds.

By the time I began describing those clouds, the machine recording Dad's pulse would show it dropping from 140 to 100.

The nurses would come over and ask me what on earth I was doing to the heart rate monitor.

My brother Stan thought I was losing my mind. "Why are you talking to him like that? He can't hear you!"

When Dad came out of his coma, he could not speak. The breathing tube the doctors inserted into his throat had so damaged his larynx that he was unable to utter sounds we could understand. But he could write. Stan and I sat next to him. I explained that I had to go home now. My young children, Aaron and Lea, needed me, and I needed them and Richard.

Dad looked very agitated. I wasn't sure he understood what I was saying, or whether it was just the heaviness of the drugs he was on. He gestured with his eyes and a slight nod of his head for the nearby white board and felt-tip pen so that he could write.

"But where will I find the clouds?"

That's when I realized that Lilienfeld had been right about many things.

A Breast Cancer Fund poster, banned in San Francisco in January 2000. The superimposed mastectomy scar is that of the author's friend Andrea Martin.

11

Doctoring Evidence

You can observe a lot by watching.

——YOGI BERRA

"Dev? There's a letter for you," my husband called from Washington, D.C. I stopped to answer my phone as I was hiking the coastal trail of the Presidio just off San Francisco Bay. It was one of those nearly perfect northern California days, the sort of weather that makes you understand why so many people move to this state with its seductive balance of balmy wind and sweet air. The sun shone through tall, glistening pines. A warm, foggy wind wafted around the point of land that protruded right into the bay.

Most couples have routines. In our marriage, my irrepressible economist husband deals very well with bicycle repairs and finances. He could have become an accountant, but he had far too much personality. The two things he's never been good at are delivering phone messages and opening mail. (We both could use a wife.) In twenty years of marriage and countless trips, he had never called about a letter. I knew something was up.

"Who's it from?" I asked. This didn't sound right. My husband had the tone of someone trying hard to sound like everything is okay. I knew he was about to lose it totally.

"There's a letter from the mammography clinic. I think you need to call them." Through the phone I heard a puff of resignation come through his closed, certainly frowning lips. A gulp and swallow followed. The unusual pause was a dead giveaway.

Or maybe we had lost our connection.

"Are you still there? Can you hear me? Hello?" I asked.

"Yes, I'm here. Did *you* hear what I just said? Do you *understand* me?" We were playing out one of those cell phone commercials where the other brand—the one you did not sign up for last week—is always the better one.

I stopped walking. I looked down at the brown earth and tree roots beneath my feet. I tried to take some of that grounding into my voice as I started to reassure him.

"Listen, honey. With these tests you know there are lots of things they find that really don't mean anything at all, especially in somebody my age. It's really nothing to worry about." I was just fifty at the time.

"Okay. Okay. OKAY. I know you're supposed to know all about this. Would you please just go and have somebody else check this out?" He resorted to that old staple of the well-rooted marriage, the borderline shout. The presumption of this raised volume, of course, is that I had not actually heard him. A louder voice is supposed to make me cave. It never works, but my husband feels he has gotten his point across.

"Listen, sweetie," I explained. "Statistically speaking, I'm really just fine. Really."

"Look, I'm not interested in your damn statistics," he blurted. "Just deal with this, would you? Stop playing doctor! Go see a real one!"

I felt trapped by his concerns into taking action I didn't think made sense. To understand why, you need to know that mammography is one of the most oversold and understudied technologies in medical history. X-rays of the breast can find small white tracks of calcium that signal cancer years before it can be felt or seen. Like many stories in medicine, it all started with a terrible loss.

Distinguished radiologist Phillip Strax had been powerless to prevent his young wife from dying of breast cancer. If only he had been able to find her cancer earlier, he lamented, she might have survived. Strax knew that as tumors start to grow they leave residues of calcium that show up as tiny white dots on x-ray. Sometimes they create telltale starburst patterns. They can also take the shape of rods, branches or teardrops. Often they look like nothing but tiny pinpoints.

A mammogram is one of the most unsexual experiences a woman

can have. Breasts are organs of pleasure. When you have a mammogram, feeling erotic is the last thing on your mind. There is nothing pleasurable about standing flush against a large metal machine with soft breasts squashed flat between hard plates.

"Don't breathe!" That's what the technician tells you after she has set the plates squeezing the breast to the maximum pressure. Breathing is one of those things we do without thinking. That's why we don't die at night; the brain keeps air coming in and out of our bodies so long as we are alive. No matter how diligent you may be on a conscious level through the experience of holding your breath during the few seconds of a mammogram, your mind is saying, "Are you crazy? I need some air and this really hurts. Let's get the hell out of here. Now!"

If you're lucky, the whole process lasts only seconds in real time. But body and brain can make it all play out in excruciatingly slow motion. You remain breathless and hold perfectly still, while the x-ray shoots through the mostly fatty smushed breast illuminating any dense areas. The fleeting discomfort of mammography is one of those things women don't usually talk about. Nobody wants to be a candy ass about going through something that yields life-saving information.

Cancer specialists had long appreciated that breast cancer, like most adult forms of the illness, appears only after years of growth. Before it can be seen on x-ray, what starts out as an invisibly damaged solitary cell of the damaged breast has to elude millions of efforts to kill or fix it. Cancer cells feed on and spew out sugar, one of the fastest fuels it can consume. Doubling times of cancer cells can vary between 100 to 400 days.[1] X-rays depict the remnants of this growth—the distinctive tracks of tiny specks of calcium that splitting cells leave—years before the disease can be seen or felt.

As one of the pioneers of the original technology, Strax put together the first large effort to see whether or not mammography made sense in New York City. It is no accident that the first trial of this slightly uncomfortable technology emerged in a city also known as a center for feminist health theory and practice. Then and now the Upper West Side of Manhattan was a hotbed of women's body politics.

Working with one of the largest managed health care systems, the

Health Insurance Plan of New York (HIP), in 1963, the study found more than 60,000 women who agreed to take part. Half were randomly assigned to get a mammogram, in order to find out whether or not healthy women with no sign of cancer would benefit. The program was stunning in breadth, bold and brash.

By 1971 the results were clear. Women over fifty who had regular mammograms died less often from breast cancer. If they developed tumors, their growths tended to be smaller and less advanced than women who had not undergone screening. But regarding those under fifty, the story was not encouraging. They fared no better with mammograms than without. After this trial ended, some scientists urged more testing before younger women were subjected to mammography radiation. Rosalie Bertell, the visionary critic, and John Gofman, the renegade physician from the Nuclear Regulatory Commission, warned that radiation incurred from regular testing could itself carry a risk of producing precisely the cancer such screening was intended to stave off.[2]

Because younger breasts are more dense, on x-ray they are riddled with lots of white spots, making it really hard to make out any tumor within. It's fortunate that most young breasts don't get cancer, since mammography is a pretty poor way to find them.

The impact of showing that mammography actually worked on older women was breathtaking. The long drought of failed promises was over. Finally, after years of promises, a life-sparing technology was at hand.

Convinced that all women would benefit from regular mammograms and caught up in the political and economic enthusiasm of combating cancer, the National Cancer Institute (NCI) and the American Cancer Society (ACS) launched a massive demonstration project on breast cancer detection in 1972 that covered women of all ages. But this national program effectively killed prospects for ever finding out whether or not mammography works in younger women.

Project leaders like Arthur I. Holleb, a senior ACS official, ardently believed in what they were doing. This would not be the last time fervent beliefs got in the way of scientific research. The campaign for mammography became wrapped in military metaphors that resonated with a public already wearied of the failed conflict in Southeast Asia.

Holleb urged putting this life-saving technology into broad operation as soon as possible.

"No longer can we ask the people of this country to tolerate a loss of life from breast cancer each year equal to the loss of life in the past ten years in Viet Nam. The time has come for greater national effort. I firmly believe that time is now."[3]

Ironically, this full-court press for healthy women to have mammograms may well have been launched to make up for delays in providing regular screening for cervical cancer with the Pap smear. By 1971 it was clear that haggling over the Pap test for cervical abnormalities—long after it had been shown to work in the 1950s—meant that thousands of women developed cancers they could have avoided. Women had always been enthusiastic supporters of the ACS. As cervical cancer rates were dropping, breast cancer was growing in importance. Here was a chance to do something about it. If screening worked for cervical cancer, and did so for women at all ages, why should mammography be any different?

Economists were not widely engaged in such matters at the time. They ask questions such as, "Are the costs of using this procedure on women of certain ages in line with the benefits?" Costs are the simple part. Benefits are harder to measure. The actual number of lives saved can be estimated, but what about unnecessary radiation and surgery to look for things that appear to be cancer and turn out not to be? What about the terror that women go through as they wait for further testing and biopsies of suspicious growths, most of which turn out to be nothing? What are the chances of mistakes being made? These questions were not easily asked at the time. If they had been, the world might look rather different.

Strax understood that a revolution in public attitudes was needed for mammography screening to work. A book he published in 1974 sent a welcome message—*Early Detection: Breast Cancer Is Curable*. Finally the ACS and NCI had a champion who spoke the four-letter word that few had dared to utter: cure. Sometimes personal enthusiasms, even for scientists, are hard to hold in check. Haunted by what had happened to his young wife, Strax became an ardent advocate for mammography. He believed it had to work. He likened the failure to get a mammogram to walking into a busy street without looking both ways. A woman who ignored the risk of breast cancer faced disaster. "Or, she may be constantly on her guard, aware that she may be affected, but with the assurance that early detection may save her life. Which course should you follow?"[4]

Some four decades after the start of the American Society for Cancer Control, leaders in cancer research were ecstatic. They felt sure they finally had a technology that could deliver on the old promise that finding cancer early saves lives. The mammography demonstration project was massive and without precedent: more than a quarter million women between thirty-five and seventy-four would be provided mammograms every year. Medical advertisers had a field day as companies began to ramp up to produce all the machines that would be needed.

As Barron Lerner has pointed out in his critical history of this era, the promotion of regular x-ray exams of the breast played into two important American beliefs. Women are charged with a moral duty to take charge of their health, and technology offered them a way to beat the odds against one of the most common causes of death in middle-age women.

The push for mammography cannot be separated from the rise of feminism at about the same time. Women's body parts and lives were named and bandied about at the time in ways that were nothing short of revolutionary. In some circles, women met in groups, equipped with mirror, flashlight and speculum, to learn how to gaze at their hidden cervix. Breast cancer also came out of the closet and bedroom. Diagrams showed women how to feel their own breasts and encouraged them to do so regularly to try to find early tumors. There has never been any evidence that regular self-exams keep women from dying of breast cancer.[5] Still, many believe that getting women comfortable with touching their bodies may eventually turn out to be useful as a way to find the disease earlier. And, unlike many breast cancer tests and technologies, this one is free and has no downside.

A further boost to mammography came from reports that First Lady Betty Ford, Margaretta (Happy) Rockefeller, the wife of Vice President Nelson Rockefeller, and Marvella Bayh (wife of Sen. Birch Bayh) had been diagnosed and successfully treated for breast cancer. Reflecting this newfound public interest in breast cancer and the expanded access to screening, the officially recorded incidence of breast cancer rose nearly 15 percent between 1973 and 1974—from 82.6 to 94.9 per 100,000 women of all ages.

By 1990, enthusiasm for the technology had begun to fade. Gina Kolata, a science reporter known for her advocacy of mammograms, reported in the *New York Times* that not all mammograms were equal. "Proponents and opponents agree that a serious problem exists with mammography: A large proportion of test sites use substandard equipment, are staffed by inadequately trained technicians and radiologists, or are rarely if ever inspected to be sure the equipment is working properly at a minimal dose of radiation."[6]

The federal government didn't even have national standards for mammography until 1994. This only happened after John Dingell, William Ford, Marilyn Lloyd, Patricia Schroeder, Henry Waxman, and a number of champions of women's health, including Assistant Surgeon General Susan Blumenthal, married to Congressman Ed Markey, and NIH director Bernadine Hardy, demanded such a program be put in place.[7] I had visited Arkansas that year, working for Jocelyn Elders, the surgeon general. We learned that in one out of every three clinics mammograms were given by persons who had no training, including office receptionists.

When I told my husband not to worry about my callback, I had solid numbers in mind. I knew that if one thousand women had regular mammograms starting when they were forty, each year seventy of them would be called back, as I was, for further testing or surgery. This could involve more radiation through magnified x-rays or a surgical biopsy to examine cells under the microscope for cancer. By the time these thousand women reached fifty, seven hundred of them would have been called back for repeated testing. Many of them would have gone through biopsies. Nearly all of them would have been just fine.

Susan Love, the famed breast surgeon, and I had written a commentary on all this in the *Journal of the American Medical Association*. We explained that the breasts of women who are still menstruating regularly contain things that can be hard to interpret. The younger the breast, the more dense it looks on x-ray. But x-rays go right through older, fatty breasts. Fat is radiolucent, showing up as dark black on x-ray images. Against the sharp black background, tiny deposits of calcium smaller than the head of a pin—the microscopic residues of calcium traces from tumor growth—tell us whether or not cancer lurks within the breast.

But dense breasts, like the ones I had at the time, tend to look full of snow, created by the abundance of epithelial and connective tissue that also leaves white tracks. They pose the classic problem that hunters know very well: you can't spot an elk behind a single tree within the forest where it's hiding. I tried to tell my husband all this. He wouldn't listen.

Like many people, he believed that any sign of growth in the breast had to come out. I knew that this simple way of thinking came about because the fear of breast cancer can be terrorizing, perhaps especially so for spouses. But I also knew that most breast tumors in women just before menopause turn out to be benign. I was willing to bet on it.

So far, I've been right.

I knew my numbers regarding regular mammography in women under fifty because of my work on the *JAMA* article. Karla Kerlikowski at the University of California–San Francisco had found that women under age fifty account for only one in five cases of breast cancer. Yet those of this age who had gone through yearly tests had more than twice as many surgeries, and nearly three times more follow-up diagnostic procedures as those over fifty. We worried about the fact that women whose lives could be saved by mammography—those over fifty—were getting fewer exams, while those in whom it made less sense were getting more of them.

We concluded our report with these words, which are repeated here, because they remain pertinent, more than a decade later:

> The public policy dilemma posed by these findings for women and their health providers in the United States today is profound. Chalmers has reminded us that "if our society had been oriented towards finding out whether new technology is efficacious as soon as possible after its introduction there would not be much left to debate more than 30 years (later)." . . . But wistful wishing cannot alter the fact that mammographic screening in women under 50 years of age does not reduce deaths, while for those over the age of 50 years it saves lives. The reasons for these results are unknown and need to be re-

solved through additional clinical studies that assess the importance of menopausal status and other factors, including breast tissue change with age. In the meantime, women must be told the truth, so they can make informed choices about their health care. And efforts must proceed apace to develop better techniques to detect early breast cancer in asymptomatic younger women, to ensure that all women over 50 years of age are provided access to the lifesaving benefit of screening mammography, and to identify avoidable causes of this major cancer.[8]

After publishing this warning, Blumenthal, the first deputy secretary for Women's Health; Deborah Axelrod, then head of breast services at Beth Israel Hospital in New York; Gillian Newstead, a radiologist at New York University; Robert Smith, head of screening with the American Cancer Society; and I organized a national workshop aimed at coming up with ways to improve mammography and promote alternatives. We held meeting after meeting and reviewed reports from the nation's top radiologists. We talked about creating master mammography readers who would provide standardized second opinions on all films, which has long been done in Scandinavia. Blumenthal came up with an innovative scheme to use technologies of the space and intelligence agencies that can read a license plate from outer space to enhance the ability to find early signs of cancer within the breast. Despite these efforts and major technical progress on several fronts, no major change in how mammograms should be read and reviewed ever happened.

Because of what was called the Defense Department peace dividend, financial allocations were being shifted to peacetime activities. The secret slogan of the breast cancer activists who campaigned for a major program of research within Defense was, "Better boobs than bombs." Given the growing numbers of women in the military and the shrewd tactics of Sen. Alfonse D'Amato and Sen. Tom Harkin, breast cancer was seen as a matter of national defense. With bipartisan support from senators whose mothers, wives and sisters had been affected by the disease, I was supposed to help run what would turn out to be a half-billion-a-year program for breast cancer research in the Department of Defense. That never happened. Instead, I got "promoted" to a position requiring full-fledged Senate confirmation. This supposedly more prestigious position didn't involve any budget what-soever. Fancy titles, rather than money, can be another way to keep people quiet, at least the ones you cannot buy.

In fact the new agency I was appointed to help run didn't even exist at the time—the National Chemical Safety and Hazard Investigation Board. While my confirmation hearing was pending before the sud-denly Republican Senate in 1994, I got a phone call from Terry Yosie. We had worked together at EPA during the Carter administration, and Yosie now worked in industry. He told me that a team of public rela-tions experts had been tracking and filming my public speaking. He was amused by what they had reported and faxed me a memo detailing my alleged extremist views, which he knew I did not hold. Mongoven, Biscoe & Duchin worked for the Chemical Manufacturer's Associa-tion, Chlorine Chemistry Council, and sent this report about me to senior administration officials in August 1994.

> Devra Lee Davis is expected to direct the Clinton Administration's policy governing breast cancer and we expect her to try to convert the breast cancer issue into a debate over the use of chlorine. As a member of the administration, Davis has unlimited access to the media while her position at the Health and Human Services (HHS) [department] helps validate her "junk science." Davis is scheduled to be a keynote speaker at each of the upcoming WEDO [the Women's Environment and Development Organization—an international group headed by Bella Abzug] breast cancer conferences.[9]

The PR flacks made me out to be a deluded zealot. I noted with some amusement that their memo had the same date as an article I'd written with H. Leon Bradlow on environmental links to breast can-cer for *Scientific American*, hardly an extremist publication.[10]

The world has changed since I was the subject of that hatchet job. But I only got wind of the campaign to undermine my limited author-ity after it had been brutally successful. I will never know whether my promotion to a presidential appointment was nothing but a way to get me out of a post where I might have had some serious influence over federal government policies and research on the environment and breast cancer.

Inside government at the time, I thought things were turning our way when the head of the U.S. Department of Defense committee on radiology funding came to me, as one of the chiefs of the environmental committee, and said, "Look, we're not coming up with anything new to do with radiology. Why don't we just give more money to environmental studies? That seems to make the most sense."

Funny thing happened the very next year. That fellow and I found ourselves far removed from any position of influence. Today mammography is booming. Digital mammography has replaced conventional as the technology most people think is better, but data to prove that this is the case are not easy to find. What is clear is that a digital mammogram machine costs about five times more to purchase and finds many more things that need to be looked at more thoroughly. What is not clear is whether this will lead to fewer cases of advanced breast cancer. Lately, magnetic resonance imaging (MRI) of the breast, with contrast dye, has been advised for women who are believed to face a greater lifetime risk of breast cancer, including those with a strong family history of breast or ovarian cancer and those who've been treated for Hodgkin's disease. The MRI costs ten to twenty times more than a mammogram and also finds even more suspicious things that need to be examined surgically and pathologically. Whether or not this leads to fewer deaths is something that should be carefully evaluated. The incentives to do so, for such a highly profitable new technology, are lacking.

Recent drops in deaths from breast cancer have been chalked up to the decline in the use of hormone replacement therapy as well as the increased accuracy of breast screening programs. No mention has been made of the possible role of the hundred-fold lower levels of cancerous pesticides and some key air and water pollutants found in the breast milk of women that has also occurred at this same time. Nor can we know whether this decline in breast cancer deaths has anything to do with a reported decline in the proportion of women undergoing mammograms in the past seven years.[11] In an ideal world, we would have the ability to track the capacity of any of these things to affect breast cancer statistics.

For a while a straightforward examination of the facts on mammography seemed possible. Committees of expert statisticians and epidemiologists were charged with taking a hard, cold look at information. Studies from Canada and Scandinavia provided messages Americans didn't want to hear—mammography screening of women without any signs or symptoms of breast cancer saved lives only in those over fifty. There could be groups of women under age fifty in whom such screening made sense, but it didn't look like a good idea for all women. The challenge was how to decide who should be screened.

That challenge never got met. Instead, lobbyists for companies making mammography machines crafted unassailable alliances with articulate, persuasive breast cancer activists, then just coming into their own. Senators and representatives rapidly got on board with what became a national demand for mammography. Lost in all the political jockeying was the ability to answer a simple question: Does regular screening make sense for younger women?

Cornelia Baines is a brave, clever Canadian expert on breast screening. Her work with Anthony J. Miller in the 1990s was one of the pivotal studies showing that mammography saves lives of women over fifty. They worried, in print at the time, that such screening might well cause deaths and disability in those under age fifty by producing cancers from radiation exposures and leading to unneeded surgery.

In 2005 she had not changed her mind at all. "A wise American epidemiologist observed that it is un-American to oppose mammography."[12] How could mammography lead to unnecessary surgery? Ductal carcinoma in situ is a common lesion that is believed not to become a cancer in many cases. It has become four times more common in recent years, largely, but not entirely, because of increased mammographic screening. When it is found nowadays, DCIS is removed surgically, often with a procedure that leaves the breast intact but not without scars, and sometimes with loss of feeling in the nipple. Some young women with DCIS choose to have both breasts removed as a precaution. Even though women who have surgery for DCIS are supposed to have radiation to ensure cell killing, in half of the cases they do not. Even though they are not supposed to have the lymph nodes from their armpits removed in most cases, many of them do.[13]

Today, new computerized systems digitize and magnify what can be seen within the breast. Some radiologists have never seen a microcalcification they can leave alone, and surgical biopsy—or the use of fine needles to withdraw cells and look at them under a microscope for evidence of cancer—is booming. Others will advise waiting half a year to look again. New machines take much higher resolution digital mammograms and find microscopic changes that could not be seen a decade ago. They cost about seven times more than conventional ones. Magnetic resonance imaging of the breast costs between $2,000 and $3,000 per procedure. Ultrasounds are also being used to image the soft tissue of the breast after any suspicious signs. Increasingly, ultrasound combined with MRI is being recommended for any suspicious breast lesion picked up by mammography.

There is no way to know whether increased testing is truly better for women's health, because, as far as I can tell, there is no independent national program for collecting information on these rapidly growing and very costly tests. For those who find a suspicious lesion on mammography and also find an excellent doctor like New York University's Kathy Plesser, the whole thing works well.

Instead of spending three weeks waiting for results of mammograms, ultrasounds, and magnetic resonance imaging, and going back and forth each time, racking up related economic and psychological costs, you can go through all of these tests in four hours if they are needed. How can Plesser do this? And how can the Komen Foundation's Ozark affiliate do it in Fayetteville, Arkansas? When so much money is made from many separate visits for separate procedures, there are few incentives to streamline.

The marketing of mammography, ultrasound and breast MRI has a life of its own, where the opportunity to conduct hard, cold analysis is hamstrung by the fabulous profitability of the business. Nobody doubts that in many cases, it all works beautifully. Still, Baines laments the loss of the ability to carry out dispassionate analysis of mammography or other more expensive and newer technologies for evaluating breast cancer at this point. She isn't alone in her dismay and has some important new allies.

The National Breast Cancer Coalition, originally an avid supporter of mammography, has joined the chorus of many breast cancer groups warning that increased screening with mammography is not the answer. Looking at all of the studies ever completed on mammographic screening throughout the world, as of May 2007, their official position on the matter is one of disappointment. Their language is stunning and clear:

"NBCC believes that there is insufficient evidence to recommend for or against screening mammography in any age group of women. Women who have symptoms of breast cancer such as a lump, pain or nipple discharge should seek a diagnostic mammogram. The decision to undergo screening must be made on an individual level based on a woman's personal preferences, family history and risk factors. Mammography does not prevent or cure breast cancer, and has many limitations. Women are told that mammography screening saves lives, but the evidence of a mortality (death rate) reduction from screening is conflicting and continues to be questioned by some scientists, policy makers and members of the public. Ultimately, resources must be devoted to finding effective preventions and treatments for breast cancer and tools that detect breast cancer truly early."[14]

Lewis Kuller is a living legend, a walking encyclopedia of public health research. The former chairman of the Epidemiology Department at Pittsburgh and an expert in women's health research for more than three decades, he's published on just about every problem there is. And if he hasn't written about a subject in public health today, then he knows all about what other people have done. But there are some issues even Kuller won't take on.

"You think I want to get killed? Nobody can question the way we do mammography nowadays. I tried to tell the health commissioner it makes no sense to be pushing mammographic screening, especially on young black women. If anything their breasts are even more dense than those of white women. If we start regularly putting them through the radiation that comes with mammography at a young age, we are just going to see more breast cancer down the road and lots more unnecessary cutting for biopsies of things that look suspicious but really aren't."

At this point, it's clear that many breast cancer activists recognize the wisdom of Kuller's views. Breast Cancer Action and the Breast Cancer Fund are among those issuing strong statements about the need for new methods to find women at risk rather than to focus on finding cancer after it's formed. Recently the Susan G. Komen Foundation added its voice to those advising against regular screening in women under the age of forty.[15]

The problem of breast cancer in young black women is especially puzzling. Why on earth should women who generally have children earlier in life and tend to have more of them develop more breast cancer? Having children is supposed to lower the chances that breast cancer will occur. Not so for young black women. National Cancer Institute statistics show that from 1995 to 2000, black women under the age of forty had twice as much breast cancer as their white counterparts. Few people are even asking why. Lucille Adams-Campbell is a researcher who's been trying to figure this out for years. A tall, trim, intense scientist, she runs the epidemiology program at Howard University. About three years ago, she produced a preliminary study that found that the main approach to predicting risk of breast cancer—the Gail model—does not work at all in black women. This model basically takes information on your family history—both mother and father, whether and when you had children and other personal characteristics—to predict the chances you will get breast cancer. Adams-Campbell asked women who were part of the Black Women's Health Study in 1995 basic questions about their lives and habits. She looked at those who developed breast cancer and learned that the Gail model correctly predicted breast cancer in eleven out of seventy-five cases. In other words, it was wrong far more often than it was right. The model missed sixty-one out of seventy-five cases of breast cancer altogether; it was predictable all right, predictably wrong. The abstract Adams-Campbell submitted to a scientific meeting on this subject was worded in the lingo of tentative science: "The Gail Model appears to be an inappropriate model to be used in African-American women to predict breast cancer risk."

An inappropriate model, indeed. This is the same model that is still widely used at clinics around the country to identify women at risk of breast cancer. It basically asks, Did your mother or sister or grandmother have the disease? Do you have any children? Did you nurse them? When did you begin having regular menstrual periods?

The man who made this model, Dr. Mitchell Gail, an experienced epidemiologist, never claimed it would work for blacks. His original work was based on a study of several thousand older white women.

Still, that hasn't stopped the drug companies from handing out neat little calculators into which this simple program for calculating risk is built and urging that it be applied to all women. Nor has it stopped clinics from using a model that is irrelevant to black women to assign them to get regular mammograms. Ask yourself why these companies would be paying for such calculators. The answer is simple and has nothing to do with science. Create demand for technologies and drugs to be prescribed and you make money. Physicians, harried with growing paperwork and office time management requirements, are among the last to question whether these free lunches, calculators and other gizmos come with hidden costs for them and their patients.

Adams-Campbell showed that the Gail model is worse than guessing when it comes to predicting breast cancer risk in blacks. If we just guessed, maybe half the time we'd be right. But the model is wrong most of the time. So, why is her work disputing the model still preliminary, still unpublished—some three years later?[16]

That's a very good question. Adams-Campbell is no slouch academically, but even she can't overcome the traditions and systems—and the billions of dollars invested in the technology—that are built around this old approach. French social scientist Pierre Bourdieu uses the term "cultural capital" to explain the social forces that propel policies of which many remain unaware. In truth, our cultural capital for predicting breast cancer isn't really that good for white women either. Half of those who get the disease have no known risk factors. They may have eaten well, given birth to children whom they nursed, and exercised, and still they get breast cancer. But by any measure, black women will fare poorly. Some people are determined to find the genes that may affect blacks differently than whites. The funny thing is that blacks in America are genetically more closely related to American whites than they are to black Africans.

But there are other parts of being black in this country that may prove pertinent to breast cancer. Does the death rate from breast cancer for black women have anything to do with the fact that blacks tend to work in the dirtiest and most dangerous jobs? While one in twelve Americans is black, one in two sanitation workers and one in three blue-collar workers are black. Researchers at the NCI have reported that when a woman moves to the United States from other countries where the risk of breast cancer is lower, something happens to change her chances of developing the disease. Within a single generation, her risk increases to that of a woman who was born in this country. The more Asian American grandparents a woman has who were born in America, the greater the chance that she will develop breast cancer.[17]

What about personal care products popular in the black community that can contain hormone-mimicking agents? Do they play any role in the increased amount of breast cancer in young women, and the greater death rate of older black women? Scientific studies of risk suggest that the longer lifetime exposure to estrogen the greater the chance that breast cancer can develop. The earlier a girl begins to menstruate and the later a woman enters menopause, the more hormones she is exposed to and the greater the odds are that she may develop breast cancer. Hormones occur naturally and regulate body functions. They tell glands and organs what to do and when to do it. The hormone amounts that a woman's body makes vary over time and depend on her body size, when she becomes sexually mature and when she enters menopause.

Something is happening to change the age of sexual maturity, and again there are differences between blacks and whites. Breast growth has become so common in girls under ten that pediatric endocrinologists have proposed considering breasts normal at ages seven and eight in black and white girls respectively. Why the age should differ between blacks and whites is one question. Why the average age at which breasts grow should have dropped by two years is yet another. One area that my colleagues and I are focusing on has to do with products that we and our families use for daily hygiene, household chores and killing pests inside and outside our homes. Many of the chemicals we use for these activities can contain chemicals that act like estrogen. Herbal remedies can also behave this way. Estrogen-like "hormone mimics" can get in the way of our body's ability to tell organs and glands what to do.

Estrogen is not just something the body makes, it's something the body can make more of, when exposed to things in food and the general environment. From studies conducted over the past three decades, we know that personal care products such as lotions, dyes, nail polishes, skin treatments, hair products, oils and creams can contain hormones and substances that act like hormones. Some hospitals sell discarded placentas to companies that put them into cosmetics that are marketed as creating especially soft skin and hair. The placenta, through which all nutrients pass from the mother to the growing fetus, is packed with pregnancy hormones.

In America and Canada, personal care products still contain chemicals that are known carcinogens, like the 1,4-dioxane we learned about in children's bath products in Chapter 9. I hadn't realized how insidious exposures of this sort could be until I met Chandra Tiwary, a distinguished, recently retired endocrinologist from the Brooks military base in Texas, where in the 1990s, he couldn't figure out why he kept seeing black babies with breasts. Girls as young as one and as old as three were showing up with breast growth and pubic hair, a condition called premature puberty. When it comes at normal ages—and those ages are also dropping—puberty can be a handful. When it hits babies, we know something is seriously wrong.

Being a methodical man, Tiwary asked the mothers about everything they did with their babies. All of them had regularly applied creams to their babies' heads to smooth their hair. These creams contained various ingredients touted for their ability to get rid of frizz. Sometimes they were labeled as containing placenta, estriol or hormones. When the parents stopped using the creams, their babies' breasts went away.

Tiwary had many of the creams from babies that grew breasts tested, those that were labeled and many that were not, and found that all of them contained hormones. He wrote to the FDA and filed what is called an adverse drug reaction report in 1996. He never got a reply.

MUCH OF MODERN MEDICINE began with exciting dreams of basic science and engineering. Just as Leonardo da Vinci had imagined in the fifteenth century, humans could be made to fly. Wilhelm Röntgen's accidental discovery of x-rays at the end of the nineteenth century found immediate applications: to look straight through the body and many other materials. Within a few years x-rays were used to find tumors years before they would be detected by physicians. When these breakthroughs first occurred, no one thought to ask whether they might have any negative impacts on human health.

The public likes to think that decisions about how our world is organized, what technologies we use to find cancer, and what substances go into baby shampoo reflect the reasoned judgment of intelligent people who systematically review all relevant information. Whether or not we allow smoking indoors, what products we place on the bodies of our children, what medical tests we put millions of people through, all these decisions have been made based on one simple consideration—tradition. Years ago Robert K. Merton showed science to be a consummately human enterprise fraught with fads and fashions. Just like the rest of us, scientists get comfortable with doing things the way they always have, even when arguments arrive suggesting that change has to happen.

Some traditions are good and honorable. We put our hands over our hearts when we sing the national anthem. We stand when the judge enters the courtroom. Nobody gets killed as a result. Other traditions are less benign. Medicine today sees itself as resting on what is called evidenced-based information. Drugs and surgical procedures are supposed to be developed and tested systematically under controlled conditions. Ideally new treatments and technologies don't become widely used until they have been shown to work.

Usually this medical proof is achieved in the context of randomized trials, where people who are otherwise similar are assigned to receive or not to receive a particular intervention. The results tell us whether or not those who got the tested treatment fared better. In fact, there are realms of medicine where evidence has never been the driving force, where tradition effectively shuts off the opportunity for cold, hard calculations. Entrenched medical practices are nearly impossible to change.

When a patient shows up with a bad headache, a physician cannot say come back in five years when I've completed my research and we'll figure out what to do. Many medical problems are inherently emergent—they require answers on the spot. Regarding how best to find cancer and what treatments make sense to try, the use of scientific information to evaluate what works is more limited than most of us realize. In an ideal world, screening tests to find early, treatable instances of cervical, breast and prostate cancer would first be tried out in small groups of people, evaluated carefully, and only put into practice on a large scale after full and methodical assessments. In the real world, that seldom happens. The story of false leads and false hopes for finding cancer is littered with exciting technologies, like mammography, that work in some groups, but get overused, oversold and understudied for all others. The result is that we miss chances to do it better until our failures become overwhelming.

There are no villains in this story in the old-fashioned sense. If anything, we are all victims of the pressures created by a disease that won't wait for answers. We eagerly turn to technologies we hope will work, and by the time we learn that they may not, we are too committed to change course.

Those who profit mightily from our dependence on these technologies have no reason to ask whether they are doing what we hope they do. We lack an independent system for evaluating the real benefits and costs of cancer detecting methods and existing treatments. The *New York Times* disclosed at the end of May 2007 that physicians who prescribe blood-boosting drugs get hundreds of millions of dollars every year in what are called rebates. "The payments have risen over the last several years," said the *Times,* "as the makers of the drugs, Amgen and Johnson & Johnson, compete for market share and try to expand the overall business."[18]

The fact is oncology is a business, as well as the grounds for trying to keep people from dying of cancer. Sometimes, its business side stands in the way of its larger, more noble goals. Those on the front lines today do not necessarily have the capacity or the incentive to be disinterested observers. In the case of these blood-boosting drugs intended to deal with the anemia so common in cancer patients, controlled supplements with inexpensive iron have been found to be as effective in many cases as more costly patented drugs in staving off anemia. But there is little effort to promote this alternative. In the U.S. today, we use three times more of these drugs than in other nations, and spend about five times more on chemotherapy, although our cancer survival statistics are not appreciably different.[19]

More times than we might like to admit, we use chemotherapy as a form of psychotherapy. All the focus on killing cancer is giving way to a new appreciation of some very old ideas. The environment of the body reflects everything that goes into it. New research on vitamin D and more nutritional supplements than I could possibly name at this point show what Hippocrates told us centuries ago: food is medicine. Another part of the equation is also coming into focus. The Susan G. Komen Foundation, the largest private funder of breast cancer research, is shifting interest to those things in the larger environment that affect cancer risk. This table, adapted from the Silent Spring Institute, condenses information on 216 different chemicals proven to cause mammary tumors in animals. None has ever been regulated for its potential to induce breast cancer.

The commitment to evidence-based medicine is relatively new. In some of its most ardent practitioners, this commitment can result in clear choices that sometimes may have tragic results. Tom Chalmers was one of the first physicians, along with Archie Cochrane, to urge the scientific study of medical procedures and drugs.[20] He was the man Dr. Love and I turned to when we wrote our article for *JAMA*. We quoted his insistence that research alone would provide the only guidance on how to design better medicine.

Chalmers and I first met on a bus when we worked together in 1980 as experts brought in to advise the government on what to do about Love Canal, a vastly polluted area. We both understood the dilemma of hazardous wastes: the environment is a mixture that can never be studied in the same way as drugs in clinical trials. I learned years later that Cochrane's first work, now little known to the world, had been on similar problems: he had shown the world that the black lung of coal miners was related to the dusts with which they worked.

A few years later, we tackled what seemed a more straightforward issue—how do you decide whether the air inside an airplane is safe to breathe. Chalmers chaired a meeting of the National Academy of Sciences committee convened to review this problem in the spring of 1984, opening with a confession: "I used to smoke three packs a day. I know how awful it is to try to quit. . . . I don't want my plane being piloted by a man in nicotine withdrawal."

Chalmers laid his cards on the table. "I want to tell you all something right now," he opened the discussion in the firm and clear tone of

Table 11-1 Chemicals Shown to Cause Mammary Gland Tumors in Animal Studies and Produced at More than 1 Million Pounds Annually[21]

Chemical Name	Air Pollutant	In Consumer Products	Food Additive	Female Occup. Exposed
1,2-Dibromomethane	x			
1,2-Propylene oxide		x	x	x
1,3-Butadiene	x	x		
1,4-Dioxane	x	x		
2,2-Bis(bromomethyl)-1,3-propanediol		x		
4,4'-Methylene-bis (2-chloroaniline)		x		
Acrylamide	x	x		
Acrylonitrile		x		x
Benzene	x	x	x	x
Chloropene		x		
Ethylene oxide	x	x	x	x
Hydrazine	x		x	
Nitrobenzene	x	x		
Nitromethane	x	x		x
ortho-Nitrotoluene		x		
ortho-Toluidine	x	x		x
Styrene	x	x	x	x
Toluene diisocyanate mixtures		x		
vinyl chloride		x		x
vinyl fluoride		x		
vinylidene chloride	x	x		
1,2-dichloroethane	x	x	x	x
1,2,3-Trichloropropane				
1,2-dichloropropane		x		
Carbon tetrachloride	x	x		x
Methylene chloride	x	x	x	x
Isoprene	x	x		
3,3'-Dichlorobenzidene		x		
Amsonic acid		x		

Source: Rudel et al., *Cancer,* 2007.

someone used to running large institutions. "There is no way that I'm going to have anything to do with a group that tries to keep pilots from smoking. There is probably something worse than nicotine withdrawal, but I couldn't tell you what that is. Nobody in their right mind would ever travel in a plane flown by a guy going through withdrawal."

What did nicotine withdrawal have to do with the NAS—an institution that most American presidents routinely tap for advice on nuclear weapons and global warming? The motto written on the gold-leafed dome of the academy's Great Hall is an ode to the powers of science: "To science, pilot of industry, conqueror of disease, multiplier of the harvest, explorer of the universe, revealer of nature's laws, eternal guide to truth."

Heading this committee of scientific experts convened by the academy's Board on Environmental Studies and Toxicology, Chalmers had been tapped to grapple with an intense debate. Sen. Daniel K. Inouye of Hawaii wanted to know whether the health of airline employees and other frequent long-distance aviation travelers, like himself, was imperiled by long flights between Honolulu and Washington, D.C. While the first manned flights had lasted a few seconds, modern aviation kept people aloft at several miles above the earth under conditions that humans had never before experienced for such extended periods of time. At the time, smoking pilots and passengers legally lit up the moment planes completed takeoff and had to stop just before they began to descend. After eight to ten hours of sitting in the cramped space of a plane, most travelers, whether smokers or not, stank like they had been in a crowded, smoky bar. The flight attendants' unions and growing numbers of passengers with lung conditions threatened by stale air were asking that smoking be curbed.

As a former nicotine addict, Chalmers knew that the worst symptoms of withdrawal hit within a few hours. The brain of a smoker doesn't work well without its regular boost of nicotine. The rates of breathing and beating of the heart, reactions to light and heat and cold, and speed of response to anything good or bad are all affected. Who on earth would want to fly in a plane piloted by a person in the throws of nicotine withdrawal, where the list of symptoms extends from irritability and anxiety, difficulty concentrating, restlessness, insomnia, tingling sensations and dizziness, and shakes?

When the NAS committee on cabin air quality announced it would hold a public hearing to gather information, the eight hundred seats of the auditorium were filled. Irate passengers resolved to hold on to their right to smoke gaped at impassioned flight crews committed to cleaning up their workplace.

Dr. Chalmers was president and dean of Mount Sinai School of Medicine during the Vietnam era and knew quite a bit about conflict resolution. A lanky, balding redhead with piercing blue eyes, he looked every inch the medical leader, a man born to fit his white coat perfectly. Chalmers told me how he had come to believe that research on the conduct of medicine was critical. "I started out as a young doctor following the practices of my teachers. You have to remember that at the time anyone who managed to become a doctor was assumed to know what he was doing. I began to keep records of what happened to my patients. I soon found out that we were recommending a type of surgery that it turned out was killing people. Nobody figured out until we began to follow up what was going on. That's how I became a firm believer that no medical practice should ever be put into place without first being studied and evaluated in a detailed and painstaking manner."

After a few days of fairly abstract discussions, Chalmers and several committee members hit on a radical idea—they wanted to go see a real airplane on the ground. The drawings the committee had to look at didn't provide enough detail. They needed to stare down into the inner workings of an airplane air handling system.

One spring day in 1985, Chalmers and I got back on a bus—this time with the entire committee. We departed from the academy's marble Greco-Roman revival building just opposite the State Department in Foggy Bottom and were dropped off onto the tarmac at what is now called Reagan National Airport. Committee members gathered under the belly of a freshly cleaned jet. John Spengler, then a young up-and-coming faculty member from Harvard's engineering department, looked disappointed. "Can't we see a regular airplane? One that hasn't been spiffed up yet?"

The airport manager looked uncomfortable. "Well, the underside of these planes when they finally land has lots of streaks on them. We wanted to show you what they look like after we get finished with them."

I looked across and spied a plane that had just emptied passengers about fifty yards away. "Let's go take a look at that one," I said, as I began to walk toward the other airplane. Chalmers, Spengler and the other committee members followed along, eliciting stares from baggage handlers who hadn't ever seen such an entourage of suits with notepads in the middle of their runway. When we got to the recently landed jet, as we looked up at its base, we could see that it was yellowed, kind of like a well-used meerschaum pipe.

"What exactly is all this brown stuff?" Spengler asked.

The airport manager was a bit embarrassed. "Well this plane just came in from California, so it had a long flight. All of our planes have air that goes through a single sock or filter. We bleed air from outside the engine in high altitudes, bring it in so that it gets warmed by the hot engines, and then mix it with the cabin air. When the air leaves the airplane, we keep recirculating it so that at any time half of the air is new and half is old. Those brown streaks are left by the nicotine from cigarettes. Actually the big problem is what this gummy residue does to the avionics—the equipment that allows us to keep these planes flying in good shape."

The committee figured out that it needed to take a good look at what was in the sock. After several hours of pushing thousands of cubic meters of air through this one central filter, that sock was full of sticky tars. This made it simple—the longer a plane was in the air and the more smoking that went on inside it, the dirtier the air would be for all of the passengers. The outside of the plane became gummed up with sticky scum that had just been scrubbed off the one we'd seen when we arrived at the airport. Of course the lungs of flight attendants and passengers couldn't be cleaned up. Ironically the airline maintenance guys were most put out by the fact that they had to keep replacing the avionics equipment; they didn't have chemicals strong enough to remove the grease and tar.

Once the costs of cleaning up smoke residue on planes had been firmly established, the facts were clear. It made no sense to allow smoke on airplanes. Contrary to what he told the committee on its opening day, Chalmers concluded that smoking should not be permitted on airplanes.

Tom Chalmers was a man who lived and died with his principles intact. Years after he shepherded the NAS committee to ban smoking, he was fighting for his life against metastatic prostate cancer. This was a man beloved and respected by hundreds of colleagues. We had a great party to salute him at Harvard in 1995. Toasted and roasted by admiring students and faculty, decked out with huge paper bow ties rendered specially for the occasion, Chalmers didn't look like a man with a few months to live. But the prostate cancer had spread to his bones, and he knew what he was up against.

We shared laughs about the sorts of things that only academics find funny at this splendid party, but there was an undeniable gravity in the air. Chalmers had already received the maximum amount of radiation to stifle the searing bone pain with which he was then living.

It was only a matter of time, we all thought. But a chance meeting with legendary cancer doctor Mitchell Gaynor left me with some hope. The former chief resident at Sloan-Kettering in New York, Gaynor saved lives that others had given up on. He was the sort of cancer doctor that cancer doctors went to when they had the disease. I asked if I could look at his files at Cornell's Strang Cancer Prevention Center to understand what he had done. At the time, he had more than twenty patients with various forms of advanced cancer who were still alive, some after more than two years. They all had gone through regular chemotherapy. None of them should have lived more than a few months.

But Gaynor had added something to their regimens—a combination of herbal remedies—that appeared to be keeping these people from dying. Many of them were doing exceptionally well. Gaynor explained that a woman who had been the chief nutritionist at Sloan-Kettering had left a year earlier with advanced breast cancer. Because the cancer had spread throughout her body, Gaynor never expected to see her again. Six months later, she walked into his office. The tumors were gone. Gaynor was stunned and asked her what she had done.

She explained that "a guy named Ralph in Wyoming had come up with this recipe for purple herbs. I had nothing to lose. You had all written me off. So I tried them."

Gaynor managed to get the purple remedy delivered to him at Cornell and gave it to patients who had been sent home to die. His results so far were amazing. Others were trying to figure out what exactly was in this mixture.

I phoned Chalmers right away, "Tom, I've found something you have to check out!" I began to explain what Gaynor was doing.

"Are you mad!?" Chalmers asked. "You must think I'm crazy! I could never take herbs that nobody has ever studied. I've spent my life studying medicine scientifically. I refuse to even think about such a thing."

"But Tom," I pleaded. "You're going to die. Why not try this?"

"Of course, I'm going to die. I know that. If there was a randomized trial, I would consider it. But unless this remedy is being studied under controlled conditions, it's out of the question."

Within a few months, Chalmers was dead. No one ever accused this man of inconsistency. His position was firm and clear; only medications derived from controlled studies, including herbal remedies, should be used in medical treatments. Chalmers had no problem concluding that smoke ought to be kept out of airplanes, because smoke left gooey, tar-laden marks that not only were unattractive but also mucked up the electronic equipment. They couldn't be good for our lungs either.

But what about substances in the environment that affect our health and our chance of getting cancer that leave no telltale signs? What are we supposed to do about exposures to tiny amounts of tars or numerous other compounds that look risky when tested in animals? What sort of action is reasonable to take when all we know is that people are exposed to these agents over a lifetime, but we have no direct human evidence that they are harmful to humans? What are good reasons for suspicion? Do we wait for people to experience harm before we act to lower the risk that such harm will happen? Ironically, the NAS committee Chalmers chaired acted to ban smoking on airplanes chiefly because evidence showed that smoke directly soiled the planes and their innards. The fact that it was likely to worsen the health of airline crew and passengers was not the principal reason for this recommendation.

The answers to the questions of what sorts of evidence are needed to justify actions to treat cancer or to prevent potential environmental harms are not simple at this point. Les Thiele is a beautiful indication of that. She is positively luminescent, with huge blue eyes, strong cheekbones, and a commanding and spirited stride that looks like she might have just come off a horse or a catwalk. Thiele saved her own life, she believes, by finding the best natural medicines to go with the surgery and radiation that removed her advanced ovarian cancer seven years ago. You won't find her in a clinical trial at this point. She's convinced that the combined effects of modern medicine, yoga, meditation, massage, acupuncture, herbs, detoxing systems and prayer keep her thriving. She may well be right. She's certainly not alone in pursuing such remedies.

Support for going outside the conventional system for cancer treatment comes from a growing number of remarkable quarters. There's nothing timid about Donna Karan. This is a woman who revolutionized fashion and kept her husband with advanced cancer alive far longer than anyone expected. She did this the way she does most things, with a focus and intensity that is breathtaking—tapping all the healing modalities you can name from aromatherapy to acupuncture, massage, chiropractic and more types of body work than most of us have ever heard of. Using the considerable reach of her influence as one of the world's top fashion designers and working with nurse educators like Susan Luck and master yoga instructors like Rodney and Colleen Yee, Karan wants the rest of the world to learn what she thinks kept her husband alive so long.

Coming up with ways to study complex approaches to cancer treatment will tax the brains of Chalmers's counterparts today. But we cannot solve the problems of the future with the technologies that have created them, as Einstein once urged. We've got to go beyond our worn cultural capital to new ways of looking and thinking about treating cancer, as well as in our efforts to prevent it.

When decisions involving products or airplanes created by multi-million-dollar, multinational industries are involved, science alone is rarely the driving force in determining what materials are to be controlled and what sorts of information are sufficient to justify such controls. In trying to study environmental hazards and new remedies to treat cancer, we face one simple fact: there is no control group. We never can find an unexposed group against which to compare the dangers of most common environmental contaminants. For new herbal and nutritional remedies against cancer, or even for new uses of the soothing sounds and relaxing smells of aromatherapy, we can't persuade patients who've been told they will die to agree to sit through a trial they may not see the end of.

All this has played into the hands of those who want to delay acting until we have better evidence, even though this will keep us waiting for what can never be obtained. Like the characters in Samuel Beckett's drama *Waiting for Godot,* we keep thinking that whatever it is we really need will show up if we just hold out a bit longer.

VLADIMIR: Well, shall we go?
ESTRAGON: Yes, let's go.
They do not move.

Clementine Szukis, of Bridgewater, New Jersey, pictured in 1988. As a hairdresser for Johns Mansville asbestos workers for twenty years, she regularly brushed white flakes of asbestos from their heads before washing, cutting and styling their hair. Her fingers show a distinct deformity called "clubbing" that occurs when asbestos-filled lungs do not distribute sufficient oxygen to the extremities.

12

The Harshest of

<u>Schoolmasters</u>

It's difficult to get a man to understand something,
if his salary depends on his not understanding it.
—UPTON SINCLAIR

HOW DO WE KNOW what we think we know? Simple facts are often culturally constructed, passed on like clothing styles or table manners. What do we do when the thing we are trying to understand is not something tangible but an idea that must be abstracted from how others experience the world? Primitive societies depend on shamans, mystics, gurus or priests to set their compass. We rely on their modern counterparts—those deemed experts. But as we have seen in past chapters, those who claim the mantle of expert often come with hidden baggage. We know enough to be skeptical of those whose paychecks come directly from a given industry. We also understand that those who advocate for the environment don't always have or care about the full facts at hand. Yet we seldom know the extent to which vested interests have shaped and spurred the development of entire fields of scientific inquiry.

The foundations of epidemiology as a science can't easily be separated from the industrial forces that decided what information got released and what questions were asked and answered. The result is that what began as an earnest effort to understand the dangers of the real world has often turned into a way of covering them up.

It's often not possible to say where truth begins and ends. Science works with one set of rules for determining what any given community considers to be facts. In public health research, studies are to be repeated with large enough numbers accumulated that the results can achieve statistical significance. Of course, what is deemed significant in statistical terms is not always important in biological terms. It is also true that what may be profoundly important biologically can sometimes elude statistics altogether.

Law, in contrast to epidemiology, works with another set of rules, which vary depending on the issues at hand. If a woman is charged with murder, she must be found guilty "beyond a reasonable doubt." The O. J. Simpson murder trial proved that, with sufficient resources, skilled attorneys can magnify reasons for doubt in the minds of twelve jurors that sometimes strike many of us as an astonishing leap of logic. For other matters, like those involving how we set environmental standards, the agencies must provide what's called "a preponderance of proof."[1] The very phrase tells us that proof is a relative thing. In civil law a fact is established if the odds it's true are just a little greater than fifty-fifty. Proof in such cases is like a playground see-saw. The tiniest bit more weight on one side than the other is enough to sway the result. Proof in physics requires odds of 90 percent (i.e., a one-in-ten chance that a result happened by accident). The biomedical sciences usually demand that the chances are 95 times out of a hundred that the results are true. This leaves us with only one chance out of twenty that we've made a mistake. Thus the level of statistical proof required in science is more demanding and precise than in law in many respects. Ideally, scientific proof is transmitted and established democratically: when a majority of scientists concur that a given fact is established as true, it is true. Until then, it's not.

These already murky waters have been further muddied by recent court rulings on what scientific evidence can be considered as having proven harm. As part of a carefully crafted and heavily financed campaign, courts and regulations are increasingly insisting that before a given exposure can be considered to have caused a specific cancer, there must be epidemiological findings linking similar exposures with particular health damages. This places epidemiology on a pedestal that it rarely merits.

In a series of surprising, well-documented articles, medical essayist David Michaels has shown that in many current instances, the absence of epidemiologic findings becomes a surefire way to postpone, avoid, or delay regulatory controls. People who allege that their poor health is due to this or that environmental hazard sometimes are well paid for their problems, providing the settlement stays secret. As a result, proof of human harm remains legally locked up. So long as things can be made out to be uncertain and unresolved, production—and profits—continue uninterrupted.[2]

Thus has evolved the well-paid skeptic who learnedly insists that experimental and clinical sources of scientific information are not the same as full-blown public health studies of affected people with clearly defined exposures. The absence of sufficient numbers of already harmed people is held up like a cross before a vampire. We are uncertain, goes the argument. We can't be sure whether what we've found in animals really pertains to people. Let's study the problem longer, before we take any action. Let's wait until we have more people to examine. The lack of statistically significant evidence on harmed humans is construed to mean that no harm has occurred. In reality, the absence of human evidence on environmental harms is chiefly proof of how hard it is to collect the information and the intensive, inventive ways designed to prolong doubt.

THE DANGERS OF asbestos are no longer disputed. Known to the ancient Greeks for its fire-resistant properties, asbestos has left a sorry legacy in every nation in which it has been widely used.[3] It is one of the best-studied workplace hazards in the world, partly because it was so widely employed as an insulator in buildings, ships, power plants and factories. But unfortunately asbestos is much tougher and resilient than the bodies of those who mine, use, transport or carry it away. It degrades into invisible, floating particles that can slip into the exquisitely fragile sacs of the lung, where they are walled off, leaving permanent scars. Chronic exposure leaves a person with less and less working lung tissue; eventually she suffocates. In 1898, Her Majesty's Lady Inspectorate of Factories singled out the "evil effects of asbestos dust" as one of the worst of numerous horrid working conditions for young boys and girls.

Specialists who looked into the suffocating lungs of asbestos workers agreed that bad luck alone did not explain why so many failed to reach middle age. The British pathologist W. E. Cooke made the connection plain in a 1924 paper entitled "Fibrosis of the Lungs Due to the Inhalation of Asbestos Dust." The dust-laden lungs he was talking about had come out of the chest of a woman named Nellie Kershaw. At age thirteen, she had gone to work in the spinning room of a textile factory in Rochdale, England, belonging to the firm of Turner & Newall, one of the country's largest asbestos manufacturers. Her job was to turn bulky raw asbestos rock into heat-resistant industrial cloth and textiles. She died at age thirty-three. The inquest ruled her death a result of asbestos poisoning, or asbestosis. The company argued that since asbestosis had not been officially listed as an occupational disease in Britain, they should not pay for her death. The firm won.

Seven years after Cook's paper reporting on Kershaw's death, the British Parliament added asbestosis to the list of compensable diseases. If Kershaw had died at age forty, her family would have received less than a year's wages.

In the 1930s it was known that inhaling asbestos fibers leaves telltale scars that could be seen on primitive x-rays as strangely scarred, a grayed matter within the lungs that looked very different from normal lung tissue. Years later, electron microscopes would provide stunning images, showing this dull, fibrous material to be the accumulated residues of asbestos dust. Asbestos fibers inflame the lung, sending white blood cells called macrophages to try to get rid of the unwelcome foreign bodies. When this fails, as is often the case with asbestos, other cells grow around the attacked area, leaving distinct scars. Lungs that contain these walled off attack zones and scar tissue don't work very well. Sometimes the heart gives out from all the extra work that has to be done just to get sufficient blood through the lungs. As the distinguished British expert E. R. A. Merewether put it, "This fibrous tissue is not only useless as a substitute for the air cells, but . . . by its invasion of new territory . . . it gradually, and literally strangles the essential tissues of the lungs."[4]

Some pathologists, and England's own Factory Inspectorate, had long urged that dusty, dirty workplaces should be cleaned up. Surviving family members, like those of Kershaw, had no doubt that their asbestos-working sons and daughters and husbands and wives had worked to death.

By the middle of the twentieth century, Kershaw had plenty of company. The technical literature contained hundreds of reports of strikingly similar disease and death in men and women who had worked directly with the dust. Many had begun working as young-sters. In response, industry argued that these findings were coinciden-tal. No matter how many individuals could be shown to have died with lungs full of asbestos, this did not prove there was a major problem. After all, these young workers had gone to work in factories because they lacked the skills or the family income to avoid it. The ones like Kershaw who succumbed were probably just more frail, or maybe they had started out with tuberculosis.

Enter that faithful friend of industry, the well-paid skeptic. The search for public health research in the very beginning cannot be sepa-rated from an effort to set the bar very high regarding what sorts of evidence could ever be considered sufficient to conclude that there was any problem with early industrial life. Industries fanned the de-mand for robust epidemiological research not because of an over-whelming commitment to public health knowledge but from a belief that such work would take a long time and would ultimately vindicate their practices. After all, people become ill as individuals. How could their workplace be more important than their parents, their good and bad habits, and all the other things that happened to them as unique beings? The idea that the workplace affected health simply did not fit with notions of individualism so key to the early stages of the industri-alizing world, nor did it dovetail with the growing confidence that in-herited defects accounted for most ailments.

In a strange way that has seldom been appreciated, before epidemi-ology could be seen as the most important type of information on workplace hazards, it was necessary for experimental research on ani-mals and case reports on individuals to be seen as happenstance. Yes, these people worked with asbestos and died young, but how can we be sure that these two facts are related? Was this really sufficient ground for controlling such an important industry? After all, experiments on animals might not be relevant to humans.

UNTIL EPIDEMIOLOGY matured as a science in the latter half of the past century, the compilation of individual cases, coupled with ex-perimental research on animals, had been the primary way of decid-ing whether a given agent was a cause of cancer. In his charming autobiography, the cancer researcher Isaac Berenblum depicted the vibrant science of controlled studies in animals that flourished in Europe and Japan in the decades before World War II. Looking back on his career in 1977, he warned of a growing threat to the field. Those who set the terms for increasingly complex, expensive and time-consuming research proposals may themselves be preoccupied with something other than the advancement of basic research.[5]

Within Berenblum's lifetime, it had become clear that free and open discourse was hardly the norm for cancer research on workers. Less clear, however, is the extent to which people wishing to conduct epidemiologic studies in the workplace have had to compromise just to get in the door. In 1930, Dr. F. G. Pedley directed the clinic at McGill University in Montreal that was completely underwritten by the Metropolitan Life Insurance Company.[6] He reported to Anthony J. Lanza, then head of efforts to study asbestos mining for Met Life. Of the miners in Quebec who were still alive and working at the time, nearly one in five had lungs that showed up on x-ray as scarred with asbestosis. On repeated occasions Pedley was denied permission to publish his findings of lung damage in asbestos workers. Because the profitability of insurance companies depended on what illnesses they had to pay for, they had a major interest in monitoring reports of work-related health problems. They did not do so with entirely benign intent. Lanza presented a highly skewed rendering of Pedley's work: the Quebec asbestos workers, he claimed, were unique. They had no evidence of asbestosis.[7]

By 1935, despite such suppression and manipulation, the literature on the harms of asbestos in workers was vast. A Pennsylvania state report listed over 125 references from all over the world. Lanza wrote an analysis for Johns-Manville in January 1931 alleging the safety of asbestos. Released four years later, this report was used to lobby against changes in the law to compensate for asbestosis in New Jersey. It revealed that half of all asbestos workers had lungs that showed classic asbestosis on x-ray. But Lanza claimed that these people, all of whom had at least three years work experience and had been selected at random, were not really disabled. They had defective lungs that took in too many fibers.

Germany officially compensated the surviving families of dead asbestos workers in 1936. Italy followed suit in 1943. The position adopted by asbestos companies when workers claimed injuries from asbestos was quite similar. They argued that these are individuals. We can't be sure that what happened to them is relevant to most other people. How could we ever know that their poor health and eventually their deaths had anything to do with what they worked with? How did industry respond to growing numbers of autopsies and sickened individuals who had worked with asbestos? They did what many industries do today: they set up secret studies in private laboratories to see whether or not animals responded in a way similar to that reported in workers. Much of this experimental work, which was done in the 1950s, only surfaced in 1978. Lawyers representing men and women who had died of asbestos pried these records from the previously secret files of the asbestos industry.[8]

THE 1950S WERE a watershed period for epidemiology. The study of the dangers of asbestos finally went beyond simply tallying cases of illness and early death. By then, so many deaths had occurred from asbestos that it was possible to take each one and compare the life experiences and histories of those who died from what looked like complications of breathing asbestos to others who were quite similar in age, shape and size but did not have such exposures. This method of contrasting cases with disease to those who were not ill appeared in Germany in the 1930s in the study of cigarette smoking and in England with the study of breast cancer. But while it was a relatively straightforward matter to compare smokers and nonsmokers or look at the reproductive histories of women with and without breast cancer, assessing workplace exposure was quite a bit more complicated. The asbestos industry bet the house on being able to show that the growing numbers of deaths were not tied with work alone but resulted from underlying deficiencies. Ultimately it lost this wager. But in the process of losing, it won decades of time during which the absence of human evidence and government regulation allowed operations to continue. Opening industry doors and records to the study of asbestos-related sickness and tallying the impact of tobacco on health basically laid the foundations for the field of chronic disease epidemiology.

The case comparison method gave epidemiology a status in science that it had previously lacked. But it raised a problem that remains unresolved to this day: Where do we find people with no exposures to

the agent we are seeking to study? Who has never been exposed to cigarette smoke today? Where are those without exposure to asbestos?

The link between asbestos and cancer was deemed proven by many
in the 1940s. Groff Conklin's 1949 report in *Scientific American* featured a graphic display of "carcinogens known to be present in human
environment."[9] Asbestos was described, along with chromates and arsenic, as causing cancer by physically damaging the body. By the 1950s
asbestos stood convicted as a carcinogen by simple "clinical epidemiology," as Thomas Mancuso, one of the government's top epidemiologists at the time, called it. By this he meant that the sheer number of
reports of men with asbestosis who also had lung cancer at autopsy
provided an overwhelming picture linking the disease with regular
workplace contact. The main indicator of this tie, consistent in every
country and in every report, was the high fraction of people dying
with asbestosis who also had lung cancer at autopsy. Sometimes the
cancer was distinctive: a tumor of the pleural sac, the diaphanous lining surrounding the lungs, called mesothelioma. Mesothelioma is believed to be exclusively associated with asbestos exposure. Often
asbestos-related cancer looked just like other lung cancers. Autopsied
asbestos workers had lung cancer ten times or more than others. The
case was effectively closed by 1939 for German compensation carriers: asbestos clearly induced asbestosis, and if you had enough asbestos
in your lungs to cause even slight asbestosis, lung cancer was deemed
occupational.[10] In the Allied countries, national security concerns
during World War II combined with industrial trade secrets to keep
this information from the public.

As laboratory evidence began to accumulate that smoking, asbestos,
coal tar and other industrial agents threatened health, industry repeatedly cautioned against confusing such experimental evidence with
proof that human harm would really take place. In a sardonic commentary in the *Lancet* in 1958, Evarts Graham admitted that absolute proof
of human hazards may never be obtained. He averred that the only way
to meet the demands for evidence then being insisted upon by the tobacco industry would be to engage in a series of human experiments:

1. Secure some human volunteers willing to have a bronchus
   painted with cigarette tar, perhaps through a bronchial fistula.

2. The experiment must be carried on for at least twenty or twenty-five years.
3. The subjects must spend the whole period in air conditioned quarters, never leaving them even for an hour or so, in order that there may be no contamination by a polluted atmosphere.
4. At the end of the twenty-five years they must submit to an operation or an autopsy to determine the result of the experiment.[11]

Graham did not realize that the preposterous experiment he had spoofed was already under way with asbestos workers. They were not confined in air-conditioned quarters but instead allowed to return to their families with deadly, invisible dust on their clothes after work. In 1954, a Turner & Newall medical researcher named John Knox teamed up with the prominent cancer epidemiologist Richard Doll to assemble information on the unwitting experiments that had been carried out on asbestos workers employed by Knox's firm. In 1950 Doll and Hill had published highly regarded case-comparison analyses on lung cancer and smoking in physicians. The company hoped that tobacco would also prove to be a major culprit for its workers. Knox and Doll confirmed that those who died decades after starting to work with asbestos had ten times as much lung cancer as men who did not work with asbestos. The company refused to allow Knox to release these findings. Appalled by this refusal, Doll moved to publish the work under his name alone.

This was not an effort for the faint of heart. As he began to prepare this work for publication, Doll and the journal editor came under pressure from the company. Just how much pressure only surfaced fifty years later when lawsuits found written requests for revisions in various copmany files. At one point Knox tried to persuade Doll to withdraw the paper completely. The editor of the *British Journal of Industrial Medicine* was visited by a director of the company and pointedly asked not to publish the report.[12]

Doll's article on the hazards of asbestos for Turner & Newell workers was ultimately published in that journal. But the publication differed in one critical way from the draft. The original paper had included a sentence which signaled that asbestos dangers might not be over: "Insufficient data are available to determine whether the risk has yet been eliminated by the improved conditions which now exist." A draft of the manuscript dated March 16, 1955, included a different, but still reserved, sentence allowing that the harms of asbestos might not yet be resolved. In a version found by the plaintiffs' attorney for Chase Manhattan Bank in 1993, someone at Turner & Newall had crossed out the following: "It is unlikely that the risk is now large, but insufficient data are available to determine whether it has been completely eliminated."[13] The bank was suing the firm for failing to disclose what it knew about the dangers of asbestos.

We don't really know what happened. All we can be sure of is that the article that was published in 1955 did not include any hint that the risks of asbestos might continue. It left the clear implication that whatever risk might persist from asbestos was not likely to be large. But size, like beauty, is in the eye of the beholder. And we now know from studies of air pollution that small risks which affect the entire breathing planet can have big impacts.

Doll is gone, so we will never know what ensued between the time he first drafted this article and when it finally appeared. But the facts that can't be disputed are these. Right after this article appeared, Doll began privately consulting for Turner & Newall. For many years he defended the company against lawsuits from some of its asbestos-exposed workforce. In 1964 he published precisely what Turner & Newall had asked of him in 1955, telling a New York Academy of Sciences conference on asbestos that the 1932 U.K. asbestos regulations might have completely eliminated occupational hazards.[14] Barry Castleman has written the definitive public history of this secret industry. He testifies regularly for asbestos-poisoned workers and their families. Castleman obtained copies of the 1955 drafts as part of a legal discovery. Years later, when asked about his changed views and the altered 1955 paper by Castleman, Doll refused to answer.[15]

It may be shocking that such a distinguished researcher worked so closely with an industry now understood to have engaged in so much disreputable behavior for so long. But it is hard to grasp the conditions under which research on the workplace had to be carried out. The only folks controlling such information and funding any studies were industries. If you wanted to understand the workplace, there was no place else to go. Some firms behaved benevolently. Many did not. History will have to judge Doll's cooperation in this context.

SARNIA, ONTARIO, is a town just miles from Lake Huron that is famed "for its breathtaking sky, blue water and beautiful waterfront parks."[16] At least, that's what its colorful promotional brochure claims. The brochure neglects to mention another claim to fame. A major industrial center, Sarnia is known for some of the highest asbestos exposure ever recorded anywhere.[17] The town also has five times the level of mesothelioma as the rest of the region. Almost 1,500 cases of the disease were diagnosed between 1981 and 2001, about one new victim a week.[18] That's impressive for a town with a population of 70,000.[19]

What's even more impressive is that you won't find a single asbestos mine in the region. Rather, Sarnia is home to 20 percent of Canada's petroleum refineries.[20] Many of the residents lived amid asbestos-lined foundry ovens and asbestos-insulated pipes, or with the manufacture of asbestos products.[21] The fine white fibers were not only common at the workplace but were carried home every day into the families and communities of Sarnia. An editorial in the *Toronto Star* reads:

> The men went to work every morning, proud to earn a livelihood for their families, and they came back each evening carrying death on their clothes. The women shook out the clothes and washed and ironed them, and were proud to be taking care of their families. Asbestos was so thick at the Holmes Caposite plant in Sarnia that shipping doors at both ends of the factory would be opened to clear the dust. It blew so thickly into the street that the traffic would come to a halt. In the park, across the street where the children played, the benches were coated with a layer of asbestos dust.[22]

James Brophy, an occupational health professional, has worked with the people of this region for more than two decades. In an e-mail to me, he described long-standing indifference:

> For over 16 years, the Ontario government tolerated asbestos levels that were described in their own documents as "some of the highest ever recorded"—we estimate over 8500 times the current legal limit.

During that period they never informed the workers, never enforced their own directives and never stopped the production process.[23]

Frank Fitzsimmons began working in a dusty factory in Sarnia in 1974, taking care of equipment that was full of asbestos. At that point the dangers of asbestos exposure were well known to scientists. Fitzsimmons's son Donald was born soon after he began this job. Frank had no idea that he was bringing deadly fibers home with him every day. Ten years earlier, in an important paper to the New York Academy of Sciences in 1964, Muriel Newhouse had shown that people who lived with or near asbestos workers could contract fatal cancers from the neighborhood. But men like Frank Fitzsimmons were not told that. Brophy, executive director of the Occupational Health Clinic for Ontario Workers/Sarnia, wrote

> When we asked the union to use the Freedom of Information Act and acquire the Ministry of Labour file on Holmes we discovered among the pages of inspection and hygiene reports an epidemiological study prepared for the Ministry. It reported that the Holmes workers were dying of lung cancer 6 times higher than the rate for the Ontario population; 4 times higher from all other malignancies, and 11 times higher from respiratory causes. This did not include 5 cases of mesothelioma all under the age of 60 with 3 under the age of 50. In the foundry itself the government measured for silica and the company was never—in over 40 years—in compliance. It was literally a killing field.[24]

Fitzsimmons's former wife, Maria LaCount, remembers, "It would be on his clothes as fine black and red dust. He would be literally black and he would pick up and hug the baby. We didn't realize this stuff was hazardous."

As a teenager, Donald was never an athlete. He'd always been short of breath. It turned out there were good reasons for this. Before he could get a driver's license, Donald was diagnosed with mesothelioma. He had never worked in a factory. His only exposures occurred through playing and hugging his daddy as a toddler. He died in the fall of 1989, barely sixteen years of age. Parents aren't supposed to bury their children. Such an unnatural loss leaves permanent marks.

"I was there when he died," Frank told a reporter. "He was down to skin and bones. The cancer had eaten him up."

For some time, Fitzsimmons didn't connect his son's death with his job. But when he finally made the connection he snapped, "I felt like I killed Donald myself because I brought this stuff home in my clothes. I feel responsible."[25]

Despite a century of evidence on its dangers, the market for asbestos is booming in India, China, Iran, Kazakhstan, Thailand and other developing countries led by shortsighted leaders determined to generate revenue today despite its lethal legacy. India has an asbestos industry with corrupt relations with government and ownership links with the media, increasing at 9 percent a year and recommended by some Indian stockbrokers as a good investment. China has its own asbestos mines. The nation is sensitive enough to market pressures to have begun building a nonasbestos brake manufacturing plant, a joint Japanese-Chinese venture financed by the World Bank. But those brakes will be for export; the domestic market is another matter. Areas of the Tibetan plateau and the arid Tsaidam basin of Tibet's far northeast are home to expanding asbestos mines.[26] The Chinese domestic housing market employs increasing amounts of asbestos cement. Regarding future deaths that this will bring, those making these decisions know but they don't care, much like the factories around China leaking benzene into the rivers.

Canada remains, at best, ambivalent on the issue of asbestos. A 1984 Royal Ontario Commission report reviewed the world literature and concluded that the dangers of asbestos were well established. It described the Johns-Manville asbestos plant in Scarborough, east of Toronto, as a "world-class industrial disaster." A three-volume compendium, the report ended on an eloquent note:

To learn from the asbestos experience is to learn from the harshest of schoolmasters—human tragedy. It is a fact that society has all too often required this stern tutelage to assimilate its lessons. This is strikingly evident in the realm of regulation. One can think of the *Titanic* and its impact on the regulation of radio communications at sea, of typhoid epidemics and their impact on the regulation of drinking water, of mine cave-ins and their regulation of mining. One can think of asbestos. This entire Report has been an exercise in learning from the asbestos experience and discerning the lessons it teaches about health hazards and their regulation.[27]

The scientific case against asbestos is no longer debated. While it is easy to portray the massive burden of lawsuits against asbestos companies as the result of overzealous lawyering, the truth is far more complex. The asbestos trials have released damning proof that the reason the record took so many years to fall into place had nothing to do with science and everything to do with the control of information. The terrific profitability of the industry, the role of asbestos in wartime, and the shifting grounds of public health research kept this hazard from being fully indicted for more than three decades.[28] In the United States today, asbestos brakes and asbestos-contaminated potting soil, insulation and kitty litter linger as hazards most people believe are long gone.

We like to think we no longer use asbestos, but today America simply imports from Mexico the asbestos-laden products we no longer make at home. In the first six years of the twenty-first century, America has tripled the amount of "asbestos and cellulose-based cement sheet, panel" shipped from Mexico, taking two-thirds of the world supply of this commodity. Mexico also provided about twice as much asbestos yarn and thread in 2005 as in 2002. Where are these products being used? One place they are not is in the twenty-five nations of the European Union. The EU has banned asbestos, as has Saudi Arabia, Uruguay, Japan, Argentina and Gabon. As of this writing, America and Canada, still one of the world's top producers, have not.[29]

Why do you suppose that today half of all cases of mesothelioma in American, French and Italian women have no known history of exposure to asbestos?[30] There are two possible answers. Either mesothelioma is caused by something else that has yet to be identified that occurs widely throughout these nations. Or, somehow, somewhere these women have been exposed to asbestos and have had the bad luck to develop this disabling and lethal illness.

DIFFERENT RULES HAVE been developed in science to decide when a specific relationship between any two things can be considered to be proven or evident, and who is responsible for informing people of a suspected hazard. At this point, the absence of epidemiological proof of harm is taken by many to mean that no harm exists. In fact, what it mostly tells us is that proof is hard to come by and may not be achievable at all. Once a hazard is identified, there are no clear rules about who is responsible for conveying this information to the public.

The ways we look at human evidence for deciding whether a hazard exists and what we are supposed to do with such information have changed markedly over time. As we have seen, well into the twentieth century pathologists and others concluded that certain things caused cancer based on their repeated observations of individuals. They tied this information to experiments showing similar associations, if such experiments had been publicly reported. This way of thinking simply follows T. H. Huxley's formulation: "Science is nothing but trained and organized common sense." But it is rapidly becoming inadmissible, even as a basis of reasonable suspicion.

The grounds for concluding that a suspected hazard is a true threat to human health have been forged over the past half century by some of the best and the brightest minds in public health. Richard Doll, of the Imperial Cancer Research Fund in Oxford, England; Hans-Olav Adami, of the Karolinska Institute in Stockholm, now at Harvard; and Dmitri Trichopoulos of Harvard's School of Public Health are world renowned for their studies of tobacco and other public health hazards. Among them they have published well over a thousand articles in leading journals throughout the world. They have written the texts and set the standards for determining what kind of evidence is deemed sufficient proof that a given exposure in fact causes a given health problem in humans. Within the past decade, professional societies such as the International Society for Environmental Epidemiology[31] and the American College of Epidemiology[32] have weighed in regarding the ethical duties of epidemiology in developing such proof. Researchers are obliged to use the best methods, inform people of findings, respect privacy, avoid coercion and specify where support for their work comes from. The last point is especially important. As one wag noted, in epidemiology you are not just known by the company you keep but by the company that keeps you.[33]

We will probably never know whether Doll's, Adami's, and Trichopoulos's ideas about how scientists should study industrial hazards were at all colored by the fact that these eminent epidemiologists secretly served as highly paid consultants for the asbestos, chemical and pesticide industries.[34] For years, Doll, in addition to his work for the asbestos industry, was paid to advise Monsanto and Dow Chemical. Adami and Trichopoulos were well compensated for their efforts on behalf of the pesticide and solvents industries.[35] Were these eminent epidemiologists just calling it like they saw it? Did industries engage them only because they happened to agree with the judgment that particular exposures to a given product, whether asbestos or pesticides or chemicals, had not been proved to cause human harm?

Typically, these scientists' work for these firms involved reviewing whatever published information was available on any given health threat. As we have repeatedly seen, getting to the point of being able to publish results on worker health was not easy. It depended on the willingness of those in industry to risk opening their records to examination. Access to such information came with considerable strings attached, including provisions that results of any industrial surveys didn't get published unless approved by the companies. That didn't happen much.

From the earliest days of industrial hygiene, advisers to industry were often asked to lay down the ground rules for how and why more research was needed to clarify various technical matters before any clear conclusion could be reached. At the same time that they were being handsomely paid for their expert explanations of why current information was lacking, these specialists also crafted the grounds that would be used by generations of epidemiologists to decide whether or not there was proof of any hazard. Their closely honed and carefully followed efforts determined what critical pieces of information were missing. They did not engage, as far as I can tell, in persuading these firms to open their records to full and complete examination by independent experts. Rather, their work furthered the manufacture and selling of doubt.[36] I have not been able to determine why, contrary to policies in their respective institutions at the time, Doll, Adami and Trichopoulos did not acknowledge much of their work for industry in publications they wrote for hire.

In all fairness, the route to independent expertise in public health has never been simple. Consider this: If researchers hadn't cooperated with industry in the 1950s, they wouldn't have had access to any information on the health of workers. At what point does cooperation become cooptation? When do those who continue to study a problem that others want to fix become part of the problem?

In epidemiology today there's no balanced seesaw for determining what's true. In studies of human hazards, the end of the seesaw showing proof of harm is generally thought to have to be twenty times heavier than the lighter end in order for a hazard to be considered established. Proof that a given condition actually results from a given exposure is considered to be established only when it has a one in twenty chance of being false. Over the past half century, Doll, Trichopoulos, Adami and other leading epidemiologists have convincingly argued that we can't conclude that a hazard exists in humans until several different studies have all shown the same thing. Even today epidemiologists are being hired by brake manufacturers to convince the courts that the data collected on asbestos from brakes and mesothelioma point to no causal connection.[37]

Look again at how Evarts Graham chided the skeptics on the hazards of tobacco smoking in his *Lancet* commentary. For agents that can cause cancer two decades after exposures begin, is it fair to insist that we must have proof that such harm has happened in many groups? Can we really require researchers to repeat studies showing that the same exposures have caused the same cancers in other groups similarly exposed before we can agree to change policies to prevent additional harm?

Irving Selikoff, a leading figure in the epidemiology of asbestos, worked, like most in the field, with some cooperation from the industry. Most of his financial support came from government grants, unions and union records, and for a while from the American Cancer Society. Over two decades, teams of researchers working with Selikoff obtained detailed information on good and bad habits of men and women in the factories. Industry leaders had hoped that smoking would turn out to explain much of the risk then believed to be tied with asbestos. They were right, but in a way that worsened their case. The chances that asbestos workers who smoked would die of lung cancer did not just add up, they multiplied. Those who worked with asbestos and did not smoke cigarettes had a fivefold greater risk of the disease than the general population; those who smoked cigarettes alone had a tenfold greater risk. But the poor fellow who worked with asbestos for twenty years and smoked a pack a day had a more than fifty times greater chance of dying of lung cancer than one without such exposures.[38]

ONE OF THE MANY ISSUES Doll weighed in on was the question of whether gases released from coke ovens harm human health. Coal tar had been one of the first agents shown to cause cancer in animals. Rodents whose skin had been painted with tar developed tumors early in the twentieth century.

Coke is baked coal, a necessary ingredient in making steel. Heated to more than 2,000 degrees in an oven, coal releases aromatic agents full of tars and compounds based on benzene. Today benzene, toluene and xylene—all cancer-causing agents produced by cake ovens—are captured and sold. But for years, they were released out of the ovens into the community in quantities nobody bothered to measure or monitor. From the turn of the twentieth century the Allegheny, Washington and Fayette County regions of southwestern Pennsylvania were the nation's top producer of coke.

For most of the century a coke oven was a pretty simple thing, a brick beehive the size of a small garage. Once started, it could not be stopped because if it ever cooled it would crack. The young men working an oven would shovel in the coal and then stack bricks in front of the entrance to allow it to reach the necessary temperature. When the coal was cooked, the oven doors were opened on both sides, sucking in outside air. The oxygen-starved coke burst into flames and was doused with massive amounts of water. What was left was almost completely pure carbon that could be used for making steel. Most health experts surmised that the heady mixture of gases released from coke ovens, full of tar, benzene and heavy metals, was unhealthy. Yet the demand that this process be shown to harm humans in statistically significant, well-designed studies was not easily met.

Carol Redmond came from a small working-class town in Pennsylvania to Pittsburgh to study mathematics. Showing an aptitude for numbers from the start, she gravitated to biostatistics, earning a master's degree at a time when the field was coming up with methods for evaluating patterns of health and disease. It turned out she also had a talent for the sort of slow, tedious work that eventually yields phenomenally important findings. Redmond began her public health career painstakingly assembling the information needed to determine whether coke ovens posed a risk to workers. Following the rules for carrying out epi-demiological studies that Doll and others had laid down, Redmond and a team of researchers in Pittsburgh worked with the unions and spent the early 1960s pulling together details on more than 70,000 workers. Where had each worked? For how long? What jobs did they have? What chemicals and dusts did they regularly use? Had they smoked cigarettes?

There were plenty of cases of lung cancer showing up in doctors' offices in Allegheny County in the 1950s and 1960s. While many men with lung cancer at the time were smokers, many were not. More black men got lung cancer, though they tended to smoke less. Many cancer victims were women who hadn't smoked. Were the ovens a fac-tor? That was what Steelworkers Union chief epidemiologist William Lloyd and many others wanted to find out.

As far as I can tell, Doll never published directly on Redmond's studies. But a letter found in the archives of Robert Kehoe, the head of the industry-funded Kettering labs at the University of Cincinnati, makes it clear that Doll tracked this matter closely for many years. Doll thanked Kehoe for making available a copy of Kehoe's private study of coke oven workers. As he did with a number of major indus-trial exposures, Doll monitored these studies of coke oven workers for years without publicly revealing their existence.

In 1956 the *British Journal of Industrial Medicine* published a study of 8,000 truly lucky men who had lived long enough to retire from working in the coke ovens: they had no additional risks from their em-ployment. Of course, this study didn't ask what had happened to those who never made it to retirement.[39] When she began looking at the patterns of death in the coke oven workers, even before she counted a single person, Redmond could tell there was something special about them. In the 1950s, and even as late as the 1980s, there was one major requirement for working the coke ovens. Those who ran the ovens were strong, often young, and most often black. At the time, nine out of every ten coke oven workers in southwestern Pennsylvania were black. Black men who sought work in the well-paying steel industry were offered only the dirtiest, most dangerous posts. They could not be machinists, carpenters, chemists or electricians; those jobs were reserved for men of British, Scottish or other European ancestry.[40] They could work atop or next to the coke ovens, where searing heat and fumes regularly burned holes through their boots. They could haul steaming molten trash from the blast furnaces and ovens.

By the time Redmond started her work, the pension funds of the coke oven workers were quite rich. They still are. Only a small proportion of those who spent their lives in this work lived to claim retirement benefits. Redmond looked at men who quit before retiring, as well as those who got their pensions. The former had twice as much lung cancer as the latter.[41] The notion that there is a typical amount of lung cancer may seem odd. But in fact, in any group, there will be a certain number of people who develop this disease. When Redmond began her studies, researchers already understood that the rate of lung cancer itself was changing fast because of cigarette smoking. The trick was to come up with a way to take into account the underlying growth in smoking-related lung cancer and figure out whether there were yet other causes. Two tools were the mainstay of such work. A standard mortality ratio contrasted deaths at different ages found in the coke oven workers, both those who quit and those who got to retire, with those that occurred in the general population of people who were the same age as the workers at the time of diagnosis. A standard incidence ratio compared new cases of lung cancer in coke oven workers with the number of cases reasonably expected to occur in those without such workplace exposure.

There were two main reasons why coming up with the right group against which to gauge the health of the coke oven workers was not simple. First, racism didn't stop with President Harry Truman's integration of the armed forces at the end of World War II or with the Supreme Court's ruling in *Brown v. Board of Education* in 1954. Some union officers and public health officials seriously asked whether black workers were inherently more vulnerable to lung cancer. Today we know that the gene pool of black Americans is more similar to that of white Americans than to that of blacks in many different parts of Africa or the Caribbean.[42] But in those days, thinking about skin color and genes was guided by simplistic notions.

A second complication was that while life is a mixture for all of us, it was especially so for coke oven workers. Their exposures were hardly limited to simple solitary agents but involved complex mists, fumes and chemicals. When researchers study a specific drug, they try to hold constant all sorts of things that can affect health so that they can contrast how the drug affects those who take it compared with those who do not. Whatever results they get, they are reasonably sure that the drug being studied is the main reason for whatever difference appears between the two groups. But trying to study workplace hazards by comparing workers with others is more complicated. What is the best control group? Where do we find people with no exposures to the things that the oven workers experience every day? We know that workers are healthier than average because they do not include those too sick or infirm to hold a job.

Redmond began her studies by contrasting the health of coke oven workers with that of other steel workers. The presumption of this comparison was disarmingly simple: let's see whether men who are working the ovens have risks beyond those of other steelworkers. The person who first thought up this contrast may not have expected to find much. After all, pollutants released into the air near the ovens seldom stayed there. But despite the limits of this comparison, Redmond showed what many had long feared. Relative to steelworkers overall, coke oven workers had between two and four times more lung cancer. When Redmond first published this finding in 1972, she was asked how she could be sure that these risks were really tied with the work. After all, the oven workers were mostly black and the other steelworkers were mostly white. Couldn't this result just be telling us that blacks have some special weakness?

It took Redmond five more years to put these questions to rest. The largest combined steel mill west of the Mississippi was Geneva Steel in Orem, Utah. Its workforce was all white and mostly Mormon. The results were clear. Mormon coke oven workers, who didn't smoke or drink much, had two to four times more lung cancer than other steelworkers.

Then Redmond began to ask some really tough questions. Does the risk end at the factory gates? When she looked at rates of death and disease in Allegheny County, where most of the coke oven workers lived, she found that retired black oven workers had about four times more lung cancer than other blacks living in the county and a seven times greater chance of dying of lung cancer than other black retired steelworkers. The reason other retirees were relatively healthy was clear: they had survived their jobs. They were obviously in better health than most of those who left before retiring. Many times, those who quit did so because they had to: their bodies had given out.

When Redmond looked at the small group of men who had lived to retire from working atop the ovens, where the fumes were most concentrated for the longest time, the full brunt of their work became clear: they had about sixteen times more lung cancer.[43] When Redmond compared coke oven workers against the entire U.S. population, she noted that, "If United States rates are used as a basis for comparison, one would conclude that lung cancer is significantly in excess in both the white and non-white workers."[44]

THE STUDIES PIONEERED by Redmond went on for nearly thirty years, carried out by more than a dozen different researchers. An impressive body of detailed material appeared in most of the major occupational health research journals. Study after study confirmed and extended the basic findings.

For quite some time, the people who lived in Clairton, Pennsylvania, in the massive shadow of the Clairton Coke Works, didn't hear about these studies. But they knew there was something different about where they lived. "We used to wake up with this greasy, black soot stuck to our cars and windows," Connie Lucas, a former resident, told me. "People would complain about what all that grit did to the paint on the cars and houses." Nobody thought much about what it might do to the insides of their bodies.

The century-old coke works at Clairton are a marvel, a complex of ducts and furnaces that runs for almost a mile alongside the Monongahela River. The plant today is one of the largest producers of coke in the world. As the steel industry declined in the 1960s, the works' owner, U.S. Steel (later renamed USX), shut down several of its huge ovens. But in the 1980s the company got permission from the Allegheny County Health Department to restart two coke batteries in addition to ten already operating.

The expansion came with a price. A number of accidents released tons of extra pollutants in the air, so much that for the first time in its history, people organized to oppose the plant. The world was becoming smarter about the alphabet soup of chemicals involved in coking. These include such known carcinogens as benzene, toluene, xylene, cadmium, arsenic and beta-Naphthylamine. This compound was once used to make dyes and rubber. A cousin of it is employed to make moth balls. At night and sometimes first thing in the morning, Joann Meier, who raised her family in the area, remembers that the air would reek of mothballs all the way to the wealthier neighborhood of Squirrel Hill.[45]

In the summer of 2004 I talked with a federal employee who had been troubled for years by what he saw at Clairton. A handsome, quiet man in his late thirties, he had things to tell me that would never be found in any records. He asked that I not reveal his name. My research assistant, Mary Katherine Nagel, and I met him one sunny day just outside a federal office building in the eastern United States in a city that I cannot name for what turned into a very long, very hot lunch hour. What came out of the Clairton mills, he said, didn't just go into the air.

"In the mid-1980s, before I began to work for the federal government, I worked as an intern one summer for an environmental assessor at the Clairton site. They were ripping out the old coke ovens and putting in new foundations. When they were digging the foundation for the new coke ovens, the groundwater filled right in to the holes."

"How could this happen?" I asked.

"All the water in the ground there," he explained, "feeds right into the river. What runs below the ground right by the plant is actually an underground plume of water just a little higher than the surface of the river. Naturally it all flows downhill, taking whatever is with it into the Monongahela River.

"We were digging to lay the foundation for the new ovens, but strange things kept happening. The water was so caustic, it stripped the paint off the excavation machinery.

"A few years later, for my first real job as an engineering pro I returned to the site as a groundwater consultant. That's when I found out that what left the plant made problems even a half mile away. There's a water treatment plant down the river. The intake filters got clogged up with toxic materials from Clairton; they couldn't just be taken to a landfill. They needed to go into an official hazardous waste disposal site, with all sorts of added costs and protections. Instead of doing this, the water district sent the clogged filters back up to the plant, where they were burned up in the ovens at night."

Lucas and her family could smell when something strange went on. "We lived right next to the mills. At night, the odors would be the worst, because they would let things out then that they didn't want you to see. If we were rocking on the porch, sometimes we could feel a hot pass of air for a few minutes in the dark, then that sickening sweet stench would make us run inside and shut the windows."

The federal employee reported that odd things happened at other times as well. "One day we ran a monitoring well in the area between the plant and the Monongahela River. We took a long piece of 40 millimeter thick polyvinylchloride pipe—the stuff that's close to indestructible. We ran a 6 inch schedule, 40 millimeter polyvinylchloride pipe in to check the groundwater. We were always looking for a bailer. This is a long tube filled with water. We had a hard time finding a bailer that wouldn't dissolve. We put our brand-new lexan pipe in. We pulled it out, and it was just a string. The bailer had been dissolved by the groundwater. The groundwater eventually dissolved the well itself."

"What could have eaten through thick plastic pipes like that?" I asked.

He answered, "Well, I can't be sure. Pretty strong acids must have been there. We also know that benzene dissolves plastics. But we will never know. Every time we would come in to sample a spot, the guys from the plant would know we were coming and would tell us where to look."

"What happened then?" I asked. "That must have been a little unnerving."

He sighed and nodded. "Yep. Well, the next day we went out to drill new wells, right where our old one had dissolved. But we couldn't. Right on top of the spot where we had pulled up empty string, someone had poured a new cement pad with a hydrochloric acid tank. Overnight they closed up the place where our wells had dissolved. This was 1987 or 1988."

"Who were you working for?" I asked.

"Officially, I was working for an environmental monitoring firm. The State of Pennsylvania had hired us to check on what was entering the river."

"What did you find?"

"Nothing," he said. "We found nothing. That was exactly what we were supposed to find.

"We used to score sites—you know, rank them—in terms of how bad the chemicals were and what techniques were being used to contain them. Well, Clairton was the worst I've ever seen. I once told someone, 'The only problem they don't have there is child labor.' I know that at night they burned things that they never would have tried in the daytime. I remember, we'd come in and find things covered with a fine layer of grit and when they would turn the filters back on the air would clear again."

I was glad that my researcher was with me. We kept looking at each other as the man spoke. He had waited a long time to tell this story.

"There was corruption there. Absolutely. I know just from talking to people at the plant. I told my friend at the EPA, 'You gotta do something about this site!' I was concerned about unmitigated discharge of pollutants in the air and in the groundwater.

"Yet the decision is hard, because Clairton was a very important job center. They are balancing, do we shut this plant down because it's the worst environmental site in the hemisphere? I can tell you this: I've been to thirty or forty superfund sites and Clairton was the worst."

"Why did you stop working there?" I asked.

"I couldn't take it anymore. I got to the point that I couldn't smell anything anymore. One day I just took the sample bottles and threw them against the side of the truck and they just smashed. I was so fed up with the site. I didn't know it, but I was getting sick myself. I went to the doctor. I was dead tired. I was very, very irritable, with massive headaches. You take one deep breath and your senses are numb—too much phenol in the air.

"The doc said that my complaints were basically in line with a mild case of benzene poisoning. He rather carefully worded it that I might be at an increased risk for leukemia.

"I was chronically exposed to neurotoxins. I would be exhausted at the end of the day. I refused to work there anymore. This was about 1988. I wasn't married then. I left and went back to grad school—Wright State, to get my master's in engineering.

"When I left, I told them they needed a health and safety plan or I am never coming back."

"What happened to your company?" I asked.

"They had had enough of me. Because I refused to go back to the site to work, they saw me as a threat. They couldn't fire me because they knew I had documented everything. There's no doubt in my mind that the whole place would be a superfund site if it ever stops running.

"I'm not naive. I know that there are serious economic issues involved. They need a happy medium. They let a lot happen that should never have happened. Like this consent agreement with the state that allowed massive amounts of toxic chemicals to flow into the water. It's cheaper to pay fines and admit you are putting things into the water than it is to fix the problem. I don't know what the amount was, but every month they would have to pay a certain fine. I probably put seventy monitoring wells that came up empty, because they were supposed to come up empty.

"Let me tell you. Back then, environmental consulting was not the place for an honest person. They're in business to make money. Many times doing the right thing can conflict with wanting the profit. Either you keep getting paid or you speak up and they'll find another consultant that will drill where they want him to drill.

"I am sure there are honest consulting firms out there, but they are not as profitable as the dishonest ones. At the time the State Department of Environmental Protection just did not want to know what was going on. They were all about defeating the monitoring process so they wouldn't get slammed by the feds.

"When I said we ought to tell the state, I was told that wasn't a good idea. They just wanted us to look like we were watching out for things that nobody really wanted or expected us to find.

"I went to school and never came back."

In 2007 I spoke with someone else who still works at the Clairton plant, which still operates right alongside the river. "Things are lots cleaner now, that's for sure," he assured me. "But hey, we all saw lots of things we know just weren't right for years. I remember one night in the late 1980s, a huge hole was dug in the ground. That same night more than fifty barrels of who knows what were thrown in there and covered up. People don't even think of trying that stuff nowadays, but back then, that's just the way things operated."

In whatever decade you ask the question, the answer seems remarkably consistent: environmental practices were shameful twenty years ago, but since then we've cleaned up our act. Except that it's never quite clear when and by what means this cleanup happened. Thus the story begun by Carol Redmond remains unfinished.

THE "MOST INFLUENTIAL Supreme Court Ruling You've Never Heard Of" is the title of an essay by a group of scholars and scientists dismayed at changes in the way the courts handle admissible evidence.[46] A radical shift occurred with the 1993 decision in *Daubert v. Merrell Dow Pharmaceuticals, Inc.* This ruling turned judges into scientific referees in charge of deciding what's fair and foul scientific information. It also played into the hands of those who insist that human harm must be proved to have occurred before we can say there is a clear connection between any exposure and a given health problem. Basically, *Daubert* relegated experimental work in lab rats to a marginal position in such matters. The court laid out four tests that judges can use in deciding whether the science presented in any case is credible and admissible:

1. Does the evidence rest on a testable theory or technique?
2. Has the theory or technique been peer reviewed?
3. Does a particular technique have a known error rate and standards controlling the technique's operation?
4. Is the underlying science generally acceptable?

Who could argue with such questions? In fact, as with much in science, where you stand on these issues depends on where you sit and who owns the chairs.

The *Daubert* case involved a lawsuit against the pharmaceutical firm Merrell Dow, brought by a mother whose child was born with a serious birth defect that she claimed arose from the use of an antinausea drug, Bendectin. The firm's expert reported that several different published studies of persons who had used the drug during pregnancy did not show such an effect. Experts for the plaintiff relied on two other types of evidence—experimental research with animals showing that use of the compound early in pregnancy produced birth anomalies, and analyses of several individual cases where such birth defects had occurred after the use of Bendectin. The Court basically threw out both types of evidence. In effect, it set itself up as the deciding official as to what constitutes acceptable science in this instance. It rejected experimental studies in animals altogether as irrelevant to human harm. In doing so, the Court ruled that the only proof of harm that could pass muster was sufficient numbers of sick, deformed or dead children to have made it into published epidemiological studies. The assembly of case reports of such harm could be rejected because it constituted an inappropriate type of evidence.

This was a stunning decision in many ways. First, the court rejected animal studies as the scientific grounds for predicting human harm and also dismissed clinical case reports, no matter how many, as lacking proof of risk in any instance. In staking this ground, the justices said that where there is epidemiological evidence, it trumps all other information. This judicial decision fundamentally misunderstands how science works and how hard epidemiologic information is to obtain. Under this view, human hazards can be ruled on only after they have been proven to have occurred in sufficient numbers with adequate documentation. The Court thus set back by decades efforts that can be undertaken within the legal system to prevent human harm.

The contrast this decision offers with past approaches could not be clearer. In the 1970s, in *Ethyl Corp. v EPA*, the Court looked at experimental evidence on the hazards of lead poisoning and reached a firm decision: if you want to prevent children's brains from being damaged by microscopic amounts of lead, then get lead out of gasoline.[47] With *Daubert*, judges were granted dominion to insist that the only clear evidence of human harm came from publications that analyzed statistically significant instances of human deformities or death.

The *Daubert* decision presumes that well-done science is like painting by numbers, displaying universally agreed on standards and methods to come up with clear, definitive facts. Since this decision was issued, a series of other rulings have all moved in the direction of ratcheting down the types of information and types of experts who are able to submit evidence that judges can look at. Not only does a person have to be a scientific expert in their own view, but they must also be considered such by a majority of other experts. If the world were a fair and just place, where all were accorded equal opportunities to develop information and publish their work, this would make a great deal of sense. But the real world is neither fair nor just, and it is growing less so by the day.

The medical world is built on increasingly fine subspecialties, often funded by those whose products they are being paid indirectly or directly to assess. Under these circumstances, what seems simple and obvious is often neither. Proof that the environment into which we are born can shape our lives and deaths and even the well-being of our children has never been easy to come by.

Consider how medicine has evolved during the past hundred years. For centuries physicians practiced what my good friend Mo Mellion, former head of the Academy of Family Practice Medicine, describes as traditional or "hand-me-down" medicine. This consisted of venerable guidance passed down from one to another physician and eventually memorialized in text books. Confronted with patients in need, doctors typically mete out what Mellion calls "the best I got medicine." Because they can't tell their patients to wait for research to be finished before figuring out what to do, they cobble together treatments aimed at alleviating pain and suffering. The practice of medicine fundamentally remains a solitary pursuit. The core of the encounter occurs between two individuals, each with hopes and expectations.

There is only one problem with this seat-of-the-pants approach to medicine, according to David Eddy, the godfather of modern medical evaluations. "In many instances, we simply don't know what we are doing."[48] As a physician-mathematician with more than the usual curiosity for following things through, Eddy did something others hadn't dared to try. Starting in the 1980s, when he was a young physician in training at Stanford, he kept track of the cases of various diseases, the treatments and their outcomes—and came up with a stunning set of conclusions. Many big-ticket, big-profit medical procedures, including annual chest x-rays, the conventional treatment of back pain and glaucoma, the use of bone marrow transplants for breast cancer, and many forms of cardiac surgery, simply don't work. This work exposed the soft underbelly of medical technology. There is little proof that a large part of the $2 trillion we spend every year on medicine actually does what we think it will—and lots of reason to think that it doesn't. A revolution in looking at medicine has begun, fueled by these unrelenting analyses.

Whether a parallel revolution can demand the creation of full and fair evidence on environmental hazards is not at all certain. But at this point we know that medicine of the future will start by looking into the micro-environment of the patient's body and the macro-environment of the world around us.

If clinical medicine rests on such unscientific foundations at this juncture, is it any surprise that fields like environmental and occupational medicine and epidemiology lag much farther behind? There are many reasons why it's been much harder for public health researchers to come up with evidence on the ways in which where we reside and work affect our health. Chief among them is one simple fact: the ability to generate information on workplace hazards has always been hampered by those who control the worlds in which those facts exist. The concept of trade secrets is meant to protect commercially valuable production methods and formulas like the recipe for Coca-Cola. But under various laws now in place, the health and welfare of workers exposed to various agents can also be trade secrets.

In 1986 William Fayerweather, then working for DuPont, made a stunning presentation to a group at the U.S. National Academy of Sciences. He had developed a computerized system for tracking the health of every worker in DuPont's chemical plants. With a few keystrokes he could tell how many had developed what illnesses and where they worked with what materials. Neither he nor his system any longer works for DuPont.

In 2003 Richard Clapp took a simple look at information about the health of men and women who produced computer chips, which a California court had ordered IBM to provide to the plaintiffs' lawyers. Clapp found that compared to the general population, more of the IBM workers died at younger ages of cancers of the breast, bone marrow and kidney. As a company that self-insured its workers and practically invented how to use computers for tracking, ordering and following all sorts of items, surely IBM had a system in place, like the one that Fayerweather had set up a couple of decades earlier, to monitor the health of its workforce? Plaintiffs' lawyers were told there was no such system. Those medical forms the lawyers got—one for each employee—with punchable circles to be filled out for machine reading were never scanned at all, the company claimed. They were just made to look like they could have been read by computers.

Clapp proceeded to do the analysis himself. He was threatened with lawsuits if he ever released his findings. Apparently under pressure from IBM's lawyers, a major international journal, *Clinics in Occupational and Environmental Medicine,* withdrew its acceptance of Clapp's first article showing that those who worked in "clean rooms" as IBM chip makers had higher rates of several types of cancer. Just this past year, another group of scientists hired by IBM produced results that showed no such harm. Their study was limited to workers at just three plants out of the dozens that IBM operates.

Three and a half years later, despite threats of lawsuits, Clapp's work finally appeared in print. He had asked whether those who had worked in computer chip manufacturing at specific plants in California, Minnesota, New York and Vermont for at least five years between 1969 and 2001 died of diseases similar to those of the U.S. population. The answer was no. Those who had worked for IBM had greater chances of dying from several forms of cancer, including those of the brain and central nervous system, kidney cancer, non-Hodgkin's lymphoma and breast cancer. Given the limits of the information available to Clapp, it wasn't possible to say which chemicals accounted for the excess rates of cancer. That information remains a trade secret.

No independent assessment of this problem can be conducted. Some nice people I know at IBM wish the company would just acknowledge the sins of the past and move on. But they say the lawyers would never allow that. Fighting lawsuits to the bitter end discourages other potential plaintiffs and saves money.

How EVIDENCE IS DEVELOPED, what information is allowed to be seen or heard, is not a simple matter when it comes to our health and its connections to the world around us. The hidden industrial sponsorship of some of the leading figures in epidemiology is deeply troubling. We will never be able to know the extent to which the creation of doubt about industrial hazards arises from real concerns or reflects the subconscious tendency of people to dance with the one who brought them.

Unmarked graves in the Smeltertown cemetery. For more than a century, toxic dusts from Encycle's now-idled smelter in Smeltertown, Texas, wafted across canals and the Rio Grande to Mexico and Sunland Park, New Mexico. Of the small Smeltertown city, only the contaminated cemetery remains. In 1998, the U.S. Department of Justice and the EPA documented that "Encycle's own business records provide compelling evidence of sham recycling. Numerous hazardous wastes with little or no recoverable metal value were mixed into Encycle alleged 'products.'" The government memo detailing sham recycling, released to the Sunland Park Environmental Group in July 2006, can be found on this book's web site.

13

No Safe Place

Why should anybody want to pay to see a play about Love Canal
when you can drive through New Jersey for free?
—DUSTIN HOFFMAN, IN *TOOTSIE*

IF THERE WAS A HEYDAY of environmental activism, it had to be the
1970s. Remember this era. There were big problems with seemingly
clear solutions. The decade opened with 20 million Americans attend-
ing Earth Day rallies around the country, joining millions throughout
the world in vague calls to protect the planet.

With rhetoric that seems strangely modern and prophetic, President Nixon claimed the mantle of the first modern environmental
president in his 1970 State of the Union Address:

> In the next 10 years we shall increase our wealth by 50 percent. The
> profound question is: Does this mean we will be 50 percent richer in a
> real sense, 50 percent better off, 50 percent happier? Or does it mean
> that in the year 1980 the President standing in this place will look back
> on a decade in which 70 percent of our people lived in metropolitan ar-
> eas choked by traffic, suffocated by smog, poisoned by water, deafened
> by noise, and terrorized by crime? . . . The great question of the seven-
> ties is, shall we surrender to our surroundings, or shall we make our
> peace with nature and begin to make reparations, for we still think of
> air as free. But clean air is not free, and neither is clean water. The price
> tag on pollution control is high. Through our years of past carelessness
> we incurred a debt to nature, and now that debt is being called.

It's irrelevant, at this remove, if Nixon was only trying to get out in front of Congress on an issue with overwhelming public support. President Nixon's speech marked the opening of a decade that saw the passage of major environmental laws, including the Clean Air Act of 1970, which included national fuel economy standards for cars. A new federal agency was created—the Environmental Protection Agency—and staffed by committed young lawyers and others with a passion for planetary defense. By 1980, the Clean Water Act of 1977 had been enacted and the Superfund law was in place to clean up hazardous wastes. The nation has not seen another period of such focused attention on environmental laws since then.

The radicalism of the project is hard to imagine nowadays. The president proposed an expansive program to protect and fund open spaces and reduce sprawl. He singled out the car as the worst polluter of the common resource of the air at the time, calling for major new laws that ensure that "the price of goods . . . be made to include the costs of producing and disposing of them without damage to the environment."

During the 1970s environmental emergencies made national headlines again and again. Highly prized oysters from the once pristine James River watershed of Virginia were decimated by sloppy discharges of the pesticide Kepone into the riverbed. Those few shellfish that remained were deemed inedible. After a month on the job, workers came down with what they called the "Kepone shakes," quivering, jittery muscles and slurred speech. Late in 1975, the once productive James River was shut down for fishing.

In 1978 Love Canal in upstate New York became a national byword for pollution. The census tract around Love Canal contained 4,897 people, surrounded by streams to the north and west and the Niagara River on the south, with a rural area to the east. President Jimmy Carter ordered the evacuation of 225 households while federal investigators combed through residues of past wastes. The nationwide scope of the problem was staggering, as one analyst noted. "By that time industry had produced—and stashed SOMEWHERE—about 100 trillion pounds of hazardous wastes, enough waste to create a highway to the moon 100 feet wide, 10 feet deep."[1] People desperately wanted to believe that whatever problems had happened at the Love

Canal and the James River, if they truly were problems at all, happened somewhere else to someone else.

It has been hard to document the burden on public health from leftover industrial garbage. Few researchers have tried to do so. National resources devoted to the problem, while never flush, have fallen to an all-time low. Like much of the effort to evaluate environmental risks to public health, the absence of a robust infrastructure to examine these matters is not accidental. Lots of people have an interest in seeing that the harms tied with environmental wastes remain understudied and underrecognized. Calls for studies by experts often amount to a way to buy time and keep things as they are. After disasters like those that hit the James River or Love Canal (or 9/11), many people just want to go home. When our lives are turned upside down by dark forces, we long to restore some semblance of normalcy. The yearning for the comfort of the familiar pushes us to set aside any hint that the havens of our homes and communities are in danger. No matter that home may be a bit shaken, it's still the base on which we come to depend, where we feel secure even when we shouldn't. Sure, the problem might be bad, but not where we live. Our place is just fine.

As the seat of national political power, Washington, D.C., is home to organizations of varying political stripes whose purpose is to issue opinions. Nicknamed think tanks, their employees get paid to ponder what-if scenarios on matters as diverse as the flat tax and whether a polar route for shipping could possibly offset sea level rise. Some are well known, like the Brookings Institution and Resources for the Future, which have been issuing weighty reports and holding schools for new congresses for more than half a century. Others, like the Environmental Law Institute, where I worked in 1980, are more obscure. ELI is the type of place where lawyer's lawyers go for advice on the intricacies of federal and state rules and regulations relating to the environment.

In the spring of 1980, ELI was asked by Congress's own research arm, the Congressional Research Service of the Library of Congress, to weigh in on a hot political topic: Did current laws compensate those who had been injured by discarded wastes? Did they prevent additional damages? Where harms had hit, did the laws allow people to recover their losses? As scientific director of ELI, I was thrilled to be working for CRS. This was a serious issue posed by a serious organization—the stuff think tanks dream about. Behind this request were two prominent senators on the Committee on Environment and Public Works, John C. Culver of Iowa and Robert T. Stafford of Vermont, a Democrat and Republican, respectively.

Our task for the CRS seemed straightforward. We were to look at a dozen or so alleged episodes of pollution that hit people's homes and neighborhoods and find out what actually happened. Was anybody truly hurt? Did serious damage occur to the local environment? Who paid to fix things? Based on our findings, we were supposed to tell Congress whether or not a new law might be needed to generate funds to clean up sites, secure public safety, and assess the human and environmental damages.

Our research team included some of the brightest young legal and seasoned public health scholars of their generation: Marvin Schneiderman, the prolific, iconoclastic deputy director of the National Cancer Institute, on leave to work with us on this project; Jeffrey Trauberman, a former biochemistry major at Brown and a recent graduate of Georgetown Law School who would later become chief lawyer for Boeing; and Leslie Sue Ritts, a Princeton honors grad who would become a partner in one of Washington's most prestigious law firms, Hogan & Hartson.

As is the custom in Washington, we worked around the clock. If our children, friends or spouses wanted to see us, they had to visit ELI's crowded, cramped offices in the apex of a triangular building just above the Dupont Circle Metro stop. Our ten-person team worked like mad to produce a solid, short and correct report within two months. Our motto was "The perfect is the enemy of the good." But we all knew we had to be damn good to survive in this environment. Luminaries from local law schools volunteered their time to review our work. I soon learned that I was wrong to have resented the intense review and multiple revisions that these heavyweights put us through.

As a sendoff for our report, *Case Studies of Compensation for Toxic Pollution,* on May 22, 1980, Sen. Culver and Sen. Stafford provided the CRS with a strong bipartisan endorsement. "The researchers undertook a painstakingly thorough analysis of the law . . . even though only six states are included in the survey, the legal and practical obstacles to recovery are representative of those which confront all victims of toxic pollution."[2]

Our report was intense and unflinching. Leftover industrial garbage, we wrote, had affected the health of people whose only crime was to live downstream or downwind of forgotten or hidden waste ponds, piles or pools. The people responsible for causing these harms in the first place were not easily found and often long gone. Polluted properties were passed among owners like hot potatoes, ending up in the hands of those least able to get rid of them.

We concluded that some people had been harmed, even killed, by industrial wastes in some regions of the country. But many instances of harm could not be proved. Sometimes people had left or died before anyone could try to determine whether their health problems could be tied to where they lived or worked. The laws did not provide remedies for their injuries or property losses. Large swaths of land remained unusable and without prospects for repair.

Within two weeks after we submitted this weighty indictment, a small bomb dropped on our group at Dupont Circle. Every single one of the hundred sitting senators received a nicely bound, impressive-looking report discrediting our work, printed on the letterhead of Covington & Burling, one of Washington's premier law firms. None of us dreamed that such a highly regarded firm would take us on unless there were solid grounds to do so. I had no idea at the time that Covington & Burling was working directly for Edmund Frost, the supersmart, well-connected general counsel and vice president of the Chemical Manufacturers Association.

I met Ed Frost shortly afterward, during the congressional debates on whether and how best to remedy the legacies of past industrial pollution. Standing six feet tall, he was always impeccably groomed, as clever and well-spoken as he wanted you to think he was, and often quite charming provided the subject fell outside the immediate concerns of his group. He lost no opportunity to let me know that we should put past mistakes into perspective. I needed to appreciate that the chemical industry was cleaning up its act. His job was to make sure they could keep doing their work without the undue interference of a new federal law that would basically tax existing companies to pay for the past sins of others. I needed to understand that the fact that someone of his obvious intelligence and skill being engaged on behalf of the chemical industry signaled a sea change in industrial practices.

The memo from Covington that Frost had thrust on the Senate was a direct hit on our team, claiming that we had been wrong on the facts, wrong on the law, biased and careless—words that cut to the quick. Eventually I learned that aside from a few typos, we had made no mistakes whatsoever. At the time I did not appreciate the old law school adage: When you've got the facts on your side, pound the facts. When you've got the law on your side, pound the law. When you have neither the facts nor the law on your side, pound the table. Covington's memo was merely a high-class way of pounding the table, but we were too naive to realize it.

As a young mother new to Washington's machinations, I was scared. I worried that this salvo from such an eminent firm was my professional death knell. I would never get another job. So when I was asked to meet with the senators and their staffs to explain what we had done, I did what my Bubbe (grandmother) always said to do: stand on the truth. I described to Sen. John Danforth and his earnest chief legal aid, Clarence Thomas, what we had found in Times Beach, Missouri, one of our case studies.

Once a vibrant locale of two thousand on the outskirts of St. Louis, Times Beach, when we began to study it, was on its way to becoming a ghost town fenced off from the world. In 1980 it was still struggling with the dismaying revelation that it was about to be officially declared a toxic waste site. Today it is the site of one of the state's newest parks, just off of America's famous Route 66. But few know its storied environmental history. The town was founded in 1925 as a tree-filled summer resort on the banks of the Meramec River, less than twenty miles from downtown St. Louis. Vulnerable to regular flooding, many of the town's first buildings rested on stilts.

The town's troubles began on May 20, 1971, when Russell Bliss, a local contractor, collected used oil from Independent Petroleum Corporation and Northeast Pharmaceutical and Chemical Company, two small industrial facilities. As he had done on many other occasions with wastes from other firms, from 1971 to 1976, Bliss applied these slick residues to twenty-three miles of unpaved dusty local roads and to a nearby horseback-riding arena. It was perfectly legal to do this.

One of the things that makes oil so useful is that it can seep into cracks and crevices not filled by light and more watery liquids, and can keep grime from swirling around. But the stuff Bliss put on the roads that day was no ordinary oil. He had mixed regular engine oil with materials taken from a company that made skin soaps based on chlorine. The oils he spread around were heavily laced with dioxin, one of the most toxic compounds ever studied.

Bliss had used the oil in his own barns and began applying it to stables in the area. Dust is not the only thing that was suppressed by this reused oil. Horses, dogs and cats began to sicken. Sixty-two horses dropped dead after one spraying at the Piatt stables in March 1971. So many dead birds fell from the sky, they had to be raked up. It later turned out Bliss had poisoned even his own land.

At first people were told they had nothing to worry about. Three months after the horses died at the stables, six-year-old Andrea Piatt came down with a rare blood disorder. Her ten-year-old sister also became ill. They both had played in an office just outside the horse show arena that had been drenched with waste oil. They got better when they stopped playing in the area. Bliss carried on. Every once in a while more horses would die. Finally, in 1979, the U.S. Centers for Disease Control came in and measured soils and collected samples of dead animals. That same year, an employee of the chemical firm confessed that the still-bottom wastes Bliss had been hauling had hundreds of times more dioxin than was then legal. CDC soon figured out that the illnesses in the girls and animals had been no coincidence. The places with the most dead animals were those with the highest levels of dioxin.

In 1983 Times Beach would become the first town in the nation's history to be bought out, for more than $30 million, and evacuated by the federal government. Considering that all this took place under a government firmly committed to doing less, the evacuation is quite remarkable.

Some twenty years after the incident, Bliss told CNN's Jeff Flock he had absolutely no idea he was spreading poison. The folks in the factory in Verona where he picked up the oil had been running around in cutoffs and sandals. He noticed a gooey residue on the floor and learned that they made ladies' face cream and soap. There was no reason to think anything could be wrong with oil left over from making facial cleanser.

Looking back on the entire episode, Bliss finds plenty of blame to spread around.

> Like I try to tell anybody, I never produced, generated one drop of this stuff. I only hauled it from one point to another. And I got 18,000 gallons of it, 6 loads, 3,000 gallons per load and I got approximately $150 a load and I and another man drove from St. Louis to Verona which is about 500 miles round trip, you won't get rich doing that . . . I hauled it away for nothing and they were hauling it to Baton Rouge, Louisiana, and paying 25 cents a gallon to dispose of it. And all of a sudden "sucker Russ" comes by and hauls it away for nothing because I thought it was motor oil.
>
> If I thought it was something bad, would I spray it on my own farm where my family is, where I have my wife buried? Would any human being do that? I don't think so. I have a boy of 40 who's lived through this whole thing. I have another boy, 25, he was born, I believe, in '72. He lives here with me and I just, no matter how good or bad you are, you're not going to hurt your family. And I had no idea, absolutely no idea at all what was in it. I still don't really know, only what's been told to me . . .
>
> . . . Business at the time was selling oil to refineries, to make heavy industrial fuel oil to smelt steel and steel foundries. If I would've known what it was . . . it would've gone to the steel mills, then it would've been burnt, and then it would've been gone. [3]

In truth, if waste oil had been incinerated in the mills, it would have been transformed but would not have been gone. Smaller residues would have spewed over a much larger area of land, leaving less detectible tracks. This would have been an unknown problem for the residents and a major loss for public health research. Where large numbers of people develop the same specific ailments, or when smaller groups have encountered similar conditions so that their experiences can be looked at together, that's when public health studies are easier. The deaths of a few animals, while troubling, don't prove that people are endangered. With hindsight, it's hard to grasp why it took so long for people to figure out what was going on. Unlike much of life today, the waste oil poured over Times Beach left small numbers of people with relatively clear and large exposures.

For several of the poorest polluted zones that we looked at, we had no idea how many had been affected. Some black communities, like Triana, Alabama, a sharecropper town about five miles southwest of Huntsville, had simply moved. Most folks who had lived there had regularly eaten channel catfish, largemouth bass and smallmouth buffalo fish from the nearby rivers. Fish are the ultimate concentrator of pollutants that enter water. The toxic site in Triana ran through an eleven-mile stretch of the Huntsville Spring Branch and Indian Creek, two tributaries that drained into the Tennessee River as it rolled through the Ozarks. For twenty-three years, from 1947 to 1970, the Olin Corporation ran a DDT manufacturing plant at the Redstone arsenal, releasing more than four hundred tons of contaminants into Huntsville Spring Branch. Regular discharges from the plant left a swath of contamination that extended downstream to Alabama's Wheeler National Wildlife Refuge, the largest and oldest refuge in the state. At the time of our report, all we could say for sure was that those folks who lived near the river had more DDT in their bodies than had ever been found before. Twenty years later, scientists from the government would mount a study to see whether the women still alive in the area had higher rates of breast cancer. They found no added risk for those who survived years of pollution. There is no way to account for the thousands who were no longer living or had moved away.

Even a half century after the factories began to operate, some new and fascinating clues have come out of Triana. It looks like everything is not okay with those who are still alive and remained nearby. Researchers at several nearby universities recently examined the genes of people who have been catching and eating fish heavily contaminated with DDT for nearly fifty years. Those who had eaten the most fish had the greatest numbers of abnormal cells and risk of breast cancer.[4]

When my meeting with Danforth and Thomas ended, the senator and his aide both assured me they realized that our work rested on

solid foundations, and that the senators would support the new legislation of Culver and Stafford to clean up waste sites. That law would
include a provision for an Agency for Toxic Substances and Disease
Registry (ATSDR), with a "mission to prevent or mitigate adverse human health effects and diminished quality of life resulting from environmental exposure to hazardous substances." It was up to the Senate
to set up a system to make it harder for people to commit environmental crimes in the future and to provide government authority to
evaluate public health threats posed by hazardous wastes. Left out of
consideration at the time was whether or how to provide funds to
those who had been harmed.

The environmental cleanup bill under consideration later became
known as the Superfund law—one of the most widely disputed pieces
of American environmental legislation ever passed. But at that point
its failures were not apparent. What was clearly visible was the desperate need to address the problems faced by people living in Times Beach
and Triana.

Lots of people we spoke with in 1980 believed that once public
health agencies had the authority to study such problems and to create
a registry of those who had been exposed, reports like ours would become unnecessary. They could not have been more wrong. At the time,
we did not appreciate that we had become part of a well-established
government tradition: studying problems as an excuse for not acting
to change current practices. There are always powerful voices asking
regulators to hold on just a few more years while the scientists complete their research before taking any hasty actions that might disturb
local businesses, especially ours.

JUST BEFORE THE Thanksgiving recess of 1980, I was called back to
meet with senators and key staff, such as Curtis Moore, the influential staff director of the Environment and Public Works Committee. A
flurry of activities arose to convince the newly elected Republican administration to accept the proposed Superfund Law. We thought the
law was set up to make those who had caused damage pay, not to pe-
nalize those who were caught decades later holding land. The law's
sponsors wanted this major new act to provide massive funding to

clean up such sites as Triana and Times Beach. The funds were to come from taxing the polluters still standing who now had the deepest pockets. Whatever objections the Covington and Burling memo had raised regarding our study of compensation for victims of toxic pollution seemed to vaporize. The law didn't pay people for their poor health or lost family members. Instead it paid them for the costs of repairing environmental damage. On the Senate floor, Sen. Danforth urged that "we have no time to lose . . . I believe the clear consensus is that we must clean up abandoned hazardous dump sites as soon as possible."

Behind the scenes, a number of senators on both sides of the aisle had advised the incoming president's new staff that this law would be better than anything they might end up with later on. After all, some people thought the government and industry should not only reimburse them for medical costs, but should also pay punitive damages for those reckless behaviors that gave rise to pollution. People got money to replace their unlivable homes, but nothing could replace their lost health and communities.

The Superfund law passed in December 1980, just before the Christmas recess. I began working at the National Academy of Sciences a few years later, in 1983, with the Board on Environmental Studies and Toxicology (which went by the immodest acronym BEST). It quickly became clear that there could never be a best way to determine what sorts of scientific information could be developed about abandoned waste sites. The effort to come up with scientific evaluations of toxic waste sites was like trying to dance with a bear. So long as you are moving and the bear is up, you don't even think of sitting down. You wait for the bear to get tired. Nobody wants to cut in.

True to the word of its sponsors, the original Superfund law had contained a provision to create the Agency for Toxic Substances and Disease Registry. But giving an agency a name, and even providing some funding, does not ensure that its goals will be achieved. Environmentalists were desperate to have this agency because they felt sure it would show who had been hurt by pollution. Ed Frost and the chemical industry also favored the creation of this agency, but they had quite different expectations. They believed ATSDR would show how little harm had come from chemical contamination. It is a matter of some irony that the ATSDR came into existence only after a very unusual team emerged. Ellen Silbergeld, the brilliant senior scientist at the Environmental Defense Fund who had already conducted breakthrough research on metals toxicology, and the irrepressible Frost joined forces to sue EPA to create ATSDR. Squeezed between usually opposing forces, EPA yielded in 1983.

It quickly became apparent why Frost had so enthusiastically championed the new agency. The first head of ATSDR, a physician who had conducted important work on lead toxicity in children, Vernon L. Houk, simply didn't believe that hazardous wastes caused any health problems except mental illness. Under his leadership, within two years, ATSDR issued reports on close to a thousand sites. All featured the seal of the U.S. Public Health Service. Most found no evidence of harm. That wasn't surprising: the agency admitted that it had no information whatsoever on health or exposure in two out of three sites. That didn't stop industry from claiming these reports as proof that no harm had occurred.

By 1986 Congress had had enough. It demanded that ATSDR start to work evaluating all the abandoned waste sites in the country. The number of such sites could be as low as 32,000, if you believed EPA, or as high as 439,000, if you went with the estimate of the congressional Office of Technology Assessment.[5] Asked by ATSDR to do so, my colleagues and I at BEST finally issued our report on health impacts of such wastes in 1991. The report took about two years to write and about two additional years to make its way through an unusually searing review process, during which some of the language was muted and the conclusions hedged. There was never any overt effort to suppress the results. It proved unnecessary. We all understood that these were different times.

Our report admitted that we often lacked information on who had been exposed to what. The NAS committee concluded, "The health of the public has remained in jeopardy at many sites long after the risks could have—and should have—been identified. Hazardous wastes have constituted a significant health hazard to specific populations at specific sites."[6]

One member of our committee, the distinguished epidemiologist Richard Remington, was dying of cancer at the time we finished the report. He insisted that we include a warning, which he crafted with unusual poignancy:

> It should be recognized that if exposure [to wastes] becomes general and almost uniform, current epidemiologic techniques will not be able to ascertain any related health effects. There is a window of opportunity to initiate studies in areas where groundwater pollution has remained high and localized. There is also an important opportunity for prevention that could forestall major public health problems in the future. The legislative mandates, policies and programs of the federal and state agencies that currently manage hazardous waste sites are inadequate to the task of protecting public health. The distribution and frequency of exposures of specific populations near specific hazardous-waste sites cannot be ascertained, because the needed data have not been gathered. Our report indicates that the nation is not adequately identifying, assessing, or ranking hazardous-waste site exposures and their potential effects on public health. We are currently unable to answer the question of the overall impact on public health of hazardous wastes. Until better evidence is developed, prudent public policy demands that a margin of safety be provided regarding potential health risks from exposures to substances from hazardous-waste sites. We do no less in designing bridges and buildings. We do no less in establishing criteria for scientific credibility. We must surely do no less when the health and quality of life of Americans are at stake.[7]

I FIRST MET TRACEY Segasti in the mid-1980s, when she was an attractive but bewildered young mother living near Brio, Texas, about twenty miles southeast of Houston. Like hundreds of others, she and her family had settled in the area eager for the new space it provided. Five years after moving to the area, she learned that the town sat on a few dozen acres right next to land that for years had been used to dump wastes from oil and gas production and chemical plants into dozens of unlined pits. By 1984, her young son was sick and not getting better no matter what medicines she applied. She found herself becoming a community activist, trying to get precisely the sort of information that industrial polluters went to great lengths to keep under wraps.

I met up with Segasti again in 2003. Remarried and living in Louisiana under the name Tracey Kuhns, she came to a meeting of the Louisiana Hunters and Fishers Against Pollution—a group Kuhns and her second husband started after she left Texas in the early 1990s. She had fled along with many others, she told me, after scientists from the University of Texas–Houston reported that babies in Brio were being born with twisted reproductive organs, and that some had died shortly after birth with monstrous defects of their heads.

Patricia Buffler, a professor at the University of Texas when reports of Brio's problems were emerging, is one of the most distinguished epidemiologists in the country. She would go on to become dean of the Graduate School of Public Health at the University of California–Berkeley, and president of several professional groups of epidemiologists. In 1994, a few years after completing her preliminary assessment of Brio, she also joined the board of directors of the FMC Corporation, a multinational chemical firm, a position she still holds today. Winfred "Buzz" Johanson was the physician who worked on the project with Buffler. Together they found that children of Brio had three to four times more birth defects than those born in other areas.

By 1994 the ATSDR had a new, earnest administrator, Barry Johnson, a career public health service officer who took the agency's mandate seriously. A courtly, amiable man, Johnson asked the group I headed up at the NAS for advice about Brio and towns like it. How meaningful was Johanson and Buffler's study? How likely was it that their findings were tied to pollution in the town? More generally, what methods could be used to examine the health impacts of hazardous wastes? What did studies in animals and modeling with computers teach us about estimating risk? We were given two years—which turned into five—to do a more complicated version of what the Environmental Law Institute a decade and a half earlier had accomplished in two months in a study for the Congress.

A second NAS committee was set up to comb through dozens of reports to state and federal officials about what had gone on at Brio. Here's what that committee's official report said about the study developed by Johnson and Buffler.

A team from the University of Texas School of Public Health (UTSPH) led by a physician-epidemiologist conducted an independent evaluation of these same data from Brio. The investigators divided the population into 3 zones of potential exposure based on proximity to the waste site and wind patterns in the area. Zone 1 was defined as adjacent to the waste site, Zone 2 was 1,460–3,000 feet (0.4–0.9 km) and downwind from the site, and Zone 3 was 2,100–4,100 feet (0.6–1.2 km) from the site and away from the prevailing winds.

The UTSPH team evaluated 652 household-response forms for various health effects. Respondents reported 121 pregnancies, of which twenty-five (20.7 percent) ended in spontaneous abortion. They also evaluated the rates of reported birth defects and used data obtained from the CDC *Congenital Malformations Surveillance Report* to estimate expected rates. Among the ninety-six live births, eighteen (19 percent) were reported to have had congenital abnormalities. The investigators attempted to correct for recall bias by using a conservative risk estimate. They assumed that the number of cases recorded in 1990 for about one-third of the area was the annual incidence for the entire population that lived in the area over the period 1983–1989, while also assuming that the medical end points were valid as reported. With this assumption, they found a lower-bound relative risk of 2.4 for congenital birth defects and 3.8 for major CNS malformations. In 181 women in the study 19–50 years old, there were 126 pregnancies, for a fertility rate of 0.7 births per woman per year. This seems very high, even in the absence of a control group, and may indicate a serious problem in the data.

There may have been biases in the ascertainment of cases and confirmation of reported congenital malformations. First, prior to the health survey, extensive media coverage about the site included anecdotal reports about adverse health effects. This may have biased interview responses. Second, the survey was conducted by volunteers and the response rate was low, leading to the possible biases. While there was some attempt to standardize the questioning procedure, there was no recording of home visits and outcomes or of attendance at training sessions. Volunteers may have been especially diligent in seeking positive responses, which would enhance the potential for recall bias. Third, the form was to be filled out by the interviewee rather than the interviewer, which could cause differences in interpretation of the questions and hence increase uncertainty in the answers. Last, no attempt was made to confirm the diagnoses by contacting physicians. Thus there is no independent verification of reported cases. Other factors also inhibited the development of an independent assessment of this problem, including the protracted nature of the dispute, the inability to gather independent information and the difficulty of obtaining validated measures of exposure.[8]

Let me break this down. Johanson and Buffler couldn't study everyone in town and had few resources. They worked with community residents and made some assumptions. First of all, they took the reported number of birth defects from one-third of the area and assumed that the other two-thirds had not a single defective birth. Even under this scenario, the rate of defective babies born in the area was at least two to four times higher than what should have occurred. If they had assumed that the other unsampled areas had rates similar to the one that they studied, the rate of birth defects would have been four to eight times higher than normal.

Johanson died of kidney cancer shortly after completing this study. None of this work at Brio was ever published.

Because public attention to the issue was high and people had been told that their children had been harmed, there was much angry rabble-rousing when the University of Texas report was first released. When people learn that their homes have been poisoned, information they provide cannot be treated as objective. Maybe the ability to remember births of defective children can somehow be distorted. No wonder why the place is now largely a ghost town.

Kuhns told me that once her neighbors heard about the University of Texas study, they began to think that having children born without ovaries and needing surgery for their reproductive tracts had not just been lousy luck. One little girl turned out to be a little boy. Several boys needed surgery to repair their penises.

"Did your family suffer from any of these problems?" I asked her.

"My grandson was born with a penis that had to be fixed, five years after we left that subdivision. I have seven grandchildren now. Some folks weren't so lucky. Some babies," she said, "had been born without faces."

I guessed Kuhns was talking about a type of lethal birth defect called anencephaly. During pregnancy, for reasons we don't usually understand, the skull and scalp sometimes do not form. In the most extreme cases, called acrania (absence of the skull) and acephaly (absence of the head), a baby will have no head at all.

More than three decades after the law was first passed, nobody wants to claim credit for the Superfund enterprise. The law provided a complex process through which those firms still in business had to pay to clean up their ancestors' trash. Interestingly, the bill didn't touch the hot-button matter of how to pay people for any illnesses they may have developed. The victims of toxic pollution were left to play the odds of toxic torts. If they had distinct damage with documented exposures, and could find a lawyer willing to take on their case, they might win the legal lottery. Like so many federal legislative efforts to redress environmental problems, it proved far easier to pass the law than to make it work.

In Brio, after a decade's wrangling over who was responsible for cleaning up which parts of the large area, most people have moved out. One set of polluters found themselves sued by their own insurance company, Lloyds of London. Superfund has been the subject of protracted wrangling that shows no sign of abating. While the law is still on the books, the federal government no longer collects funds from industry. Every once in a while one hears rumors of its complete demise.

OTHERS HAVE WRITTEN tomes on the merits and demerits of a law intended to make corporations pay to clean up waste left by those who are long gone. Justice seems the last consideration in the search for blame and remedies. The fact that locales that turn out to be riddled with pollution often look superficially normal feeds the idea that maybe there's not much to worry about after all.

Sometimes neither the land nor the people show any trace of their toxic encounters. In the border region of El Paso, Texas, and Juarez, Mexico, the landscape itself, a badlands of sandy volcanic mesas and steep, narrow canyons, is so unrelenting that you'd think nothing could leave a mark on its vast and expansive canvas. The actual border is nothing but a great empty concrete ditch. Today the cities of El Paso and Ciudad Juarez form the largest bi-national region in the world, with over 3 million people.

Before the ditch was set in concrete, the Rio Grande threaded through a narrow valley that allowed travelers to pass between the Mexico mountain range and the Franklin (Rockies) mountain range. Its riverbed would shift after rare heavy rains, taking the boundary line with it and causing disputes over which country owned what land or animals. In a massive industrial resolution of this problem, the United States built the concrete channel in 1964, creating a permanent boundary between the two countries. Nowadays a combination of diversions for golf courses upstream and drier weather leaves the channel dry much of the time.

The name El Paso conjures up images of the Wild West, of gun battles and outlaws. The town became famous in environmental history in the way that the gun slinging outlaw Billy the Kid was famous. But the only shootouts in the past few decades have been between lawyers representing the town's aged lead and copper smelter and those working with the town and its residents. Today an 828-foot brick smokestack stands silent in the city's center.

In 1972 El Paso sued the American Smelting and Refining Corporation (ASARCO), alleging that it had violated the federal Clean Air Act by releasing a steady rain of sulfur on the surrounding communities. When copper is smelted with limestone and silica, the raw rock in which it can be found is essentially baked at 1,200°F, releasing gases of sulfur and other impurities. These contaminants don't go far because they are heavy metals. Tiny traces of these toxic metals can be tracked into homes and schools on the soles of shoes or on bare feet. Small children, then as now, play in dirt and dust. In 1970 thirty-five children in one small area of El Paso were hospitalized with brain damage and other signs of metal poisoning.

Investigations by Philip Landrigan, a pediatrician then in his second year as a public health officer for the Centers for Disease Control, discovered that many of the children of the region were not well. Three out of every five living within a mile of the stack, in an area named Smeltertown after its dominant business, had dangerously high levels of lead in their blood. This was a public health emergency. Landrigan told his bosses and anyone else who would listen that there was no doubt that this contamination came directly from ASARCO's smelter. His 1975 article in the *New England Journal of Medicine* detailing these conditions forced examination of every other smelter in the country. Studies showed that levels of lead that did not immediately sicken children still dulled their brains and nervous systems.

ASARCO's answer to this crisis was straightforward. Smeltertown families were booted out of their homes. Only the dead remained. The small Smeltertown cemetery of marked and nameless graves was covered with blackened, windswept sand. Longer stones or slabs of poured concrete presumably indicate adults, and smaller ones outline those who died as children. The name and short life of Guadaloupe Carmona, 1925–1927, are handwritten on a poured slab. Mounds of small gray stones mark other nameless graves.

In the Environmental Law Institute's report for the Library of Congress in 1980, we described El Paso as a well-established case of mostly historic interest. We knew that the lawsuit against the company had been settled and that the land surrounding the smelter had been bought by ASARCO for less than half a million dollars. The purchase was made on the condition that all the residents were to be removed so that their former home sites could be used to store acid tanks and railroad cars.

When I visited the region in 2003, I learned that some environmental solutions, unlike love, are not forever. El Paso's problems are not nearly as well resolved as I had believed. In fact the story has taken a strange turn. In May 1992, ASARCO set up two CONTOP (continuous top-feed oxygen process) furnaces. These hot-burning ovens never slept. All day every day, they burned tons of toxic wastes at 90 percent efficiency. This meant that just 10 percent of what they tried to burn ended up intact. Still, 10 percent of hundreds of thousands of tons of wastes fired over several years left enough metal poisons in the region that the furnaces were put out of business by the U.S. Department of Justice (DOJ) after operating just seven years. DOJ found that the burned wastes had "little or no recoverable metal value" and that the recycling had been a sham. In March 2005 the inspector general for the U.S. State Department reviewed records on the health of workers at the U.S.-Mexican border and concluded that many of them were sick and unable to get independent medical care in this region.[9]

Memos released from EPA during the Clinton years show that so long as the CONTOP furnaces were running, the company told the world it was recycling materials. Think back to the waste oil that Russell Bliss distributed or took to be burned in mills. If this waste is laced with dioxin or heavy metals, then when it gets burned, thousands of tons of toxic agents are finely spewed back into the air over large regions. Recycling thus becomes a neat redistribution system, taking measurable solid wastes and turning them into immeasurable, ultra-fine air pollutants.[10]

Pollutants do not need passports. The residents of El Paso and Juarez know this, because they are joined by more than a century's worth of leaden soils and plumes that have freely crossed back and forth over the U.S.-Mexican border and left many zones uninhabitable including some areas of Sunland, New Mexico. Commerce, of course, crosses borders as well. In 1999 ASARCO was bought for more than $1 billion and today is a completely owned subsidiary of Grupo Mexico. They have declared their intention to reopen this century-old facility.

What happened to the hundreds of millions of dollars that ASARCO had set aside to pay for cleaning up El Paso? In a stunningly cynical move, Grupo Mexico was granted permission to use that money to pay down corporate debt. Not a penny has been spent to remedy the damage from this longstanding pollution.

At this time, ASARCO faces bankruptcy because of its responsibilities to clean up dozens of Superfund sites. Of an estimated $6 billion in cleanup costs for old ASARCO areas throughout the United States alone, the firm has set aside less than $100 million. The Steelworkers Union in Dallas used the Freedom of Information Act recently to unearth an EPA memo warning that any sampling of metals in El Paso could show that the smelter had burned illegal wastes for years. Many locals suspect the plans to reopen the rusted old smelter are just a ploy to keep the plant from being declared a Superfund site. So long as the company declares its intent to operate, it can't be prosecuted for having abandoned the area.

ONE NOT TOO SULTRY day in the spring of 2003, long after Superfund had become a hobbled earth-moving bureaucracy, I came face-to-face with how far we still had to go. Some people say we've found all the really big problems of heavy industrial wastes and nailed those that can be nailed to clean them up. Others argue that the problem of waste is just a matter of definition. Some pollutants, like PCBs in the Hudson River and mercury or arsenic residues in southwestern Pennsylvania, are so widespread that practical solutions are nearly impossible. The law originally covered discrete geographic zones of presumably manageable size. A national list of the priority hazardous waste sites was generated, setting in motion a process whereby those companies with the most money left would pay to fix up past harms. But what happens when the problem doesn't fall within narrow boundaries but extends through an entire river basin or across major parts of a city?

In an effort to see firsthand some of those nagging pollution problems that seldom get national attention, in the spring of 2004 I drove with Florence Robinson, a Heinz Foundation Environmental Awardee, on a toxics tour of emptied towns in the corridor between Houston, Texas, and Baton Rouge, Louisiana. Robinson is one of three children, from a family that includes people who crossed over to be white. Her immediate family includes a number of teachers and at least one lawyer. Before the Civil War a thriving Maroon culture of blacks who refused to be slaves set themselves up as free people in the bayous, living off raccoon, alligator and fish, and holding on to significant parts of their African culture. They understood what plants signaled health in their community. Today, Florence is a scientific expert in this tradition.

One of our first stops is Bayou Sorrel, just south of Baton Rouge and south of Iberville Parish near Calcasieu Parish. In 1978 a twenty-year-old truck driver named Kirtley Jackson dumped a truck full of hazardous waste directly into an open pit that ran into the bayou. That's what people did back then. Nobody can be sure what happened. Perhaps he had disposed of one of the materials known to burst into flames or release toxic gases when it hits water. Maybe he combined chlorine bleach with ammonia, releasing chlorine gas. Whatever it was, something in the wastes he dumped reacted with those already in the pit and Jackson was overcome. He died instantly. Public notice of his death was one of the first things to bring attention to the need for some kind of law to protect people from toxic trash.[11]

Florence and I are joined by Dean A. Wilson, who offers ecotours of the wilderness around Bayou Sorrel, not your usual hook-em and shoot-em trip to the swamps. We cruise slowly over the water in his swamp buggy. A gigantic, buzzy air propeller pushes us over waters that are sometimes just a few inches deep. Mists roll off the waters, giving the place an ethereal, primeval cast. It feels primordial, like the set of a movie where the dinosaurs are about to return. The temperature is not hot. Fish, frogs and birds keep mosquitoes in check. We are covered by an ancient arboreal canopy of shade. The boat putters close to shore amid mangroves and huge banyans, sometimes called walking trees because they put down roots from upper branches that may extend fifty feet from their central trunk.

Robinson explains the land's predicament. "Louisiana is under the Napoleonic code and one of the consequences of it is that when water flows over land, instead of that land then going back to the state or federals it is still owned by private owners. Here people own wetlands or swamps, so they can cut down some of the oldest cypress in the world. There are four types of trees that can survive repeated flooding: the willow, the cypress, the tupelo and the water locust. Cypress are prehistoric trees that date from even before the dinosaurs and they are related to Sequoia—other trees that can live a thousand years or more."

As we motor through the swamp, we pass Delta Downs, Starks Lookout Tower, Gum Cove Ferry, Black Bayou Ferry, Ellender Ferry, Southwest Louisiana Boys Village and the Indian Bayou Ferry. The most industrious animal in the area today is the beaver, whose dams attract birds, fish and other wildlife. Otters, minks and raccoons all feed off the world that the beaver builds. Louisiana has some very primitive fish—the bowfin, the garfish, the sturgeon and the paddle fish. These have evolved from ganoid fishes—one of the most primitive fish left in the world.

Their modern descendants are not edible, not because of their biology but because of where they've been. We pass by a posted fish advisory, tattered and worn, stuck on a wooden stick in the middle of the water. In big red letters it reads: WARNING—Contaminated Zone.

Wilson explains how ineffective the signs are: "Look. See where the warnings are posted? That's just where folks would set their pots to try to collect crab and other things that creep along the mucky bottom where all the poison settles. Some warning system."

The Atchafalaya basin is unique among Louisiana's basins. It's one of America's central bathtubs, where rainwater and other fluids that roll off of land eventually flow. Its problems are not merely local but national. Nearly one-third of the runoff in America ends up there, bringing with it all the things that enter water as it moves through farmland and industrial zones. Fish and the muskrats and raccoons that eat them end up with some very interesting contaminants in their flesh, everything from heavy metals like mercury and cadmium to organochlorines made in any number of distant factories and refineries.

FURTHER INLAND, Robinson and I drive through Iberville Parish visiting places that had once been vibrant small towns. Few have ever heard of Reveilletown, Louisiana. In 1987 thirty families, in what was then a poor black community next to Georgia Gulf's flagship chemical plant, sued the company because their land was packed with toxic contamination. The company responded by buying up the entire town, paying the residents for their homes and leveling the entire neighborhood. Some of the local environmental and community advocates protested that this solution removed the people but did not remove the risk.

The region we toured had its own nickname—Cancer Alley. In the late 1980s, Florence told me, the Oprah Winfrey show brought some attention to the sick children and other residents of the area. The CDC couldn't say why babies born to people living near the Marine Shale Processing plant had ten times more neuroblastomas—a rare and usually fatal tumor of the brain in babies.[12] The local citizens thought it might be tied with the fact that this recycling plant basically took in toxic waste and turned out an ash of pellets of supposedly harmless stuff for playgrounds, in the process discharging various toxics into the air and water. It all was perfectly legal.

Oprah asked one of the medical experts why many people from the area, including many babies, were so ill. She heard the usual litany—you smoke and drink too much, and you have too much dirty sex, whatever that was supposed to mean. In response to this reply, she held up an infant with its swollen head and said, "Yes, this little baby must smoke about three packs a day, doc."

A few years after Oprah's visit, the town of Mossville in Calcacieu Parish also was wiped off the map. Living downstream of several major chemical facilities, locals got used to what was called "sheltering in place." Della Sullivan grew up in the town and told me how this felt. "A big boom would go off, rattling the house and everything in it. Sometimes windows would crack. Running out in the middle of the night in this swamp can be scary, especially for little kids who grow up looking out for swamp monsters."

I asked her, "Come on now, did you really believe in swamp monsters?"

With a deadpan look, she answered, "Of course there are swamp monsters. What do you think a water moccasin or an alligator really is? We grew up knowing things to stay away from. Nobody in their right mind goes into a swamp at night in their bedclothes unless they be scared out of their head.

"Let me tell you, when we were little we thought sure we saw a bull with the fire coming out of its nose. A bull chased us with be-red eyes and be-fire coming from his nose.

"O course, my siblings and myself, we all saw the same thing. It was just like dark, dark, figure tall, tall like the devil. Big, big old eyes, and he'd stare down at you. He'd scare you, but he wouldn't do nothing you know. We would all holler and stuff about it. The really strange thing is we only told this tale to one another when we got grown. Everybody thought we had seen the same thing, even though we never talked about it."

Swamp monsters were not the only things in the area that didn't leave clear tracks. Della explained, "After the plants started up, you couldn't hang your clothes on the line. We would put them on there because we didn't know, you know, just really where it was coming from. We'd put the clothes on the line, and when we'd take them in just black spots all over it. And even in your house, in my house, I could take and wipe the windows, take a white rag and wipe it, just black, like black smut."

Of course, black spots appearing on white clothes hanging in the air and all over windows had to be mysterious to folks who had always been able to be outside without any worry. It was clear that the swamps had been home to lots of unexplained phenomena.

"Lots of strange things kept happening. Even the squirrels went crazy on the back streets chasing people," she remembered.

I was incredulous. "The squirrels went crazy?"

Sullivan nodded her head. "Sure enough. They kept running after people, instead of staying away like normal squirrels."

I pressed her, "Any other things you noticed with the wildlife?"

She replied, "Well, the snakes got really strange. We were fishing one time and some water moccasins came up like a horror picture and they swiveled like this." She shook her wrists quickly as though they were about to fly off.

"One snake actually started chasing us. These things were mean and angry, not like normal snakes at all."

The residents of Mossville shut the doors and windows of their homes to smoke and fumes, but couldn't shut their bodies from pollution that entered their water and food. For fifteen years scientists from ATSDR have been testing and retesting people in the region for residues of dioxin—one of the nastier toxic byproducts of industrial life. They found that area residents had three times more dioxin in their blood than the average American. The older the person, the higher the levels. At the time of this writing, the official federal report from the Agency for Toxic Substances and Disease Registry on the dangers of dioxin remained under revision. The studies of the blood of those still alive from Mossville continue. In 2005 CDC issued a report noting that older persons from the area had more than four times the amount of dioxin in their blood than older persons in the rest of the country.

All that was left of Mossville when Robinson and I visited two years before the hurricanes of 2005 leveled the region was a solitary white painted board with a slogan painted in black:

In memory of workers and citizens who have paid with their lives for a toxic environment. Our fight for a clean environment is for you, our families and our future.

Surrounding this statement were more than fifty hand-lettered names. At the time I took a photo of this memorial, only Daisy Jones and Flora Nelson were still legible; Della Sullivan told me of many others. "Most of my family you find listed on the sign, from Sheila

Ruth Maden, to Fred Sullivan, to Gwendolyn Sullivan Fonteneau, to Adam Sullivan, to Bessie Rignathen Sullivan, just a number of them. Wilda Mae Bennett Sullivan and I can go on and on because most of my family was wiped out, you know, through bad health and cancer."

As we walked along what was left of Mossville, we found the remains of small cement-block foundations. Tall grasses claimed the space of what had once been a vibrant hunting and fishing community.

In Mossville, the lucky ones who were still alive collected money from Conoco and Condea-Vista before they left the town. But there was one catch, as an investigator told me. "There was a clause in the agreement they had to sign to get paid that no matter what pollution, no matter what illness or any other adverse effect ever came up in the future, from no matter what chemical, no matter what source of what chemical, they were no longer going to be allowed to sue the chemical companies if they got sick later on."

I haven't been back to the area since the big hurricanes hit in 2005—first Katrina broke the backs of New Orleans' levees and swamped other coastal lowlands, then Rita sucker punched what was left less than a month later. In the ocean, as hurricanes build and move across the surface, a train of lee waves is produced. Behind them, a large zone of upwelled water rises that sweeps over whatever it finds, until it runs out of steam. Jerome Ringo, head of the National Wildlife Federation, comes from Mossville. He told me that a wall of water more than twenty feet high swept through what was left of the small town. When it receded, it spread toxic sludge and waste broadly.

Nobody has dared ask what this may mean for the health of those who still try to live off the waters. But a recent report from a surprising source in China may be sadly pertinent. The official government newspaper—there are no others—reported on May 22, 2007, "Many chemical and industrial enterprises are built along rivers so that they can dump the waste into water easily," Chen Zhizhou, a health expert with the cancer research institute affiliated with the Chinese Academy of Medical Sciences, told the newspaper. "Excessive use of fertilizers and pesticides also pollute underground water. The contaminated water has directly affected soil, crops and food."

Of the ten most lethal illnesses recorded last year in China, cancer was number one, followed by cerebro-vascular and heart diseases.

ABOUT TWENTY-FIVE YEARS AGO, Florence Robinson came back from Baton Rouge to live near Devil Swamp, a place she'd fallen in love with as a child. As we looked around the area, I asked what had first attracted her to the place. She beamed as she reminisced.

"This was a real grand honest-to-goodness swamp. It was beautiful, full of oak trees draped in mosses. It was a true wild area. It's on the banks of the Mississippi River—a genuine overflow swamp. I had always wanted to live really 'in the country.' I found my little dream house on the edge of the Devil Swamp, found it at night, because I was actually commuting every day from Monroe to Southern where I was teaching chemistry, which is about a 180-mile drive. So I did that for about two weeks and then I found this house and had to wait the thirty days to move into it. I moved into it and it was heaven. And at the time I didn't realize that all the factories were around it, because you couldn't see it, it was a heavily wooded area, and you couldn't see all the industrial activity."

I asked, "What kind of plants were near your home?"

"All kinds of petrochemical plants, chemical plants. There was no refinery directly in that neighborhood, but there were several chemical manufacturing plants, a lead smelter, Calcine coke plant. Then there was something they call the Union Tank Car Company. This was basically a business that washed out the bottoms of tank cars and dumped a lot of that stuff into the little creeks that ran into the main bayou that ran into the Mississippi River."

"When did you start to wonder what was going on around you?" I asked.

"We used to go out in the swamp, my son and I, we used to just go walking, and I remember the first year I was there I found a pond that had salamander eggs in it, and that is so rare to find down here, and I was just so excited about it. The next time I went to that pond it was covered over with an oil film. We kept seeing things like that. You'd go out and then you start seeing a little bayou that was covered with this nasty looking oil film, and I got to the place I just stopped going out there.

"And it was probably a good thing for me. Probably the best thing that ever happened, NOT to keep going out there."

I asked, "You think this kept you from having as many health problems yourself?"

She replied, "Of course, it took a lot of time for me to even think there could be something wrong for us. It was really not until the mid-eighties it began to dawn on me that there are some real problems out here. I mean, we'd always had extremely bad odors. It's a funny thing, swamps always smell a bit from natural decomposition. You absorb these odors and then you get to the point you don't really notice them.

"Things were happening to me that I wasn't aware of. For example: At my house, I never woke up in the morning feeling refreshed. I never woke up in the morning feeling, 'Oh boy, it's time to get up and get moving.' I always woke up feeling like I'd been shoveling coal all night. And it didn't matter what time I went to bed. Interestingly enough, I did a lot of traveling in those days; when I would go other places I'd wake up in the morning fresh and ready to go. Literally bounce out of the bed. I had convinced myself I was simply not a morning person. That's not the problem. The problem is, I was in a really bad, polluted area. And I would get away from my house and, of course I had eight o'clock classes every morning and I felt so awful, I was dragging through the eight o'clock classes. By noon I would start waking up. And it was just that I was getting rid of my toxic load, and then I started to feel better in the afternoon, and I'd stay at school until late, come home, maybe have dinner, usually go back out, I had to do something else, and then get back home and work and go to bed, and wake up the next morning feeling exhausted."

I asked Florence about her son, who had grown up there. "Had he ever had any problems with his health?"

She sighed. "Sure did. When he was in high school, his first year in high school, they called me at school one day, said he was sick. He'd had a severe headache. He'd vomited. He went blind."

I wasn't sure I had heard her. "What? He went blind in school?"

Robinson went on. "Yes. I got to the school and he was lying down. And of course I'm terrified. Oh my God, my kid's got a brain tumor. Something of that sort. By the time I picked him up he could see again. I took him to an ophthalmologist, who said it was a classic migraine. He said anytime you feel something like this coming on, you take these pills. Michael took those pills all the way through high school. Let's see, he stayed high school, did one year in college and then he left the area. He hasn't taken any of that medicine since."

Robinson lived close to another area with unexplained health problems that we had looked into as part of our NAS report: St. Gabriel, in Iberville Parish, Louisiana. Today it's home to one of the largest chemical plants in the country. In the 1980s, a local pharmacist named Kay Gaudent noticed she was dispensing medications for what seemed like lots of miscarriages and other women's health problems. A team from Tulane University came in to study the issue and found what they believed were high rates of failed pregnancies. By the time others were brought in to review all this, the high rates had gone away. Florence remembers vividly what happened.

"The people who were brought in to review what had been done changed the way they averaged the rates; they only looked at so-called confirmed miscarriages, when a woman had surgery to remove embryonic remnants. They manipulated the data in one way or another and eventually reported a small increase in miscarriages, and the other problems of reproduction went away."

"What happened to the pharmacist after that?" I asked.

"She lost everything. Her family had owned the pharmacy for several generations. This is an industrial area. Most of the people either work at the industrial facility or know someone who works in the industry or in the service companies. The local industry stopped using her pharmacists for their needs, and so their customer base dried up."

With such a load of pollution in Louisiana, I couldn't resist asking, "Why does the state's license plate say Sportsman's Paradise?"

Without missing a beat, Robinson replied. "That's easy. Because it *used* to be. People loved to hunt and fish. It was a renewable resource–based society. The Cajun people were really in sync with the environment. They lived off the land. They knew that if they didn't do something to ruin the land that it would take care of them. And it did. But they've traded the old way of life for new ways of death. Things that have been unheard of before these generations that we're now in."

It's a wonder that people like Florence Robinson keep on in their efforts, given the way the situation is stacked against them. But every once in a while, they win on an issue. Working with students and professors from Tulane University, Robinson kept a new polyvinyl chloride plastics plant from being built in an area that already had more than a hundred different chemical facilities. No reward could ever be as sweet as knowing that you've made the world a bit less dangerous by standing up to and standing down forces that would rather see you go away.

I asked Florence one last question, "What keeps you going after all these years?"

"I have been running on anger for fifteen years. I got angry all over again yesterday, when you were talking to those folks about Brio and whatnot, even though I've heard these stories a million times. When you look at them in a certain broad sense, they're all the same. It's little people who have been so terribly exploited and put upon who are suffering so gosh-awfully terrible and the government and the industry is treating them like dirt. And it makes me *so angry.* When I first started, I went as an innocent American believing in my government and believing the government was there for me, and then the government showed me what an asshole they were."

More than two decades since the Superfund law was passed to do something about the problem of toxic wastes, lots of people have scars to show for their battles to make the law work as it was meant to. Others have scars from fighting the opposite battle. One wintry day when all this was a faint memory, I was skiing with my daughter, Lea, at Arapahoe Basin in Colorado. In Summit County, A-Basin is less glamorous than the Keystone and Copper Mountain resorts, with the reputation of a skiers' ski place—no frills, just glorious mountains, old-fashioned lifts, and vistas of unimaginable beauty. Folks who ski there tend to be serious about the sport. As I was skiing to the lift (there are no lines at A-Basin), I heard someone call, "Dr. Davis!" I squinted and recognized my old nemesis on the ELI study. Expertly shussing down the slope toward me was Ed Frost, the former general counsel of the Chemical Manufacturer's Association, the very fellow who had commissioned Covington and Burling to trash our study.

Frost swooshed to a stop, neatly dusting my skis with the spray from his own. I had never imagined that someone who always wore the best suits would be dressed for warmth rather than power. Folks who ski A-Basin do not wear fancy ski duds. Frost fit right in, garbed in a plaid shirt atop layers of frayed long johns, navy blue bib ski pants that looked as though they had been in somebody's closet for years, a well used black neck warmer and a beige wool stocking cap.

"Mr. Frost?" I asked, somewhat incredulous.

"Yep," he replied. "Don't be so shocked. I love this stuff, too."

I looked at him askance, not quite sure what to make of all this. But he had started this exchange, so I waited.

"Listen," he said. "I gotta tell you something."

This was going to be good, I told myself. Part of me wanted to say, "Listen, you jerk, you cost me months of my life frantically running around defending my intellectual integrity against that slimy attack you had somebody put together on my work." But I didn't, because I had long ago come to understand that his assault had done the work and me a big favor: I got to meet interesting and powerful people in Washington whom I might otherwise never have made contact with.

So I waited. He turned to the attractive young woman next to him and explained:

"My daughter here is in graduate school," he said. "She is studying environmental management." He looked genuinely proud. Why was he telling me this?

"Listen, I really owe you an apology. My daughter has convinced me that we need to do better on these things. I know that we should not have been so heavy-handed in going after that report you all did on toxic pollution in those communities. You guys did a pretty decent job on a hard topic. But that was just how the game was played, you know."

It turned out that Frost had undergone a conversion of sorts and had left the chemical industry to join a group called Clean Sites, a joint effort of industry and others to do something constructive about the problems of toxic pollution. He had decided to stop denying reality and fighting to keep things as they were and align with those who were trying to do something practical. I could not resist a smile at this unexpected development.

He turned out to be a damn good skier as well.

So what does all this have to do with cancer? Let me tell you. The case studies we pulled together in our contested report for the Congressional Research Service all involved little towns, where most of those exposed were either dead or moved away. The number of those who survived was often so small that nobody could demonstrate whether what happened to them had anything to do with where they lived.

Years later, in our NAS report for the ATSDR, we tried to pull together all the studies we could find on toxic pollution in communities. This wasn't an easy thing to do. There are scientific journals dedicated to organs and cells, with names like *Heart, Lung* and *Brain*. There are even journals with the word *environment* in their titles that provide solid technical reviews of how to measure pollution and what can be done to clean it up. But no regular journal is devoted to technical reports on the complex, messy world of environmental pollution. So we combed through government reports from state and local officials of lots of small towns where exposures to common contaminants had occurred.

With Remington and others, the academy had formed a Committee on Environmental Epidemiology to tackle the job. The committee started this review of what was called the gray literature (local city, county or state reports—things that never appear in regular journals). We carried on this effort for four years and got nowhere. The reviewers kept telling us there were too many scientific uncertainties. People living in different communities cannot be combined. You cannot lump together apples and oranges. We were told to abandon this approach because it wouldn't be scientifically rigorous. That was undoubtedly true on some level. Yet it was also beside the point.

Apples, oranges, bananas and hot dogs and sauerkraut, all are eaten. Sometimes they can and probably should be combined. Our failure to reach agreement on the controllable causes of cancer is not merely due to the insistence on keeping things separate, with the notable exception of the belated and finally successful effort to limit smoking, and the failure to deal with the complexities of real life. Our effort against cancer—with its singular focus on treatment and its neglect of the other things that cause it—is not working, or certainly is not working well enough.

It is not simply that cancer is one of the diseases that afflicts the survivors of these polluted towns or those along China's poisoned rivers in disproportionate numbers. What afflicts them more is that the very place they lived in—the air they breathed, the ground they walked on—was toxic. The real failure of the Superfund law, like the failure of the war on cancer, has only a little to do with bloated bureaucracies or scheming lobbyists or unfortunate yokels with trucks full of the wrong stuff. Ultimately it's a failure to look clearly at what's right in front of our faces.

The way knowledge is packaged on environmental hazards is hardly a matter of scientific happenstance. Whenever a public health matter is raised that affects billions of dollars in sales of some set of goods, the burden of proof that is imposed in reviewing this threat can become so arduous that it can never be met. As Edward Bernays advised more than fifty years ago, the best public relations efforts appear in the guise of objective scientific opinion. Scientific uncertainty is easier to purvey than most of us ever dreamed.

Putting on Makeup Shouldn't be like Playing with Matches

Which cosmetics company do you trust with your daughter ?

www.safecosmetics.org

14

Chasing Tales

All scientific work is incomplete. . . . All scientific work
is liable to be upset or modified by advancing knowledge.
That does not confer upon us a freedom to ignore the
knowledge we already have, or to postpone the action
that it appears to demand at a given time.

—HARRIET HARDY

THE 2000-YEAR-OLD MAN, as channeled by the comic Mel Brooks, had some strange opinions about human accomplishments. When asked to name the greatest invention of the last two thousand years, Brooks' very old man replies with a thick Yiddish accent, "Saran Wrap! You can make a big Saran Wrap, a small Saran Wrap, you can see right through it!"

Incredulous, the interviewer asks, "Saran Wrap? What about fire?"

"Hmmm. That was also good," Brooks answers pensively.

One of the miracles of modern chemistry, Saran Wrap was discovered, like many chemical breakthroughs, as an accident. In 1933 a college student named Ralph Wiley, who worked cleaning glass in the Dow Chemical labs, could not get one vial to come clean. It was covered with a film made of polymers of vinyl chloride that would become the basis of Saran Wrap. Fire, ironically, would prove to be the product's undoing. When burned, this clear plastic wrap released dioxin. As of July 2004 the original product no longer exists. It's been reformulated based on a less toxic plastic, low-density polyethylene.

Like many modern chemicals, vinyl chloride was invented by German researchers in the nineteenth century. It was originally produced in small amounts by cumbersome methods. As chemistry advanced in its ability to meld simple, shorter molecules into longer ones, vinyl chloride became a workhorse of modern industry. When boiled at temperatures of more than 500°C, hot enough to melt sand, ethylene dichloride breaks apart to produce vinyl chloride and hydrogen chloride—a clear, colorless acid so strong it's used to clean metals. In industrial practice, the searing hydrochloric acid released when vinyl chloride is produced easily mixes with oxygen and then reacts with more ethylene on a thin copper wire catalyst to generate even more ethylene dichloride and water. This elegant, simple interaction created a revolution in the chemistry of plastics.

When assembled into chains, or polymers, vinyl chloride becomes polyvinyl chloride (PVC) and is nearly indestructible—as strong as cement pipe, as flexible as rubber and cheaper and lighter than many of the things it replaced. Benzene is one of the few things that can make it disintegrate.

As a slightly sweet smelling gas placed under pressure in metal cans, vinyl chloride can be used as a propellant for all sorts of liquids. At one time it whipped heavy cream, vaginal deodorants and hairsprays into the air. More than half of this gas is chlorine, the stuff that was used to suffocate enemy troops in World War I. It was presumed that vinyl chloride gas lacked the poisonous properties of chlorine, but this was one of many assumptions that were later proved wrong. Vinyl chloride turned out to be slower-acting than chlorine but just as lethal.

Why did it take so long for the dangers of vinyl chloride to be widely accepted? Some ways of thinking reflect nothing so much as the willing suspension of disbelief.

JUDY BRAIMAN SEEMS like an improbable revolutionary. She is a small, strong woman, with the tough edge of a grandmother who knows how to throw a football. Considering what she went through, her toughness is hardly surprising. She played a pivotal role in informing the public that vinyl chloride is dangerous.

In 1966 Braiman was a thirty-year-old mother living in upstate New York. She was married to a successful attorney and had three children in diapers. Her hair had thinned quite a bit after the birth of her daughter in July 1965. These were the days when Annette Funicello, Sandra Dee and other movie stars wore teased hair that looked like inverted bushel baskets atop their heads. The wife of an attorney, Braiman was expected to look stylish. Her hairdresser told her about a new hairspray that was sure to give her lots of volume. Several times a day in her small bathroom, she sprayed and teased her hair with lacquer, perfecting a smooth, puffy, stylish look.

By the time her daughter was six months old, Braiman had a cough that would not quit. She couldn't walk upstairs easily. Her ribs ached. She had lost eight pounds and was spitting up blood. X-rays showed that her lungs were laced with lesions. "Even though it's been forty years, I will always remember those pictures," said her doctor, William L. Craver, now a retired thoracic surgeon. "They showed patchy, fluffy infiltrates scattered throughout both lungs." Dr. Craver told Braiman to get her affairs in order. He explained that she had what looked like choriocarcinoma of the lung, an exceptionally rare and very fast-growing tumor connected with pregnancy. Few ever survived.

Braiman prepared for the worst. Her lung was to be removed, along with a rib, some muscle, and whatever was found to be growing throughout her chest.

Her internist told her there was a fifty-fifty chance she would survive the operation. She woke up from the surgery with a scar that ran from just below her heart around to her back. Still hazy from the anesthesia, she heard the surgeon say, "The good news is that you do not have cancer."

"I looked around and stared at him," she remembers. "I had spent this time getting ready to die. To tell the truth, I thought I had cancer and they were lying to me because they wanted to make me feel better. So when they said things were okay, at first I did not believe them. I made them show me all the slides."

In fact, Braiman had neither cancer nor tuberculosis. But her lungs contained about sixty different small, round deposits of what turned out to be hairspray lacquer. Each one had set off a local lesion called a granuloma, a small abscess surrounded by a zone of acute inflammation. On the x-ray they looked just like tiny tumors.

"The doctors were pretty curious. They had never seen anything like it. They asked me to bring all my aerosol products to the hospital."

In a report on her case, Dr. Craver noted that Braiman's lungs had been speckled with small spheres of fatty globules. Of the seven spray products Braiman used at the time, the only one that contained fat was Bonat hairspray, the product she had used several times a day in her tiny bathroom. Whether Braiman's many lung deposits would have formed without vinyl chloride can't be known. That they arose from that hairspray seems beyond dispute. But the particular chemistry that created her lesions remains unknown.

"My doctor thought this was pretty important and very unusual. He shared what he had found with other surgeons in town."

Her lungs returned to normal within a year, but Braiman felt like she had gotten a second life. She was determined that whatever happened, she was not going to keep quiet. Talking to the press, she decided, was precisely what she needed to do. Columnists Jack Anderson and Les Whitten wrote about her story the next year, noting that aerosol sprays can contain chemicals that "may sear the eyes, damage the lungs and weaken the heart."[1]

For Braiman, that was the beginning of what turned into a long campaign. From that point on, she became an activist on environmental contaminants in the home, school and community. She wrote letters. She corralled congressmen and anybody in a position of authority. She spoke often and vociferously, on radio and television, about the need to reduce the toxic materials found in personal care products. Her efforts to bring public attention to vinyl chloride and its dangers did not go unnoticed.

"After I began talking to the media about what I'd gone through, the company threatened me and said that if I discussed this publicly, they would sue me for slander."

In a letter from the Bonat company, dated December 17, 1973, the company raised the ante on her actions: "Please be advised that should you carry out your threat to allege on your next television show that your personal injury was caused by a product of ours, you will be held accountable for such allegation and suit will be instituted against you personally and against the association for all damages which might be sustained as a consequence."

In 1973 a can of many widely used hairsprays was one-half vinyl chloride gas. By that time, in response to growing numbers of reports about the dangers of this compound, the hair products industry had begun quietly discouraging use of the gas. Companies rushed to put out aerosols that they could advertise as free of it. That year the Clairol Corporation told an Associated Press reporter, John Stowell, that it was no longer making Summer Blonde Hairspray with vinyl chloride. Stowell was outraged to learn that just before it shut down production, Clairol had manufactured an entire year's supply of vinyl chloride–containing Summer Blonde in under a month.[2]

Thanks to these and other tactics, millions of cans and thousands of gallons of vinyl chloride remained on store shelves and in people's homes while the industry scampered to get the stuff out of its products on the one hand, and challenge efforts to compel it to do so on the other. Driven by consumer pressure and increasing reports of grave health problems in workers who made vinyl chloride, EPA and the FDA first tried to ban the sale of such products on October 8, 1974. Following the predictable array of legal delays, the ban did not go into place until four years later, June 1978.[3]

Another person who learned firsthand of the penetrating powers of chlorinated hydrocarbons was my friend JoAnn Pacinelli. Fresh out of college with a chemical engineering degree, in 1975 Pacinelli landed a job as a chemist in a plastics factory in Malvern, Pennsylvania. "I spent many hours working with no hood, no gloves, and lots of intriguing modern chemicals. We knew we were engaged in important work on important things. We were formulating polymer films and foams and used lots of solvents, things like free isocyanates and chlorinated hydrocarbons. We had absolutely no idea that our own health could be at risk."

A competitive rower and serious weight lifter, the twenty-two-year-old Pacinelli suddenly found she tired easily and had a persistent ache in her chest. X-rays showed her lungs full of what the doctors thought were small tumors. She was told she had lymphoma and had six to nine months to live.

Pacinelli's dad refused to accept that grim diagnosis. He found a lung surgeon who agreed to look at what was growing in her lungs before deciding to remove the entire organ. A small piece of Pacinelli's lung was taken out. It was full of granulomas—walled-off areas containing irritating materials she had inhaled. She left her job, relocated to a first-floor apartment and spent much of the next three years tethered to an oxygen tank. She never returned to the lab bench.

A year after she quit, Pacinelli visited the factory where she had worked. She asked the company's technical director whether there were many cancers in the young men who worked on the production lines. "He admitted that we had a lot of cases of cancer then. He told me, 'It's cheaper for us to pay people for their disabilities or pay their families for their deaths, than it is to change the way we make things around here.'"

COMMERCIAL PRODUCTION of the solid form of vinyl chloride—polyvinyl chloride or PVC—skyrocketed from a million pounds a year at the start of World War II to more than 27 billion pounds in 1973. Well before Braiman and Pacinelli survived their brushes with modern chemicals, workers in plastics factories grew familiar with another strange effect.

Stunted fingers had been reported in individual production workers in Italy and France as early as the 1940s, but it was a rare condition and had not been linked to vinyl chloride. Before 1965 only seventy-two cases of dissolved bone, or acro-osteolysis, had been recorded in the world literature. Most of these were in families that shared a genetic defect. In France and Italy in the 1960s, what had once been a rare defect of the bone was reported in groups of vinyl chloride workers.[4]

A 1964 letter from Rex Wilson, a plant physician where several cases had been found, to his counterpart at a vinyl chloride plant in Ohio, asked that workers not be told of these concerns, but be looked at "as quietly as possible." The language of this letter tells us just what the priorities were in this matter.[5]

November 12, 1964
Dr. J. Newman
33880 Lake Road
Avon Lake, Ohio

CONFIDENTIAL

Dear Doctor Newman:

   We have recently observed several cases of hand disabilities in one of our plants, the cause of which is as yet unknown.

We would like to determine as quietly as possible whether similar disabilities might exist in the Avon Lake Plant, and for this reason I would like to have you casually examine our employees' hands as a part of any other medical service you provide to them.

Specifically, the disability that we have seen is characterized by soreness of the tips of the fingers. Roentgenography of the hand has demonstrated bone changes, particularly bone resorption, in the distal phalangeal joints. In some cases this is quite marked. Blanching of the skin from decreased temperature is prevalent. Several of the cases exhibit fibrous rope-like dermatological changes of the hands. The disability has been described by several consultants as Raynaud's Phenomenon. In one instance it was called Scleroderma. The presenting symptoms do not clearly fit the classical description of either of these diseases.

I would like you to make careful observations of the hands of our employees as you see them for any reason. If you observe any soreness in the tips or any symptoms similar to those that I have described, will you please make a careful notation of the job assignment of the individual as well as his work history with the Company. As yet, we have no firm opinion that the disabilities we have seen are occupational in origin. We are investigating this matter and hope to eventually resolve this question. A logical avenue of investigation is to obtain as much information as we can about our own employees.

I would appreciate your proceeding with this problem as rapidly as possible, but doing it incidentally to other examinations of our personnel. We do not wish to have this discussed at all, and I request that you maintain this information in confidence. Will you please advise me by January 1, the approximate number of hands that you have seen and of any positive findings.

Sincerely yours,
KHW/js Rex R. Wilson, M.D.
cc: Q.F. Backmeyer
A. Vittone
W.W. Baughman

In 1967 a group of British general practitioners published a report in the *British Medical Journal* describing a man whose fingers had become stunted while he was working with levels of polyvinyl chloride believed to be perfectly safe.[6] At the time one could still look on such incidents as singular occurrences. The man's employer took the position that he must have a genetic defect or some rare disease. But within a few years, many similar reports appeared of men who had worked with vinyl chloride and suffered stubbed fingers. In one case a man's jaw dissolved. The notion that these strange cases had some common cause became hard to reject.

By 1967, the results of Dr. Newman's quiet survey at the Avon Lake plant were clear. Thirty-one cases of this strange syndrome—one-third of all known cases in the world at the time—occurred in a group of three thousand workers.[7]

The vinyl chloride story is not a shining moment in the history of occupational medicine. If the compound looked problematic in men who worked with it, the companies decided, the next logical step was to see how animals respond. In 1971 the Italian scientist Paulo Viola first reported finding cancers of the skin, lungs and bones in rats exposed to high levels of vinyl chloride.[8] Aware of this work before its publication and fearful of what it could mean to the plastics industry, a group of companies, including Montedison, a major Italian manufacturer of vinyl chloride, had earlier commissioned Cesare Maltoni, a toxicologist based in Bologna, to begin a series of studies of vinyl chloride gas.

Maltoni's work would change the way the world looked at this compound and would also set the stage for rethinking the way such research should be done. For two years, four hours a day, five days a week, five hundred lab rats were subjected to various levels of the gas. Those in the highest-dose group got 10,000 parts per million (ppm)—a level that nearly anesthetized them. Those in the lowest-dose group got 250 ppm—an amount that consumers could easily encounter when using aerosol products such as Judy Braiman's hairspray or that workers might inhale during the manufacture of PVC. Another group of rats was raised without any such exposures.

Two hundred fifty ppm is a tiny amount. If you made a stack of pennies the height of the Empire State Building, a single penny would be one ppm. One ppm is also equivalent to a single minute over two years. Two years, or one million minutes, was the usual time that rats or mice were exposed to test compounds. Maltoni's innovation was to allow the rodents to live beyond their million minutes to their natural lifetime, another half million minutes during which he was able to look for tumors that occur in the last third of life. Basically he let the animals live after retiring from their routine toxic exposure, something human workers aspire to enjoy after they end their jobs in factories.

The levels of vinyl chloride gas Maltoni used were small and the time periods long, much like what workers experience in real life. After the exposures ended and all of the rats—those exposed and those not—lived out their natural lifetime, they were splayed open and every organ of their bodies was weighed and examined for signs of disease.

No government knew of these private industry studies while they were under way.

The results were horrifying. About one in ten of the exposed rats had an exceptionally rare tumor of the liver, angiosarcoma. Not one of the unexposed animals did. In animals or humans, angiosarcoma is a death sentence. It starts in the lining of the blood vessels of the liver, and once the malignant cells break through into the bloodstream, they inevitably spread throughout the body. There is no known treatment. In Maltoni's experiment, those rats that got the highest doses had the greatest numbers of tumors, but some rats developed this malignancy even at the lowest dose.

By the fall of 1972, Maltoni had no doubt that vinyl chloride was a serious health threat. At first, he followed the rules of his contract with the companies and kept quiet. Montedison assured him it would release his findings at an appropriate time and place. Maltoni expected that this would occur when the manufacturers met with officials of the Italian, French, Japanese or U.S. governments. To his dismay, the meetings took place, but disclosure of his work—or even mention of its existence—did not.

Maltoni was furious. Disgusted by industry refusal to release his results, in 1974 he violated his agreement and published the results in *la Medicina del Lavoro,* a medical journal.

A decade earlier in 1964, John Creech, a physician in Harlan County, Kentucky, had noticed that the fingers of a man who worked in B.F. Goodrich's vinyl chloride plant had become shortened and stubbed. The bones had literally dissolved from within. Within a few weeks, Creech had seen three more cases in the plant. "If four people doing the same type of work in the same room, the same department, come down with a bizarre situation like this, it doesn't take a rocket scientist to link it to industry—to their workplace."[9] By 1973, four men who had worked with vinyl chloride at the same plant had died of angiosarcoma. Each year, fewer than two dozen cases of angiosarcoma were expected in the entire United States. To have four from the same county, let alone the same factory, within a span of two years was either an astonishing coincidence or no coincidence. Creech's report on this appeared on February 9, 1974, in the Centers for Disease Control publication *Morbidity and Mortality Weekly Report*.[10] With this report, Maltoni's work with animals became indisputably relevant.

The evidence continued to mount. More cases of men with angiosarcoma were found and reported to a number of other industrial governments. Peter Infante, a young public health researcher for the state of Ohio, told reporters for the *New York Times* in 1975 that babies born to people living near factories producing vinyl chloride from 1970 to 1973 had higher rates of birth defects. The wife of one worker remembered that since her husband had begun working with vinyl chloride, she'd had three miscarriages.[11] Before that, she'd given birth to two healthy children.

By then, however, millions of dollars were at stake. To the manufacturing companies, it made sense to fight any effort to restrain production. From the very first reports that vinyl chloride could dissolve the finger bones of workers, cause cancer in animals and deform babies, the industry had a simple response: more research is needed. Let's keep studying whether there really is a problem, while releasing enough information that people would feel assured the problem if it exists is trivial. It's a lot cheaper to set up laboratories to evaluate chemicals; it takes a lot of time to get things right. Nobody can be opposed to serious scientific investigations.

It's pretty heady for any scientist to be told that her work is so important that we need lots more of it. Funding for research on vinyl chloride and benzene began to flourish. This was not because of an inherent fascination with the problem, but rather because industry figured one thing out. If we keep providing funds to study a problem, researchers become invested in keeping their studies going. More research is always needed. Science always finds more to do.

Just as with tobacco and asbestos, the attempt to weigh the hazards of vinyl chloride had to overcome a series of carefully crafted, well-honed efforts to obstruct information and intimidate those who questioned what was going on. First it was argued that case reports of injured workers or of women with lumpy lungs weren't sufficiently rigorous. The accumulation of sick people with detailed, well-established histories might be troubling, but these alleged harms shouldn't be deemed sufficient proof to require major changes in vital national production systems on which the military and industrial foundations rest.

The only way to determine if there is a true problem is to conduct detailed, painstaking experimental studies. But as soon as experiments with rodents revealed serious health problems, they were challenged as not relevant to humans. Finally, enough time passed that public health studies could be carried out on large numbers of people who had well documented exposures. These in turn had to overcome numerous statistical and procedural objections, in addition to facing court challenges on the grounds that studies of large groups of people were "mere statistics." In the meantime, thousands and sometimes millions of people continued to be exposed to conditions that had been known decades earlier to be dangerous:

Pete Gettelfinger started working with vinyl chloride at the B.F. Goodrich plant in Harlan, Kentucky, on September 9, 1954.[12] In the fall of 1973, as his wife, Rita, recalled, Pete's first test for cancer came back showing something wrong. They dealt with this news as hard-working people often do, hoping that if they didn't talk about it, perhaps it would just go away. Pete and Rita both kept going to work at the plant and did not say a word to each other or to their kids.

Early the next year, the family was sitting together after dinner watching a television news broadcast. Some workers in Pete's factory had cancer, the reporter said. The room got quiet right after this report.

RITA: . . . So we were sittin' here watchin' the news and one of the children turned around to their daddy and says: "Are you one of 'em?"

PETE: Well, I didn't feel bad. At that time, I didn't feel bad. Two of my liver function tests did not look good, so they decided to get me out of the vinyl chloride building. And Dr. Creech said, "Ray," he said, "I'm going to put you in the hospital." The last day I worked was the 22nd of February, and they put me in the hospital on the 23rd.

RITA: He went to work then on a Monday, and he came into the office where I work about ten o'clock and told me they were going to have to take a look at his liver. And I think it's one of these things—no matter how well you think you're prepared . . .

On March 1, surgeons performed an exploratory operation on Pete and found angiosarcoma.

PETE: I made up my mind, before they took me to the operation room, whatever will be will be and I will accept. I will accept whatever they might discover. I prepared myself for it bein' real good and I prepared myself for it bein' real bad. It turned out, in the eyes of most, to be real bad. No, I wouldn't say that. I'd say it turned out, in the eyes of most, to be impossible. In my opinion it was real bad. But I still think I've got a chance.

RITA: And I mean no matter how you tell yourself that I'm ready for this—I sort of thought, you know, swell, I'm going to be like the Rock of Gibraltar no matter what comes. Well, the Rock of Gibraltar sort of crumbled.

Even with the knowledge of how dangerous Pete's job was, he still supported it.

PETE: I took pride in the kind of work I was doing. Definitely. Definitely. Definitely. 'Cause I can say this myself: from the raw material on the loading stations, all the way to sending the finished product out, I have done all those things. I've seen it done good and I've seen it done bad . . .

Every one of the men who's died so far, I knew. All of those who died were friends. Two of 'em I worked with from the first day I went to work there. They had angiosarcoma. I worked with them half my life.

RITA: It sort of sends a chill up and down your back. I mean, every one of them he had either been to their funeral or to the funeral home to see them.

PETE: I think about this a lot—it's helped me a whole lot—the fact that we got 6,500 guys in the United States makin' a livin' workin' with polyvinyl chloride in the form I was usin' it in. And I'll bet you we've got a million that are makin' a livin' in plastics. And I feel that our industry must survive. And I'll cite to you an example. The whole petrochemical industry is now maybe fifty years old, just since they made their first little thing. And the plastics industry, you might say is thirty or thirty-five years old. And it has economically given millions of people things that they couldn't have. If it wasn't for plastics, now, the price of wood would be so expensive that the average man couldn't afford to have a rockin' chair like this one on where I'm settin'. And yet that's killin' people. It may kill me.

But look how much safer it's made an automobile or the wiring in your home. So far now they only got twenty-eight dead [worldwide]. And they got two or three they don't know what's goin' to happen to them. But the industry is tremendously important. That's why it's so important for the industry to survive—for the employees that work with it. You've got to look at everyone's viewpoint. I can't say that I'm one of the lucky guys, but I must say that as long as we have put products on the market that has helped the average person economically, that must be weighed against all that's bad too.[13]

Gettelfinger was born in 1931. He was diagnosed on March 1, 1974, and died just about one year later, March 11, 1975. Too weak to work when he gave the interview but believing he would be the lucky one, Gettelfinger never imagined he would die half a year later. We have to ask, What did the company know about the dangers of vinyl chloride at the time Gettelfinger was working? When did they first know it? Information from various companies' files compiled by Gerald Markovitz and David Rosner from legal records assembled by the attorney Billy Baggett Jr. in Lake Charles, Louisiana, and by Larry Agran from interviews with workers and public health experts, make it clear that some large companies—B.F. Goodrich among them—understood the risks of vinyl chloride by the early 1970s.

In fact, the control of vinyl chloride became a bona fide regulatory victory. In April 1975, the official OSHA limit for polyvinyl chloride dropped from 500 ppm—a level that industry knew was hazardous—to one. Emissions of the gas that had previously been released into the community were captured. Contrary to dire warnings that this would mean the end of the plastics industry, business boomed.[14] But the deceptions continued.

More than twenty years later, with *Deceit and Denial*, their detailed history of the rank and sordid history of vinyl chloride, Markovitz and Rosner got the attention of twenty of the biggest chemical companies in the world—including Dow, Monsanto, B.F. Goodrich and Union Carbide. The chapter of their book that especially rattled these companies was called "Evidence of an Illegal Conspiracy by Industry." The title came directly from a 1973 memo by the Manufacturing Chemists Association's lawyers, warning that concealing evidence of the connection between vinyl chloride and cancer "could be construed as evidence of an illegal conspiracy by industry if the information were not made public or at least made available to the government."[15]

Their book detailed the cynical machinations of the companies in keeping the lid on public reports of the rare and deadly deaths affecting the men who worked as pot cleaners, chipping off fresh polymer residue from deep within the inside surfaces of six- by ten-foot tanks, with only a two-foot-square opening for air at the top.[16] The sources of information that these historians used included hundreds of thousands of pages of internal corporate documents. The National Science Foundation underwrote their efforts to organize the information. For their labors, Markovitz and Rosner found themselves at the center of a major lawsuit.

In an effort to quash public discussion of the book, these firms filed a "slap suit," or strategic lawsuit against public participation (SLAPP). This is a lawsuit filed for the purpose of rattling people that forces them to spend time and money defending themselves. The suit charged that these two academic historians had committed fraud. They had sullied the reputation of honorable companies. How had they done this? By reconstructing negotiations, using the industry's own documents, showing the three-decade struggle to keep the public unaware of, or confused about, the dangers of vinyl chloride. In the academic world, a charge of fraud is comparable to one of manslaughter. Defending against this accusation in the university world can cripple anyone's ability to function. This sent a strong message to anyone writing books on the duplicitous history of the chemical industry: Mess with us and we'll mess with you. Subpoenas from the industry demanded not only the records used in writing this book—many of which came directly from the plaintiffs' own files of company documents—but those of the University of California Press, which published the book, the book's academic reviewers, and the Milbank Memorial Fund, a nonprofit research organization that had supported the authors' work.[17] At this point, the suit has become moot, as the facts that Rosner and Markovitz reported have become clearer. Despite the intimidating intention of the suit, the authors prevailed and the case was dropped. The history of vinyl chloride is just as sordid as the authors indicated.

In 1979 top scientists were brought to Lyon, France, to review public information about vinyl chloride for the World Health Organization's International Agency for Research on Cancer. Amassing all that was publicly known, the committee declared that "vinyl chloride is a human carcinogen. Its target organs are the liver, brain, lung and haemo-lymphopoietic system . . . there is no evidence that there is an exposure level below which no increased risk of cancer would occur in humans."

Less than a decade later, in 1987, the IARC conducted a second assessment, extending the first. This report looked at more recent information and concluded that besides fatal angiosarcoma of the liver, vinyl chloride caused a number of other tumors in workers, including other types of liver cancer as well as cancer of the brain, lung and bone marrow.

Recognizing that this was a shot across its bow, the chemical industry sought the best counsel it could afford—one of the world's top epidemiologists, Sir Richard Doll of Oxford University. The next year, Sir Richard published his own analysis of the compound in the *Scandinavian Journal of Work and the Environment* in which he argued that angiosarcoma of the liver was associated with vinyl chloride, but that other, more common cancers of the brain and liver were not.[18] This paper pulled together information on worker health from four studies and ignored others. Although this limited analysis found more cases of brain cancer than expected (29 observed versus about 20 expected), Doll dismissed this additional risk as "not statistically significant" and concluded that there was nothing to suggest that these added brain cancers are occupational in origin.[19] The paper did not mention that Doll had performed this work as a consultant[20] to the Chemical Manufacturers Association.[21]

As a result of this analysis by an eminent authority, workers who developed the more common tumors of the brain, liver, and lung after exposures to vinyl chloride, whether in Italy, America, France, Japan or elsewhere, were not able to get compensation for them. Not until 2000 did it become known that Doll's efforts on vinyl chloride had not been the independent musings of a disinterested expert. A letter found after his death in 2005 indicated that Doll had served as a consultant to Monsanto since at least 1979, at a fee of $1,500 a day.[22]

What is the regulatory status of vinyl chloride today? Some nations, like Sweden and Thailand, have banned it entirely. The European Union is trying to phase it out. But business is booming in China and India. In the United States, the Vinyl Institute (VI) awards member companies for improved environmental performance. Its website proudly notes that in 2006, the CertainTeed plant in Sulphur, Louisiana, and the Shintech plant in Freeport, Texas, won VI's Environmental Excellence Award for the fifteenth consecutive year. The 2006 award is based on outstanding performance in 2005, under the U.S. Environmental Protection Agency's national emission standard for hazardous air pollutants (NESHAP).

EPA regulations require PVC resin and vinyl chloride monomer (VCM) plants to control, measure and report vinyl chloride emissions from production processes and products, including air and water releases from residual VCM stripping operations, VCM emission points and reactor openings. In addition, plants must identify and repair leaks from equipment and pipelines in their facilities and maintain records on those repairs.

"The vinyl industry is strongly committed to keeping our workforce safe and our environment healthy, and demonstrates that by investing heavily in training, technology and process improvements," said Tim Burns, VI president. CertainTeed and Shintech, in particular, have been models of safety and stewardship for 15 years, and I am proud to honor them again."

The nineteenth-century English scientist Michael Faraday is best known for divining the laws by which electricity works. In one of his lesser-known moments, in 1825, he extracted something he named "bicarburet of hydrogen" from a gooey mixture of oil and gas. A few years later, the German chemist Eilhard Mitscherlich distilled this same compound and named it *benzin*. In 1845, an English chemist, Charles Mansfield, found a way to pull benzene out of gobs of coal tar on a large scale.

Sweet smelling, seductively intoxicating hydrocarbons can be alluring, the stuff of dreams. Benzene proved close to miraculous at dissolving tars and other sticky industrial residues. It sliced right through tough, greasy grime and made paints that dried quickly. For some years scientists couldn't actually figure out how the molecule hung together. After years of pondering how benzene could exist, Baron August Kekulé von Stradonitz awoke with a vision. According to an apocryphal story, one wintry night in 1861 a snake wrapped itself around the sleeping brain of this Bohemian noble. He awoke with the idea that benzene took the structure of a serpent swallowing its tail.

From the start, scientists understood that benzene had unusual properties, both good and bad. It remains one of the most effective solvents ever found. This same capacity to penetrate and dissolve much of what it comes into contact with makes benzene one of the more toxic agents of modern life. Oils and greases are not the only things it can take apart or slip right through.

The danger of benzene, like that of vinyl chloride, remained out of sight for years, despite ominous warnings from some of the first people to work with it. Most folks in the nineteenth century just didn't live long enough to get cancer. By the twentieth century, industrialization had developed so broadly and cancer had become so prevalent, that the idea that benzene and cancer might be related was rejected by most people in positions to do so.

Like many modern hazards, the story of benzene, like that of vinyl chloride, has played out behind doors that have remained shut for

nearly one hundred years. In many ways these histories are a sophisticated game of scientific hide and seek: the complexities of science become a shield against the suspicions of consumers and workers. No
one in her right mind could ever oppose efforts to study a problem.
You want to be careful not to act too quickly, especially when the economic risks are much graver than purported public health concerns.

In 1862, not quite two decades after it was first synthesized on an
industrial scale, benzene was described in the *Lancet* as a "new domestic poison." People soon learned that this was not a material to be trifled with, domestically or in any other setting. Benzene could stupify
those who breathed too much of it for any period of time. Early reports of any industrial hazard generally ascribed blame to the foolish-
ness of those trying to use it.

The tools for depicting the ways that any exposure affected human
health were themselves changing. When benzene first emerged as a
mainstay of industrial production at the turn of the twentieth century,
most modern statistical methods for assessing hazards had not yet
been developed. The solid clinical case report was the chief way that
medical experts exchanged information on such matters.

Like most pioneers in medical research at the time, Alice Hamilton studied in Germany, after completing her medical studies in
America. In Germany she came in direct contact with the great manufacturing revolution underway for coal- and petroleum-based
chemicals. Huge new factories on the Rhine and elsewhere were
producing an array of compounds, including benzene, toluene and
other aromatic hydrocarbons.

In an important series of articles published in 1916 and 1917,
Hamilton provided the first comprehensive account of the health effects of benzene. Her work is peppered with references to case reports of benzene poisoning found in German, French and other
national journals, along with American cases reported at Johns Hopkins in 1910. She depicted, without comment, studies on Hungarian
prisoners who were given benzene as an anesthetic. With small doses,
most of them felt dizzy, nauseated and short of breath. Larger doses
caused sleep and anesthesia, followed by nausea, vomiting, headache,
dizziness, depression and drowsiness. As with many hazards, workers
already knew what they were up against. Those employed in dry-

cleaning establishments where benzene was used talked about the "naptha jag," which Hamilton said resembled being mildly drunk.

Hamilton recounted that other physicians had figured out the power of this widely used solvent. "A physician in a town where there is a large rubber factory told me of such a case in his practice. The patient, a strong man, had been dipping wooden forms in a tank filled with a solution of rubber in benzin, to make seamless surgeons' gloves. He felt dizzy and ill and left work, but on his way home he staggered and would have fallen, had not two men helped him. Later, when in bed, he lapsed into unconsciousness and when the physician saw him he was comatose, very pale and almost pulseless. He recovered completely.[23]

Some weren't so lucky. Assigned to clean out vats and tanks that had held benzene, workers would often crawl inside. Overcome by the fumes, sometimes they never got out. Young girls proved especially vulnerable. Hamilton reported that in Sweden in 1897, nine young women who worked in a bicycle tire factory in Uppsala developed bruised spots on the skin and a tendency to hemorrhage, sometimes excessively, from the nose, mouth or gums. One had uncontrolled menstrual bleeding. Four girls died after being sick less than a month.[24]

In 1917, Hamilton visited forty-one plants for the U.S. Bureau of Labor Statistics where she investigated the production of explosives—a matter of vital national importance. Hamilton found problems much like those the European doctors had reported.[25] People working with benzene-based compounds sometimes ended up poisoned. Sometimes they bled to death from massive hemorrhages.

At first, benzene was one of the by-products of making coke from coal, tapped off of the oven fumes directly. After the Second World War, the need couldn't be met by this old technology. Plastics, including polyvinyl chloride, demanded so much benzene as a precursor chemical that another technology was required to produce it. Today, most benzene is reformulated by cooking petroleum hydrocarbons with hydrogen gas under pressure. Between 1919 and 1940, at least thirty-three publications advised replacing benzene with safer solvents. Many were issued by the National Safety Council and Dr. Hamilton.

In 1948, the American Standards Association, a group of industry experts, maintained that a person could safely be subjected to 100 ppm of benzene over an eight-hour work period. It's not clear where they got this notion. But it is clear that it made no sense even at the time. Whoever was running the American Petroleum Institute understood this, and offered a radical statement in the *API Toxicological Review* the same year: "Inasmuch as the body develops no tolerance to benzene, and as there is a wide variation in individual susceptibility, it is generally considered that the only absolutely safe concentration for benzene is zero."[26]

MARVIN LEGATOR BEGAN working in the petrochemical industry in the 1950s with a small firm that got bought out by Shell. He left that post to work for the Food and Drug Administration, believing this would be the heyday of toxicology. He found himself full of ideas and short of funds.

In 1969, along with Alexander Hollender, Samuel Epstein and Joshua Lederberg, then at Harvard, Legator founded the Environmental Mutagen Society. EMS was set up to promote the study of the ways in which the environment caused genetic damage. The group also weighed in directly on such matters as why the country needed a federal law to control toxic chemicals. At one of the early meetings of the EMS at Brown University, Legator remembers, "I bumped into Paul Calabrese, who was head of the cancer group at Brown University. Suddenly I find myself needing funds for research. We had a meeting where Jack Killian, medical director of Dow, talked all about how we really needed to monitor workers on a continuing basis. I was really impressed by this. I thought, 'Boy, we can really find out whether benzene is getting to the bones of these guys!' Well, sitting there listening to this with me is Sam Epstein. Sam was furious. He turns to me and demanded, 'How dare the greatest fetus killers in this country tell us how they are going to save lives?'"

The question of how best to study workplace hazards was not fully resolved at that time. Studies in Italy had shown that printers and shoemakers exposed to benzene had twenty times more leukemia than men without such exposures. Peter Infante, starting with the

Ohio Health Department and continuing with various agencies of the federal government, kept finding evidence that workplace hazards, whether benzene or vinyl chloride, didn't just stay within factory walls. Children born to workers had greater numbers of birth defects. Some of them even had higher chances of developing cancer within the first five years of life. Groups of men with well-documented exposures showed up with well-documented cases of leukemia.

In 1970, Bernard Goldstein, a young investigator working at the Institute of Environmental Medicine at New York University, was asked by the American Petroleum Institute to examine the world literature on benzene, including all experimental and workplace studies and case reports. His review reached a straightforward conclusion: benzene caused leukemia. After that, Goldstein says, "API refused to fund us. They had all sorts of reasons, but basically, that was the end of their funding. At no time did anyone ever say that because you've found this, we can't fund you. I got no pressure to change my publication. Of course, I did get lots of questions to justify it. And I got no more funding from them to do this work." At the time, Goldstein's analysis of the accumulated scientific record was one of the most valuable records available on a public health hazard.

Despite hearing stories like Goldstein's, Legator always saw the best in people. He didn't share Epstein's cynical attitude at all. In his work on genetic damage, Legator believed that solid experimental work would show the way toward resolving matters of environmental risk. He found the folks from Dow Chemical to be enthusiastic and genuinely interested in basic research in workers.

"When Killian made me the offer to work with him, I jumped right to his defense. I thought he was a regular guy. Sam called me the enemy. Jack offered me a really well-funded consultantship to Dow Chemical to do toxicology at the University of Texas Medical Branch. I leapt at the chance."

For a few years, Legator was in high scientific cotton, with plenty of money as his team carried out studies of chromosomes in exposed workers. They literally wrote the book on how to examine DNA for evidence of structural damage. Legator remembers this period fondly, up to one critical point.

In life, sisters borrow things from one another. But when this happens within chromosomes, it can lead to troubling results. "Sister chromatid exchange" occurs when related segments of a chromosome cross over and produce unhelpful duplications of things that would be better off not produced at all. Legator showed that sunlight, x-rays and benzene could all induce this sort of genetic damage.

The Dow study had begun collecting blood from men when they first began working. Eight years later, researchers took additional blood and looked at the amount of chromosomal damage that had occurred in the meantime. The results were stunning.

"We were doing fine studying all these benzene workers for several years, until we found what we were looking for. We showed that benzene really zapped chromosomes. At that point, Dow pulled the plug. Our funding ended.

"So, the next year, I'm at an EMS meeting and there's Sam. He had heard all about what happened with our work. He sits right down next to me and without missing a beat, let me have it. 'You dumb bastard, it took you eight years to figure out what I told you.'"

"What could I say? Epstein had been perfectly right all along. I was the naive one." Legator laughed as he told me this story in 2004. At the time, he was riddled with cancer. The disease was no surprise to him. "What has happened to me is no shock. I spent my youth awash in chemicals that can slice right through you. I know what they mean."

THERE'S A WIDESPREAD BELIEF that the Occupational Safety and Health Administration regulates chemical exposures in the workplace—and there was a time, under Eula Bingham in the late 1970s, when it actively tried to do so. You will even hear people complain that OSHA is overzealous. The truth is that it's been a paper tiger for decades.

Nowhere is OSHA's failure clearer than in the cases of benzene and vinyl chloride, where well-heeled efforts are under way to roll back existing standards even as this book is being written. Under Bingham in 1979, OSHA took the radical step of setting a benzene standard of one part per million—a level then believed to be as low as feasible. In 1980, the Supreme Court ruled against the standard, arguing that the agency had failed to prove that this low level would provide measurable health benefits in workers. Basically, the court threw out the idea that any exposure to a cancer-causing substance increased the risk of the disease. It demanded evidence that only time and money and epidemiologists can produce. The Court insisted that sufficient numbers of sick or dead workers had to be assembled to provide proof that harm had already happened before allowing the agency to act to prevent further harms. This decision presumed that the previous standard of 10 ppm had been set by some coherent, defensible method. It had not.

Starting in the middle of the twentieth century, standards for exposure to workplace contaminants were determined by informal agreement between industrial and government experts. As lawsuits began to mount from injured workers alleging harms due to various workplace dusts, expert groups arose to recommend appropriate exposures. The American Conference of Government and Industrial Hygienists (ACGIH) was launched with the laudable goal of crafting guidance about how much exposure to any given agent appeared sensible. In coming up with recommendations for such standards, industries shared unpublished information from their own files about airborne hazards of the workplace under their purview. Who could object to this generosity? For healthy workers, there are presumably threshold levels of exposure, below which they will not suffer harm over a working lifetime.

But who decides what these thresholds are? How are standards set? How do they factor in mixtures? Here again, we rely on experts. Of the more than six hundred different chemical threshold limit values (TLVs) of the ACGIH, one hundred rest solely on the opinion of company experts. Many of these are more than three decades old. TLVs are supposed to be "health-based recommendations derived from assessment of the available published scientific information from studies in exposed humans and from studies in experimental animals." One of out every six workplace standards is based on no studies whatsoever. What do you do where there are no published studies or even private reports on which to hang your hat? A 1988 expose, *Documentation of the Corporate Influence on Threshold Limit Values* by researchers Barry Castleman and Grace Ziem, found that many TLVs rested on "important or total reliance on unpublished corporate communications."

Most people have no idea that OSHA is a ghost and has been so for years. Nowhere is the agency's chimeric status clearer than in its dealings with workplace air contaminants. In January 1989, OSHA made the ACGIH the official arbiter of workplace standards, by the willy-nilly adoption of all of ACGIH's TLVs as enforceable standards for the workplace. In July 1992, a court declared this move illegal, basically reestablishing the even older ACGIH standards of 1968 that were originally adopted when OSHA was established by Congress.

In its twenty-five years of existence, OSHA has set standards for an average of one carcinogen a year. But many of these were developed under the administration of Bingham, or even earlier, more than a quarter century ago. As to how closely OSHA today works with industry, a recent agreement between the agency and the American Chemistry Council, a trade association consisting of large chemical companies, makes it quite clear. OSHA and the ACC are working together to:

> Promote membership in each other's program to potential
>   companies and program auditors.
> Provide expertise in the development of training and education
>   program for VPP [OSHA's Voluntary Protection Program]
>   evaluators (including VPP Regional staff) and Responsible
>   Care auditors regarding the similarities and differences in the
>   two programs and communicate such information to
>   appropriate individuals.

SINCE THE EARLY 1980s, Yin Song-nian and colleagues at the Chinese Academy of Preventive Medicine have been conducting a massive epidemiologic study of workers exposed to benzene in 12 Chinese cities. In 1989, they reported a statistically significant excess of leukemia and lung cancer, along with possible increases of liver and stomach cancer, and lymphoma. The Yin report states that "leukemia occurred in some workers with as little as 6–10 ppm of exposure."

The U.S. National Cancer Institute is now cosponsoring further research in China and expanding its scope to include shoemakers, spray painters and workers from many other manufacturing industries. Richard Hayes and Martha Linet are co-principal investigators on the project. The average exposure of these workers is 8 ppm. Paint in China still contains benzene in the range of 7–8 percent, a level comparable to 7,000 to 8,000 ppm. The numbers don't tell the full story. Paints manufactured in China have been shown to sicken people within hours and increase their lifelong risk of fatal anemia and leukemia.

In fact, studies led by the National Cancer Institute, conducted by researchers in China and the University of California at Berkeley and published in 2005, have finally provided what the Supreme Court asked for two decades ago. Looking at the workplaces and health of thousands of workers in more than 700 factories in 12 cities in China, Martyn Smith and colleagues have found a number of disorders of the bone marrow, including cancer, clearly worsened in those with greater exposures. At first, results focused on those with exposures to benzene higher than those permitted in industrial countries today—40 to 10 ppm. Recently, these researchers have found that even at exposures of 1 ppm, Chinese shoemakers with certain common genetic traits suffer a loss of bone marrow function. They have lower white blood counts, a condition that can lead to anemia and leukemia.

These studies could end up being ignored because some were all too willing to create uncertainty that might very well lead to the setting of higher levels of acceptable exposure.

MYRON MEHLMAN WAS one of Marvin Legator's frequent collaborators. A 1957 graduate of the City University of New York, Mehlman's first job was with the Army Chemical Center at Edgewood arsenal, where he ended up in charge of the lab. In 1964, he earned his doctorate at MIT. As the director of toxicology and manager of Mobil Oil's Environmental Health and Science Laboratories from 1978 to 1989, Mehlman directed the work of dozens of top ranked industry researchers in epidemiology and toxicology, in the United States, Japan and globally.

From where he stood, Mehlman was watching an epidemic in slow motion. He had a simple idea about benzene. From inside Mobil, he argued that gasoline, which could contain benzene and hundreds of other known and suspected carcinogens, should be labeled as a carcinogen. He made these statements based on research conducted by

Mobil scientists with animals and with workers that consistently showed increased risks of leukemia at levels commonly encountered by workers in refineries and gas stations.

In September 1989, Mehlman went to Japan, where he was surprised to learn that gasoline in that country could be as much as 5.7 percent benzene. Shocked, Mehlman told his Japanese counterparts, "This is extremely high and very dangerous. You have to do something about it." The managers responded that they couldn't do this because to do so would cost their refineries hundreds of millions of dollars. Mehlman retorted, "Then you shouldn't be selling it." This turned out to be one of his last official acts as a Mobil employee. That same year, Mehlman claimed that the company in the early 1980s had falsely reported his toxicological studies to company officials and to outside agencies.

When Mehlman returned from this trip, he was fired and barred from setting foot on Mobil property. He wasn't allowed to clean out his office or talk to his staff. He was charged with improper usage of postage and personnel as part of his work with Princeton Scientific Publishing Company, an activity he had been operating for several years. On May 3 of this same year, Mobil's vice president of research had nominated Mehlman for membership in the National Academy of Sciences. On September 14, he had been given a $12,800 raise. Two weeks later he was a former employee.

Mehlman did not exactly go meekly. He estimates that in fighting the lawsuit he filed for wrongful termination, Mobil spent about $20 million just for copying and filing documents. As part of the New Jersey Conscientious Employee Protection Act—the whistleblower law—he was awarded nearly $10 million under his countersuit, an amount that was upheld despite company appeals.

Even today, the battle on benzene is not over. Part of it has moved to China. Unhappy with the efforts under way by the NCI and others in China, five major oil companies, including ExxonMobil, British Petroleum, Chevron-Texaco, ConocoPhillips and Shell, are betting $27 million that they will be able to "contradict earlier claims that link low- and mid-levels of benzene to cancers and other diseases from exposure to benzene." The *Houston Chronicle* reported that a research proposal submitted to Marathon Oil stated that "the [Chinese] benzene research was expected to provide scientific support for the lack of a leukemia risk to the general population, evidence that current occupational exposure limits do not create a significant risk to workers and proof that non-Hodgkin's lymphoma could not be caused by benzene exposure."[27]

Whatever the role general exposure to benzene may or may not play for cancer overall, it's clear that two groups have unusual risks of leukemia that have not been generally appreciated until quite recently. Those who smoke or who live with smokers will incur higher benzene exposures from burning cigarettes. We now also understand that those who have the bad luck to live in homes where old underground storage tanks with fist-size holes in their rusted shells have let gasoline seep into their basements face important health risks. This is not a trivial problem, as there are more than half a million such leaking structures around the country—many in urban areas. In a five-year study of an unusual group, Evelyn Talbott and other researchers at the University of Pittsburgh found that folks in eastern Pennsylvania whose homes had fumes leaking into them from rusted storage tanks have about four times more leukemia than others without such exposures. What's extraordinary about this work is that the researchers were able to find a group with an unusual and well-documented exposure and a relatively unusual disease. Much of the time the prospects for finding such a problem are minuscule.

MYRON MEHLMAN CONTINUES to speak out on the problems of cancer-causing ingredients of fuels in this country and internationally and the need for preventive policies. Judy Braiman is now leading efforts to keep pesticides out of schools and promote local organic agriculture. Joanne Pacinelli is the chief operating officer of an international green technologies company that is working to replace toxic everyday products with safer alternatives.

Donald Rumsfeld was the Chief Executive Officer of the Searle Corporation when FDA approval was granted to market aspartame in 1981 despite objections of scientific reviewers.

15

Presumed Innocent

When yesterday's "triumph of modern chemistry" turns out
instead to be today's deadly threat to the global environment,
it is legitimate to ask what else we don't know.

—DENIS HAYES

As THE CHIEF FINANCIAL OFFICER of one of the largest medical
centers in the world, John Poretto knew that Houston had a really big
problem. Less than five years after it was built in 1977, the nursing
school at the University of Texas Health Sciences Center at Houston
was officially declared a sick building by the State Department of
Health. Fumes and odors riddled the place. The facility had become a
management nightmare.

Sick building syndrome is not about buildings that are ill, but about
what buildings can do to affect the health of those who spend time in
them. Most of us spend nine out of every ten waking hours indoors.
People at this huge Texas facility were leaving work early with blinding
headaches, unrelenting coughs, dizziness, bleary eyes. A few had chest
pains, pounding hearts and tremors. A structure intended to house
schooling in health care had itself become an environmental danger. It
would take more money to cool, heat and fix the structure than it had
cost to build it. For Poretto, this was a tough lesson. "I learned the
hard way that we had to stop what we were doing. You just can't use
the tools of the past to solve the problems they've created."

Today we understand that volatile gases can seep out of some plastic, bendable building materials. Formaldehyde and other fumes can

be released from compressed fiberboard furniture and synthetic carpets. Irritating leaks can issue from disinfectants and compounds intended to make our workplaces cleaner. Many of these agents are known to cause cancer in animals. Some, like formaldehyde, are proven to cause cancer in humans. In their daily use of these compounds, even at low doses, those working to keep the facilities free of infection were coming down with severe and sometimes disabling symptoms, placing themselves at risk of cancer and other chronic ailments years later.

What happened to that building, and to many other facilities around the world, provides striking proof that the way we build and run buildings can affect our health. But what about other modern conveniences on which our lives rest? Technologies that radically change daily life quickly become necessities. We use cell phones and other electric-powered devices to do things that once only happened in science fiction. We rely on drugs to treat learning disabilities, esophageal reflux, bladder spasms and newly created psychiatric diagnoses that didn't even exist a few decades ago. How can we find out whether these transforming technologies bring any risk to our health? If enough of us are floored by some hacking, spasmodic cough at the same time and in the same place, as can happen with sickening buildings, we can then figure out that the place itself has induced our health problems. But many modern hazards wreak slowly accumulating damage from the combined effects of technologies or medications that we can't imagine living without.

Unlike sick buildings, there are some dangerous conditions or technologies on which we depend that can't be torn down and built over. We are like the puzzled man in an old cartoon holding a box labeled "toxic rubbish," standing by three mail slots: local, out of town, and far away. We'd like to send our garbage as far away as we can, but every day the lonely planet we inhabit seems to get smaller. There is no safe place for some of our trash. Hermaphroditic polar bears and gender-confused Beluga whales are being found in the Arctic Ocean with levels of pollutants in their fat that would qualify them for burial in a toxic waste site. Children with autism or other brain defects, tumors or leukemia, and young men with testicular cancer, lower sperm count and testosterone are increasing everywhere. Fewer baby boys are being born in many industrial countries. Are these things coming to light only because some people are looking for problems? How do we figure out whether other parts of the modern world are placing us at risk of coming down with health problems that we could avoid?

To get out of its sick building, Texas first had to understand how it ended up in trouble. Poretto was not the only one to grasp the paradox of what had happened. Brian Yeoman had worked in the Texas medical system for more than a quarter century, ending up as vice president for campus development. Like many who work in large institutions moving tons of materials around, Yeoman had not connected the dots between the ways that decisions he made every day about what to buy directly affected the environment. After a trip to the backcountry of Canada left him wondering about the environmental impact of decisions he was making at the time, he began to ask questions. He learned that carpeting which the medical center was about to throw out would create a mountain of waste so high it could be seen from outer space. All of it would end up in the local dump.

"I went to Sheldon dump in Houston. It's in an African American community." Yeoman stood there and watched bulldozers pile up old box springs, TV sets, carpeting and other waste. "It put me into very intense guilt and reflection. I began asking, 'What are we doing?'"[1]

Under Yeoman's and Poretto's leadership, the UT Health Sciences Center began its own quiet revolution. It gave up using Styrofoam containers and vacuum cleaners with bags, both of which can take centuries to degrade. "We had successful recycling programs for all the easy stuff and for much of the hard stuff like construction demolition materials," Yeoman says. "We were laying down more than 6,000 gallons of persistent organic chemicals a year. When we moved to integrated pest management, we reduced that to 15 gallons."[2]

The green building revolution makes clear the power of the purse to create markets. In Texas, at the University of Pittsburgh and throughout the Kaiser Permanente system in California, those in charge of buying materials and keeping buildings running have become the next generation of environmental champions. The world's largest firm, Citibank, requires all its facilities to reduce their ecological footprint and adopt the Equator principles—a mandate for the corporate world to promote sustainable finance and operations.

These companies are doing this because they save money by avoiding cleanups and repairs while keeping their staffs healthier and more productive.

Green businesses are looking hard for environmentally safer investments and purchases. Where they don't find them easily, they are helping invent them. An entire new industry is making building materials from natural products and recycled materials. Old blue jeans can be shredded into fluffy cotton insulation. Used paper can be compressed to create work surfaces held together with natural plant resins. Floors can be made out of cork or natural rubber, or other fast-growing plants and trees that can be harvested and replanted. Walls can be painted in naturally based paints that don't leave painters or workers with red eyes, parched throats or asthma attacks. Hospital gowns can be made from recycled paper.

At Poretto's urging, the University of Texas figured out that the only way out of the large, financially draining building was to build an entirely new nursing school in a totally new way. "The finance arena is not the most likely area from which to expect creative approaches to saving the environment," says Poretto, who began his career as an accountant at the university in the late 1960s.

Poretto and Yeoman were also influenced by the Swedish oncologist Karl Henrik-Roberts, inventor of The Natural Step. Working with other public health, environmental and cancer experts in 1989, Roberts realized that physicians like him worked in separate silos that kept them from asking why so many more people were getting cancer. Doctors, Roberts said, needed to think of the entire phenomenon of cancer as like a tree: they needed to stop focusing exclusively on the diseased leaves and begin working on the roots and trunks that determined the leaves' health. They needed to consider the entire organism from its inception to its later growth. This means seeking balance and repair rather than domination and destruction. It also implies a basic commitment to changing the conditions that give rise to cancer, and doing so long before cancer can come about. And it means coming up with ways to prevent cancer from coming back by removing hazards known or believed to increase that risk.

UT's sick building was eventually torn down. The one that replaced it in 2004 has become a mecca for those of us working to create green health care facilities around the world. It uses about half the energy of a conventional building. Rainwater and waste water are reclaimed and recycled. The nurses and staff who work in the building love the natural light, the plants that soak up pollutants, and the grand sense of space. The largest green building in the Southwest, the $42 million structure features a bookstore with café, a large auditorium, and state-of-the-art modular movable labs and teaching space for close to a thousand people.

As one of the largest such buildings in the country today, the eight-story school of nursing gets one-fourth of its energy from the hottest free source in the solar system—the sun—and requires less than half the energy of a conventional building overall. Set in a region with a near-desert climate, the building also includes waterless urinals and low-flow toilets, and smart-glass coated windows that allow maximum light to come in with minimum heat. Each of these urinals can save 40,000 gallons of water a year, which translates into saving several million gallons annually for the building as a whole. The cafeteria offers organically raised local foods, uses plastic cutlery made out of potato starch or corn, and returns waste back to the growers for use as compost.

This building even looks different from the usual cold, nondescript medical center facility. It's got character. The outside is made up of a number of recycled materials: wood siding made of sinker cypress trees that lay on the bottom of the Mississippi River, panels of recycled aluminum and cement columns made of flyash (a toxic waste product of burning coal). The sensor-driven windows include adjustable louvers, tensile fabric and heat sensors that automatically determine how much light to let into the building, reducing demand for heating and air-conditioning. With many of its windows facing west, the building gets less of the searing Texas heat. Its metal exterior glistens in the afternoon sun. People enjoy walking the inviting, naturally lit stairs. Two of the three sets of stairwells are covered and outside, close to a small park. Daylight also falls into the center of the building through two large open atriums. "Throughout the building, we can be inside and feel like it's outside," said Kim Nuñez, a thirty-nine-year-old nursing student.[3]

In 2006, the American Institute of Architects Committee on the Environment designated the nursing school one of its Top 10 green projects for the year. Yet there's still a ways to go. As of March 2007, the Denton Cooley Building, across the street at the same institution, included a hospital-based McDonald's complete with french fries, hamburgers, cheeseburgers and other unhealthy offerings. One green building does not an entire revolution make. Still, it's a start.

The University of Texas and Citibank are part of a growing national and international movement. Until 1996, in many industrial countries the burning of medical trash was not controlled at all. Mixtures of discarded plastics and hospital garbage were routinely burned in local boilers. When polyvinylchloride plastics are heated they are transformed. In addition to routine gases like carbon dioxide, burned plastic can give off the first gas to be banned in chemical warfare, chlorine. These burning gases then engage in a molecular dance to regroup into new compounds. When these new molecules contain two oxygen atoms, they are called di-oxin, a chemical nickname for a complex family of seventeen distinct compounds. Not all of them are highly toxic. But one form of dioxin is the most toxic known by-product of modern life.

Most hospital administrators had no idea they were creating environmental hazards. As has often happened in modern health care hazards, a smart nurse figured it out. Charlotte Brody found herself appalled by the fact that places of healing turned out to be major uncontrolled sources of pollution. Hospitals and doctors were inadvertently breaking the basic precept of Hippocrates—*primum non nocere*—first, do no harm. Brody, a gifted organizer, working with Gary Cohen in 1996, brought together a small group with the disarming idea of persuading health care institutions that they should protect and promote the health of their local and global communities as well as that of their patients. The group first tackled the burning of medical waste in local incinerators—then one of the largest sources of dioxins—then moved on to address major toxins in health care worldwide. Today, Health Care Without Harm is a global coalition of 443 organizations in 52 countries—all focused on getting toxic products out of hospitals and shrinking the hospitals' ecological footprint.

For reasons too obvious and complex to require mention, women are often the ones to figure out that one person's trash can be another's treasure. Global Links was started by three women who were appalled at all the usable medical supplies that were being tossed out after surgery. Sterile sutures, scissors and swabs were routinely put in the garbage. Tapping volunteers throughout southwestern Pennsylvania, the group sends unused medical supplies to developing countries, where these simple items can make the difference between being able to do surgery or not. Lives are being saved with materials that would have otherwise ended up in dumps. Each year, eighteen tons of supplies are kept out of the local landfills and more than $150 million of goods are shipped to more than seventy countries. Reducing costs of shipping wastes is but one part of the greening of medicine. As Poretto points out, "Those who are the stewards of the public's money and trust must abandon the thinking that there is never enough money to do a thing right the first time, when somehow we always find money to fix the problems on the back end, where the costs are far greater."

Today, Poretto and Yeoman are enthusiastic advocates for green building well beyond the medical world. Poretto is overseeing the construction of green, hurricane-proof homes on one of the outer banks of North Carolina. Yeoman heads up a research and education group for the National Association of Educational Procurement, helping higher educational institutions reduce their wastes and maximize their cost savings while reducing the amount of natural resources they use up.

If Poretto or Yeoman had ever suggested to the University of Texas administrators in the 1970s that they ought to spend a bit more money on construction so that they could lower their costs of waste handling and repair, and avoid illness in their staff, they would have been looked at with disbelief. The bottom line for all businesses is to keep yearly costs low and income high.

A growing number of planners understand that there are times when it makes sense to spend more now to save more later on. "Life cycle costing" refers to the need to think beyond a single year to the full life of whatever material we are purchasing or using. It may be cheaper to buy a conventional lightbulb, but it costs six times more to use it and makes the local air dirtier. This is why Australia, Cuba, Canada, California and Venezuela, and many firms like Wal-Mart and Home Depot, are moving to ban inefficient conventional lightbulbs.[4]

Today at Harvard, a focus on life cycle costs is changing much more than lightbulbs. One of the world's most prestigious and richest universities, Harvard is full of impressive, decaying stone buildings that are always in need of various repairs and restorations. Recently resigned President Lawrence H. Summers came to the campus as an economist and former Treasury secretary with a reputation for turning around the economies of developing countries. He believed environmental concerns were for the rich. Like many development economists, he also felt that the best way for a country to improve its environmental record is to grow the economy to the point where people feel they have enough money to spend on a cleaner environment. If in the meantime the brains of children became permanently damaged by lead poisoning or the country's beaches were lost to erosion, these were seen as problems that could be somehow fixed later on, when more money was available.

Some engineers and architects are forcing us all to think beyond the traditional concept of buildings to understand their fit with the human ecosystem. Jack Spengler is no ordinary engineer. As director of Harvard's Environmental Engineering Program, Spengler proposed that the institution spend money to save money by creating a green fund. Money would be used to purchase technologies to reduce pollution and lower operating costs. Summers was deeply skeptical. Harvard is a wealthy school, with an endowment of well over $30 billion. There is no shortage of people with ideas about how this money should be spent. Running such a large operation, Summers looked at green issues as wasteful extravagances. In truth, some of the early ideas for promoting less polluting approaches were developed by folks who had never needed to balance a checkbook. But Spengler and others convinced him that some of the ideas made financial sense.

Today, nowhere is the new face of greening clearer than at Harvard. Within a few years, the university was setting aside $10 million each year to underwrite technologies to reduce electricity use with motion-activated sensors, capture waste heat, and recycle rain, wastewater and carpeting, with a return on investment of more than 30 percent. The first adopter of many of these technologies was none other than Harvard Business School—a place not known for taking financial risks but for teaching others how to maximize their earnings while reducing their vulnerabilities.

Some five years later, as he was leaving the institution, Summers told Spengler that the green fund had been one of the best investments Harvard had made. Today, other institutions, including my own, are following Harvard's lead. The University of Pittsburgh's Green Action Loan Fund is showing that spending money to change valves and install sensors in water systems, put timers and motion detectors on lighting and electrical equipment, use computers to set peak flow loads, and capture rain water and reroute gray water to irrigate lawns pays off in real money. But it also pays off in lowering the size of the physical world that we need to rely on in order to conduct our ordinary business—our ecological footprint.

MANY OF TODAY'S health hazards are not at all like those of the massive sick building in Texas or the inefficient old buildings of Harvard and the University of Pittsburgh Medical Center. Chronic ailments do not conveniently become evident when people suddenly succumb to symptoms neatly in one locale. Instead, cancer and neurological disorders arise over years of time as a result of assorted triggers in our lives that may reflect where we have worked, what we have eaten, and where we happen to live now and in the past. The difficulties of unraveling the complex factors that contribute to current patterns of disease cannot be overstated. But their complexity does not mean that they need to remain unresolved mysteries.

We are heavily invested in doing things as we always have, working out of places that have just been that way forever. Entire costly systems are built on wireless technologies and other electric-powered advances. We know they work extraordinarily well. We live with them, depend on them and can't imagine life without them. We hope they don't harm us. We can't imagine not using cell phones, not ordering diagnostic radiation tests to resolve medical problems or not taking greater numbers of drugs to treat disorders that didn't even exist years ago. At this point, we can't easily know whether any of these breakthrough technologies carries any long-term hazard.

Does the absence of agreed-upon proof of these potential hazards mean that they are not dangerous? There's got to be a better way to build our world than waiting for enough bodies to drop or sicken before we decide we've got a problem. We've got several looming health problems that require fundamentally different solutions than the technologies that gave rise to them. Why are more children developing cancer and learning problems? Our ability to know whether unexplained patterns of disease are linked in any way with modern technologies and medications is severely hampered by a closed system that leaves us no independent means to evaluate such growing public health problems. We are only asked to do so after a pattern of disease has become so overwhelming that it makes network news.

PHONES AND OUR CELLS

Cell phones transform and save lives. Their risks are not easily thought about, given how intimately they have redefined normal life. Wrenching calls left on answering machines by victims of 9/11 gave some a chance to connect at the last moments of their lives. Today much of the world relies on cell phones rather than land lines. The growth in the use of these phones and the towers needed to connect them remains explosive. And, the discussion of their possible risks is ominously absent.

There's no debate that driving, biking or conducting any other mechanical physical activity while talking on a cell phone is a bad idea. That's why several states and some countries have already banned such practices. But what about the long-term risk to our health? Do the towers that transmit cell signals or the phones themselves convey a risk? Does living close to high-power electric switching stations affect the blood and brains of children? What about those who repair or build such towers? What about switching stations where high-voltage electricity is stepped down and sent throughout our workplaces and homes?

Then think of Ronald Reagan and George Bush Sr.'s political adviser Lee Atwater, General Electric's Jack Welch, Dan Case, the high-powered brother of AOL founder Steve Case, Calgary business leader Clark H. Smith, writer Bebe Moore Campbell and other heavy users of the first generations of cell phones when they were first introduced. Each of these brain cancer cases spent hours with some of the early cell phones next to their skulls. So did Maryland neurologist Christopher Newman. After developing a rarely survivable brain tumor—an anaplastic astrocytoma-grade III, he filed an $800 million lawsuit against Motorola, Bell Atlantic and others. That suit was dismissed in 2002 on the grounds that science did not prove harm.

One of the problems with studies of cell phones is that the issues they are trying to understand are inherently complex. Science works best examining one thing at a time, as we do with drugs in clinical trials. The problems posed by cell phones in the real world are like huge simultaneous equations—mathematical formulas of relationships between multiple unknowns. How can you determine the role of one factor, such as cell phone exposure to the skull, when others, like diet, workplace conditions and local air pollution, are changing at the same time and at different rates? The science that was invoked in Newman's case was not the work of lab researchers conducting experiments in test tubes under highly controlled conditions, changing one condition at a time to see which triggered the most serious or severe effects. Instead, it was the ever-more perplexing studies of epidemiologists, who are forced to make sense of the real world with all of its complexity and ambiguity as it integrates the effects of multiple risk factors all at once.

Studying brain cancer is one of the toughest jobs in epidemiology because it is a rare disease, takes years to decades to develop, and impairs the very systems that might give us clues, a person's ability to recall and describe past activities and exposures that might have put them at risk. What happens to moms and dads where they live and work and from what they eat and drink can have an impact on whether children develop brain cancer. But, because the disease can take forty years to develop in adults, and because most adults with brain cancer often develop problems of speech and recall—either from the disease itself or from the treatment—and usually don't survive, we often have to interview their remaining family members about their life histories and try to figure out what could have led to the disease. Few of us really know all the good and bad things we've dealt with in our lives, let alone those of our relatives.

When it comes to sorting through the risks of cell phones, we have lately been assured that there are none based upon reports from what appear to be independent scientific reviewers. For example, researchers from the Danish Cancer Society reported in the *Journal of the National Cancer Institute* in 2006 that they found no evidence of risk in persons who had used cell phones. Headlines around the world boasted of this latest finding from an impeccable source published in a first tier scientific journal. The press coverage of this study tells us a great deal about what journalists and the rest of us who depend so heavily on these phones would like to believe.

"Cell Phones Don't Cause Brain Cancer"
—*Toronto Daily News,* December 10, 2006

"Cell Phones Don't Raise Cancer Risk"
—Reuters, December 6, 2006

"Big Study Finds No Link Between Cell Phones, Cancer"
—*San Jose Mercury News,* December 6, 2006

"Study: Cell Phones Don't Cause Cancer"
—*Albuquerque Tribune,* December 6, 2006

"Study: Cell Phones Safe"
—*Newsday,* December 7, 2006

"Cell Phones Do Not Cause Cancer"
—Techtree.com, India, December 7, 2006

But let's look at what the researchers actually studied.

They reviewed health records through 2002 of about 421,000 people who had first signed up for private use of cell phones between 1982 and 1995. A "cell phone user" in the study was anyone who made a single phone call a week for six months during the period 1981 to 1995. The study kicked out anyone who was part of a business that used cell phones, including only those who had used a cell phone for personal purposes for eight years.

This research design raises a lot of questions. Why did they not look at business users—those with far more frequent use of cell phones? Why lump all users together, putting those who might have made a single cell phone call a week with those who used the phones more often? Why stop collecting information on brain tumors in 2002, when we know that brain tumors often take decades to develop and be diagnosed?

When you are looking at a large population to find an effect, generally the more people you study, the better your chance of finding something. But if you merge a large number of people with very limited exposure together with a small number of people with very high exposure, you dilute the high-exposure group and so lower your chances of finding any effect at all. It would be better to compare the frequent users with non-users, omitting the limited users altogether. Lumping all these various users together is like looking all over a city for a stolen car when you know it's in a five-block radius. Perhaps you'll find what you're looking for, but the chances are greater that you won't. It would be far more effective to limit your search to the five-block area. The Danish study was designed to look definitively thorough—421,000 people!—but in fact it was biased against positive findings from the start. Given how broadly cell signals now penetrate coffee shops, airports and some downtown areas of major cities, it is very difficult to find any truly unexposed groups against which to compare results. Because cell phone use has grown so fast and its technologies change every year, it is as if we are trying to study the car in which we are driving.

Another study that was well publicized in 2000 found no increased risk of most types of brain cancer in cell phone users; but the average length of use among participants was less than three years.[5] Still, the study found that those people who had used phones for even this short period of time had twice the risk of a very rare brain tumor—neuroepitheliomatous cancers, the kind that wraps itself around the nerve cells of the lining of the brain, right at the locus that cell signals can reach.

Of course, epidemiologic research is the research that works best when we have solid information on the nature of the use or exposure we are trying to understand. All of us have cell phone bills that provide detailed records of our use, and most of these can be accessed online. These were not used in this study, nor in any study of the industry to date. A gold mine of data lies untapped that could enable researchers to distinguish non-users from low frequency users from high frequency users, thereby increasing the validity and sensitivity of studies.

Underlying this whole body of research is clear evidence that cell phone signals penetrate the brain. As the Danish researchers admitted in their own study, "During operation, the antenna of a cellular telephone emits radio frequency electromagnetic fields that can penetrate

4–6 cm into the human brain."[6] What the research seeks to determine is what this means biologically.

We know that cell phone signals can warm the side of the head, where the auditory nerve is located. An earlier Swedish study, used in Dr. Newman's case, compared more than 1,400 people with brain tumors to a similar number without the disease between 1997 and 2000. Tumors of the auditory nerve were three times more frequent in persons who had used cell phones for more than a decade.[7] In 2004, other Swedish researchers found that long-term cell phone users had significantly more tumors on the auditory nerves than nonusers.[8]

The study of chronic health problems is hardly as simple as we often presume.[9] We notice that most people with lung cancer have been smokers, or that many women over sixty who get breast cancer have used hormone replacement therapy. We deduce that a single condition gave rise to a single outcome, even though we know that life is not so simple.

George Carlo is an epidemiologist who once directed a multi-million-dollar, multicountry study of cell phones that was overseen by the U.S. government and funded by the industry starting in 1993. He was fired or resigned, depending on whose story you credit, and has continued to work on the issue ever since.

The study Carlo never completed for the industry began as a series of projects looking into whether cell phone signals disrupted cultures of animal cells growing in the laboratory. Some of the work done in laboratories clearly showed that wireless signals could affect the ways cells talk to one another to stay under control—what is called gap-junction communication. Under healthy conditions, cells send messages through proteins and enzymes that keep things in order and tell badly behaving cells to get in line or die. Wireless signals disturb this ability. Cells that can't communicate well are prone to grow out of control. In essence, wireless signals promote a kind of social breakdown among cells. Unfortunately, the implications of this work were never completed.

The human health component of the study of cell phones remains unfinished, and it may well be unfinishable. A major international study of brain cancer in wireless phone users is still underway, headquartered at the International Agency for Research on Cancer (IARC) of the World Health Organization in Lyon, France. The large study was designed to combine more than 3,000 cases of brain tumors from around the industrial world and was supposed to release its results in 2006. In Canada, Daniel Krewski, a respected epidemiologist who heads that country's national study of cell phones, receives much of his funding from the industry. Some have asked whether this constitutes bias. Krewski is also part of the IARC study.

The former director of the IARC, Lorenzo Tomatis, is concerned about the lack of independence of this important work. He complained publicly in 2004 about the close cooperation that was developing between the cell phone industry and those who were studying brain cancer that could be associated with cell phones' use. When Tomatis returned to the facility to meet with colleagues with whom he had worked, he was treated like no other former director: he was ordered to leave and security guards escorted him from the building.

More than a year after the IARC study was to have ended, it's still "under way." At the time this book was in final editing, in May 2007, the chief of the IARC study on cell phones reported that she did not have any idea when the work might be published. It is now in its tenth year.

At the core of the IARC project is a major effort to learn from brain cancer patients whether they used cell phones more frequently than did others. The limits of the work are easy to grasp. The ways to overcome them are not. Still, some German findings published just last year are disquieting.

The German study captured information about the daily lives of people in Mainz, Bielefeld and Heidelberg. What did they have for breakfast regularly? Where did they live? How often did they use the cell phone? For how long? On which ear? These are the sorts of things epidemiologists like me hope you remember. This work contrasted the life experiences and reported cell phone use of 366 people with deadly tumors of the brain called gliomas and 381 with slow-growing, usually benign tumors of the membranes that cover the spinal cord, against some 1,500 people between the ages of thirty and sixty-nine who had better luck and did not have brain tumors. When asking both groups about their past and current uses of cell phones, they did not find any increased risk in those who used phones for less than a decade. But, those who reported having used cell phones for ten years or more had twice the risk of coming down with gliomas.[10] This is a tumor that begins in the glial cells of the brain, the cells that support neurons and hold them together. The growth of gliomas can be silent, with symptoms that mimic flu or a headache. But eventually, they become undeniable. People lose speech, sight, movement or hearing, depending on where the tumor starts and where it ends up.

It should be obvious that looking at people with a fatal illness and asking them to try hard to remember what they did up to forty years ago is not easy. With all the information governments now assemble to combat terror, including library and cell phone records, what would it take for those powers to be directed toward learning whether our use of cell phones places us at risk from a disease that could be averted through better design and technology?

That's not a question likely to get much attention at this moment, but it is well worth asking. The studies to date that have not found a general, clear and consistent risk from cell phones need to be understood as tentative. They have for the most part looked at older technologies over short periods of exposure. None is asking about the impact of cell phones on the brains of children and teenagers—one of the fastest growing groups of users in the world today. The governments of England, Israel and Sweden advise that those persons under eighteen should not use cell phones at all. American toddlers learn to play with toy versions of them.

What makes this especially troubling are the results from several other studies that have looked at more recent regular users. After a decade of heavy use, cell phone users have double the risk of brain cancer. The tumors tend to occur on the side of the head that the user typically favors.[11]

Another, entirely different set of data on electromagnetic fields, exposures of which cell phone signals are but one type, comes from looking at an illness even more extraordinarily rare than brain cancer—breast cancer in men. The total number of cases of male breast cancer in the United States today is thought to be less than 4,000, but some 1,400 new cases are reported each year, according to the American Cancer Society. Studies of men who work with electromagnetic fields in radio and television or in assembling cell towers have found that they have much greater risks of breast cancer as well as cancer of the brain.[12]

The table on pages 408–409, from the National Institute of Environmental Health Sciences, shows the relative risks found in studies conducted on breast cancer in men (and some women) working in jobs involving regular contact with electromagnetic fields. These risks contrast the amount of cancer found in those with known exposures compared to those without such exposures. Of course, there can be no completely unexposed group for comparison purposes in the workplace. Both men and women have greater rates of breast cancer if their jobs involve heavy exposure to electromagnetic fields.

Men typically do not get breast cancer, and when they do, the disease is often much more difficult to treat. Still, this table shows that for many professions involving work with electronics, men have between two and four times more breast cancer than those without such experiences.

Electricians, railway workers, telecommunication line workers—in a striking number of industries men have greater risks of breast cancer. How can we know whether electromagnetic fields are responsible for these differences? Perhaps they all work with solvents or other chemicals that are also associated with giving men breast cancer? As with all workplace hazards, we can only draw conclusions from the totality of information at hand.

What about the rest of us? What about children who live near power lines or cell towers? What about the growing number of young children and teenagers who have their own cell phones, despite the fact that Scandinavia and many other countries warn that children should not use cell phones at all? We hope that today's digital technologies are safer than the older analog phones and other wireless systems. The Cellular Telecommunications and Internet Association reports that in 2007 there are more than 180 million subscribers in the United States, up from 110 million users just three years earlier. Experts estimate that by 2010 there will be 2.2 billion subscribers worldwide. Cell phones are becoming so ubiquitous that soon there will be no control group against which to compare their risks.

With respect to the risks to children from living close to electromagnetic fields from power stations or high voltage transmission lines, some well-respected researchers, like Dan Wartenberg of Rutgers and others from the Karolinska Institute in Sweden, believe a growing

Table 15-1 Risk of Breast Cancer and Workplace Exposure to Electromagnetic Fields[13]

Reference, Country	Cohort Description	Exposure Classification	Males		Females		Comments
			No. of Cases	RR (95% CI)	No. of Cases	RR (95% CI)	
(Tynes et al., 1992); Norway	37,945 male workers (aged 20–70) followed 1961–85 in the Cancer Registry of Norway	12 electrical occupations	170	2.1 (1.1–3.6)	NA		SIR
		Electric transport work <ISCO codes 631, 632, 641, 693)	4	4.0 (1.1–10)	NA		
(Guénel, et al., 1993); Denmark	All actively employed Danes (aged 10–64) in 1970 followed 1970–87 in the cancer registry. 172,000 men and 83,000 women in jobs exposed to magnetic fields were compared with reference workers in unexposed jobs.	Jobs with intermittent field exposure	23	1.2 (0.77–1.8)	1526	0.96 (0.91–1.0)	SIR relative to economically active subjects

(Floderus et al., 1994); Sweden	Male railway workers (aged 20–64) in the 1960 census compared with all employed men in 1960–69 (940 person-years in 1960–69 and Cancers for 1960–69 and 1970–79 obtained from Cancer-Environment Registry.	Jobs with continuous exposure	2	1.5 (0.16–4.9)	55	0.88 (0.68–1.2)	SIR adjusted for age. All cases occurred in 1960–1969 follow up period (none in the 1970s)
		Engine drivers	2	8.3 (2.0–34)	NA	
		Conductors	1	2.7 (0.4–20)	NA	
		Railway workers (and station masters, dispatchers & linemen)	4	4.3 (1.6–12)	NA	

Source: Christopher J. Portier and Mary S. Wolfe, eds., "Assessment of Health Effects from Exposure to Power-Line Frequency Electric and Magnetic Fields (NIEHS Working Group Report)," June 1998. http://www.niehs.nih.gov/emfrapid/html/WGReport/WorkingGroup.html.

body of evidence shows that there's a serious problem.[14] They aren't the only ones suggesting that electromagnetic fields be considered a possible human carcinogen. The World Health Organization officially recommends that power line siting decisions should consider ways to lower exposures and keep people out of the line of high voltage electricity and has classified EMF as a possible human carcinogen as has the U.S. National Institute of Environmental Health Sciences.[15]

The debate over electromagnetic fields and cell phones takes place on a playing field that is not at all level. Much of the research funding is provided by the telecommunications industry just as much of the research funding on more general electromagnetic field research was provided by the electric power industry. It may not surprise you to learn that the highly publicized Danish Cancer Society study that exonerated cell phones and the yet-to-be completed IARC study are directly funded by the industry.[16] Whether this affects the design of the studies and their outcome can't be determined. One group will have an answer in 2009, after they complete a long-term animal research project. What are we supposed to do while we wait for those results?

DIAGNOSTIC RADIATION IS another modern miracle we have come to depend on. The Nobel Prize in Medicine or Physiology was awarded in 1979 to Godfrey N. Hounsfield and Allen M. Cormack, the engineer and physicist who invented the system for creating three-dimensional images of the human body. Computerized imaging technology is now such a large, profitable industry that it has its own futures market. Seven times more CT scans are conducted today than just ten years ago.[17] The leading manufacturer, Cardinal Health, is one of the twenty largest companies in the world, with revenues of more than $81 billion a year.[18]

New government regulations in the United States are shutting down what had been highly profitable ventures in which physicians would prescribe tests on machines they themselves owned. When offered a three-dimensional look inside an old set of knees or cranky stomach, a patient does not usually ask whether this remarkable test might increase her long-term risk of more serious ailments.

When my then eleven-year-old daughter was given a CT scan of her abdomen to see if she had a ruptured spleen, I was just like most parents with a child in the emergency room. All I wanted to know was that she would be okay. But when I asked the young woman radiologist if there was a way to shield my daughter's chest, she looked at me skeptically and asked, "Why? She doesn't have any breasts." I explained to her that we know that radiation exposure to the chests of girls before puberty increases the chances that breast cancer will develop later on. The woman looked at me as if I were slightly insane.

Many physicians have no idea how much radiation their patients are exposed to from regular diagnostic procedures. A major pediatric journal reported that one in every three procedures ordered for children was not appropriate.[19] Infants are at greatest risk, because the impacts of radiation have that much more time to become evident.

My colleague Dwight Heron, Vice Chair for Clinical Affairs at the University of Pittsbugh Medical Center Cancer Centers and Chairman at the Department of Radiation Oncology at Shadyside, spends his days diagnosing and treating cancer pataients with the help of modern computerized diagnostic scanning systems. I asked him what he thought about the current enthusiasm for CT and PET scans of healthy people.

Heron says, "It's a big problem. Radiologists appreciate that we could be creating more cancer in young people by what happens in emergency rooms all over the country today."

Heron referred me to a surprising new advocate on this issue. In a 2007 white paper on radiation in medicine, the American College of Radiology noted that in the past quarter century, the amount of radiation the U.S. population receives each year from medical imaging has increased fivefold. A single computerized scan of the stomach today can give half the dose that was shown to induce cancer in those who survived the atomic bomb blasts in Japan. The ACR advises that "the current annual collective dose estimate from medical exposure in the United States has been calculated as roughly equivalent to the total worldwide collective dose generated by the nuclear catastrophe at Chernobyl."[20]

Let me translate this. Modern America's annual exposure to radiation from diagnostic machines is equal to that released by a nuclear accident that spewed the equivalent of hundreds of Hiroshimas across much of Russia and Eastern Europe. In 2005 the Chernobyl Forum, an organization led by the International Atomic Energy Agency and the World Health Organization, estimated that about 6.5 million people were exposed to 5.6 Röntgen per second (R/s). This is equivalent to 20,000 Röntgen per hour (R/h). A lethal dose is around 500 Röntgen over five hours, so in some areas, unprotected workers received fatal doses within several minutes of radiation from the Chernobyl explosion in 1986. Conservative estimates are that as a result of this massive explosion there will be 30,000 to 60,000 more cancer deaths that would not otherwise have occurred.[21]

Concerns about unnecessary medical radiation in young children today are now ricocheting throughout the medical community. A group of Yale researchers, looking at current patterns, estimates that in one year, 700 people will die from cancers associated with head CTs and 1,800 will die from radiation-induced cancer from abdominal examinations carried out when they were infants.[22] Reduced brain function, learning problems and lowered IQ from such potentially unnecessary and inappropriate exams is not easily calculated, but it cannot be trivial.[23]

Most physicians and the rest of us are unaware of the dangers shown in Table 15-2 from a major radiology journal.

To put these dosages into perspective, even a properly calibrated CT scan of a child's stomach can be equivalent to six hundred chest x-rays, while one of an infant's head can be equivalent to a few thousand. Imagine a lifetime of emergency room visits, with repeated scans, and it becomes clear that these risks could create a major cancer burden of the future.

Emergency physicians have not yet gotten the message. A survey of emergency room doctors at a major medical center found that none of them was aware that some of the diagnostic procedures they were ordering increased the risk of cancer for their patients thirty years later.

The risks of radiation, unlike many other cancer risks, are not based on theoretical models or experimental research with lab rats. The numbers we use to estimate the chances you will get cancer from radiation come from real people who survived the atomic bombings that

Table 15-2 Radiation Risks of CT Scans.[24]

Exam Type	Machine Setting	Relevant Organ	Approximate Equivalent Dose to Relevant Organ (mSv)	Equivalency in Chest X-rays* .15–.01 mSv
Pediatric Head CT Scan	Unadjusted[a]	Brain	60	400–6000
Pediatric Head CT Scan	Adjusted[b]	Brain	30	200–3000
Pediatric Abdominal CT Scan	Unadjusted	Stomach	25	166–2500
Pediatric Abdominal CT Scan	Adjusted	Stomach	6	40–600
Chest X-ray (PA/lateral)	n/a	Lung	0.01/0.15	.01–.15
Screening Mammogram	n/a	Breast	3	20–300

*Chest-x-ray equivalency based on NCI estimates in this table

[a] Unadjusted: refers to using the same settings as adults

[b] Adjusted: refers to using settings adjusted for body weight

Source: Society for Pediatric Radiology and National Cancer Institute, "Radiation & Pediatric Computed Tomography: A Guide for Health Care Providers," Summer 2002, http://www.cancer.gov/cancertopics/causes/radiation-risks-pediatric-CT.

Table 15-3 Effective Radiation Doses for Common Medical Procedures[25]

For this procedure:	Your effective dose is:	Comparable to natural background radiation for:
Abdominal Region:		
Computed Tomography (CT)-Abdomen	10 mSv	3 years
Computed Tomography (CT)-Body	10 mSv	3 years
Intravenous Pyelogram (IVP)	1.6 mSv	6 months
Radiography-Lower GI Tract	4 mSv	16 months
Radiography-Upper GI Tract	2 mSv	8 months
Central Nervous System:		
Computed Tomography (CT)-Head	2 mSv	8 months
Chest:		
Radiography-Chest	0.1 mSv	10 days
Computed Tomography (CT)-Chest	8 mSv	3 years
Children's Imaging:		
Voiding Cystourethrogram	*5–10 yr. old:* 1.6 mSv	6 months
	Infant: 0.8 mSv	3 months
Women's Imaging:		
Mammography	0.7 mSv	3 months

Source: http://www.radiologyinfo.org/en/safety/index.cfm?pg=sfty_xray.

ended World War II in the Japanese cities of Hiroshima and Nagasaki. Of the estimated 600,000 people affected by the atomic bombs, fewer than 270,000 are alive today. There is a possibility that this record has created an interesting bias. Those who died of radiation sickness relatively quickly were probably weaker than those who survived the blast. As a result, the grounds on which we base our estimate of radiation-related cancer may tell us what happens to healthy survivors but not their far weaker neighbors who succumbed to the blasts.[26]

Of the more than 10 million cancer survivors in this nation, those who underwent extensive radiation to treat or find their disease, like Elizabeth Edwards, wife of Sen. John Edwards, and Tony Snow, press secretary to the White House, or those who had the disease as young children, face lifetime risks of other cancers as a result. Other studies show that the risks of cancer from radiation in cancer patients treated for Hodgkin's disease could even be greater than those of the atom bomb survivors. This apparently greater vulnerability of the weakened to the damaging effects of radiation is something that researchers like Alice Stewart and Rosalie Bertell warned about nearly half a century ago.[27] The world is catching up with them.

Stewart's work on the dangers of radiation in England was simple and powerful. She visited every county and county borough health department in the country, handing out questionnaires that asked mothers of children born between 1953 and 1955 about things that happened to them when they were pregnant. Within a year, she had determined that the mothers of leukemic children were three times more likely to have had routine x-rays of their abdomens during pregnancy. These results, published in the *Lancet* in 1956, flew in the face of assurances from obstetricians that the practice was harmless. Stewart's findings also upset those advocating the continued use of nuclear weapons and testing. The year 1956 was the peak year for above-ground nuclear testing and radioactive fallout. Obstetricians and nuclear weapons advocates alike maintained that small doses of radiation were harmless. In fact, Stewart's findings showed that a single dose of diagnostic x-rays early in pregnancy more than doubled the child's risk of leukemia.[28]

In 1960, Richard Doll and William Court-Brown published a study of patients who had been treated with x-rays for ankylosing spondylitis (a spinal deformity) and concluded that medical radiation had been harmless. Because of that work, medical x-rays for this condition and for prenatal evaluation were not officially discouraged in Great Britain until 1985 by the National Radiological Protection Board.[29] In the United States the American College of Obstetricians and Gynecologists warned against routine diagnostic x-rays of pregnant women in 1980.[30] The public understanding of the perils of radiation shifted as this book was being completed. Janette Sherman detailed some of these recent revelations:

The death in November 2006 of the Russian dissident Alexander Litvinenko, a British citizen, made headlines around the world as he succumbed to the alpha radiation given off by a minute amount of Polonium 210. Polonium is in the same family of elements as sulfur, selenium and tellurium and goes to those parts of the body that normally take up those chemicals. . . . The radiation damage to Litvinenko was relatively brief before it killed him. Death from radiation exposure is rarely as swift. If the result of exposure is cancer, sickness and debility can extend for years. In the interim, treatment can involve surgery, pharmaceutical chemicals and ironically, more radiation. Still there is no guarantee of cure, and the cancer plus the "treatment" can severely disable or kill the person. . . .

The U.S. Atomic Energy Commission and its successor, the Department of Energy (DOE), denied any harm from bomb fallout for years, but in 1997 the National Cancer Institute released a report that showed doses of Iodine-131 (I-131) more than 100 times greater than earlier government estimates. The report had been completed in 1992, but five years elapsed before the Secretary of Energy, Hazel O'Leary, released it. The massive 100,000-page report estimates exposure to I-131 from the Nevada bomb tests of the 1950s and 1960s. The data are broken down according to place of residence, birth date, gender and milk consumption. The DOE admitted that sufficient radioactive iodine had been released from nuclear tests to account for between 11,000 and 212,000 Americans developing thyroid cancer. In 2001, Joseph Mangano of the Radiation and Public Health Group (RPHP) received a call telling him that collected, but untested, baby teeth had been discovered in storage at Washington University in St. Louis. They were ultimately transferred to RPHP, and preliminary publicity about the teeth resulted in several hundred contacts from people who had donated teeth as children. Many reported cancers in themselves and in their children. The most common type was thyroid cancer, which is strongly linked to bomb test fallout. But with no funding available to test the remaining St. Louis teeth, an opportunity to evaluate the impact of radioactive fallout in Americans was lost. With the nearly four decades that have passed since the study ended, and with the ability to obtain health information and death records via internet contacts with tooth donors, it is the perfect time to complete the study.[31]

Most physicians today don't know that a typical CT scan of the chest can be equivalent to four hundred chest x-rays.[32] This is not an inconsequential number. Your risk of dying of cancer from a chest CT is comparable to the chances that you will die in a car crash by driving 4,000 kilometers.[33] While the risks of driving are well known and the subject of major public relations campaigns, those of CT scans remain secret, and those of nuclear energy or the uses of depleted uranium remain masked in mystery.

ANYONE WHO HAS ever struggled with a learning disabled child understands the urge to manage the problem with medication. Children can literally bounce off the walls and furniture of their homes, injuring themselves and their caregivers. Teenagers with untreated learning disabilities tend to fare poorly in school and in the community. Often they end up in legal trouble or in jail.

Some people have asked whether the rise in these problems is tied to exposure to heavy metals, such as lead, to modern chemicals, or to endocrine-disrupting chemicals that can affect the brain. The brain doubles in size in the first two years of life, and dulling metals can impair irrevocably the ability to see, hear, think and learn for the rest of people's lives. Rats exposed prenatally to just a single injection of some pesticides and other toxic chemicals are more excitable and less focused than others.[34] There is even evidence of lifelong damage to the brains of children exposed early in life to lead or methylmercury— compounds that are by-products of industrial activities, including coal burning. Other work has shown that mothers with higher levels of some chlorinated chemicals in their blood when pregnant bear children who develop many more problems learning, paying attention and growing up than those without such exposures.[35]

Whatever causes learning disabilities, it's clear that prescription drugs have become the principal treatment. America today uses most of the Ritalin consumed in the world. In some school districts, more than 10 percent of all children are on it at some point in their lives. They take Ritalin to help them focus.

We may have a much bigger problem than the ability to focus. Several papers have come out indicating that Ritalin may pose extraordinary risks to our genetic makeup. One paper, from a team of researchers at the M.D. Anderson Cancer Center at the University of Texas in Houston and the University of Texas Medical Branch at Galveston, studied the blood of a dozen children who were eight years old before and after they were placed on Ritalin. They looked at three different measures of damage to chromosomes that can be tracked. Their results were so stunning that they halted the study and took the children off the drug.

On each measure, after just three months on the prescribed dose of Ritalin, the children had between three and four times more genetic damage than they had when they started out. This damage was measured by examining the shape, size, and structure of their chromosomes. Coming from the two Greek words meaning color, *chromo,* and body, *soma,* a chromosome is a large building block of genes and proteins that determines our capacity to repair and survive various threats. Healthy cells are able to repair all sorts of injuries that occur just from being alive. The white blood cells of these children showed a major loss in this ability. This was not some incidental dink to genes but the sort of defect in genetic operations that has been found in adults to lead to an increased risk of cancer.[36]

This study followed only a few children. But each of them showed the same damage to their chromosomes after being on Ritalin for just three months. Whether these risks occur in only a small subset of all children is not known at this point. It could be that there was something special about these children that suddenly caused a deficit in their bodies' ability to repair themselves.

If this finding of increased genetic damage were the only indication of a problem, it would be troubling enough. But in fact, the National Toxicology Program studied Ritalin in mice and rats and found a significant increase in liver tumors in male and female mice, though not in rats. Other work that the Texas team has heard of but not been able to see has also apparently found evidence that Ritalin has lingering effects.

The Texas group's senior investigator, Randa A. El-Zein, told me that shortly after their report appeared in Cancer Letters, the team got some unusual visitors.

A group of government experts from FDA, EPA and the National Center for Toxicology Research all flew down to talk about what they had found. They were told by the scientist leading the group, "The data are very interesting. We can't turn our eyes and say we don't have an effect. We will get back to you quickly." This sounded perfectly appropriate. Given the importance of Ritalin for many children and the disturbing implications of their findings, the responsible thing to do is to study more children soon and see whether this is a broader problem or some unknown condition peculiar to the children in their study.

Chemotherapy drugs can induce hyperactivity because cancer-treating drugs often frazzle the nerves. Many children who are on chemotherapy are also given Ritalin. A number of chemotherapy agents, including radiation, are known to increase the risk that other forms of cancer will develop. El-Zein worries about the combined impact of Ritalin and chemotherapy. As far as she knows, nobody is looking into this.

So far, El-Zein and her group have heard nothing. They keep hearing rumors that other groups are studying the problem, in Germany perhaps. One study of 28 children placed on Ritalin tracked 8 of them for a few months and found no evidence of chromosomal damage.[37] With 10 million prescriptions filled each year, you would expect a major effort to evaluate the long-term impact of this widely used drug. The FDA has announced no plans to require additional information. The federal grant El-Zein and her colleagues were encouraged to submit proposing an expanded study of more children was turned down.

By now you can probably persuade most people that cigarettes aren't worth the risk. But what about artificial sweeteners like Aspartame? Now one of the most widely used food supplements in the world, this chemical was first approved for limited use for diabetics, for whom glucose, the usual form of sugar, can be life-threatening.

In January 1977, FDA Chief Counsel Richard Merrill made agency history. He formally asked the U.S. Attorney's office to convene a grand jury to decide whether to indict the major producer of aspartame, G.D. Searle, for knowingly misrepresenting "findings, concealing material facts and making false statements" in aspartame safety tests. That this investigation never happened speaks volumes about the difficulty of acquiring independent information on commercially valuable products.

Two decades after this indictment had been filed, I spoke with James Olney, a research neurologist and psychiatrist at Washington University in St. Louis, about aspartame's early history. What he told me was hard to believe. In 1969, Searle asked the researcher Harry Waisman to study aspartame in seven infant monkeys. After a year of drinking milk flavored with the stuff, one was dead and five had suffered severe epileptic seizures. In the spring of 1971, Waisman died and his research was never completed. Olney's own studies showed that aspartame paired with the food flavoring monosodium glutamate produced brain tumors in rats.

In 1973, reviewing what Searle did include, Martha Freeman, an FDA scientist, determined that the information submitted on the safety of aspartame was not adequate. She recommended that aspartame not be allowed on the market.[38] Events eventually made her advice irrelevant.

Shortly after the proposed investigation was announced, Searle hired Donald Rumsfeld, who had just left office as Defense Secretary, to be its chief operating officer. He started in early 1977. That July, the chief attorney in charge of the grand jury, Samuel Skinner, resigned and went to work for Sidley & Austin, the law firm representing Searle. The person who replaced him, William Conlon, would eventually join Skinner at the same firm.

They had their work cut out for them. That August the FDA released its official report on aspartame, called the Bressler report, that included much of the information that formed the basis for the grand jury probe. It depicted a stunning number of irregularities. In one study of nearly two hundred animals, half of them weren't autopsied until long after they had died. Have you ever found a dead mouse in a trap? After a week, what's left is dried out, shrunken and stiff. Imagine trying to figure out whether that mouse had cancer and which organs were affected. The report noted that some rats that were recorded as having died later sprang back to life.

Immediately after the Bressler report was released, the FDA formed a task force to investigate the authenticity of research done by Searle regarding the safety of aspartame. A senior FDA investigator,

Jacqueline Verrett, looked into these allegations and seconded the Bressler report's findings. A seasoned toxicologist, Verrett testified to the U.S. Senate in 1989 that the work she had reviewed on aspartame ten years earlier did not pass muster:

> At this point it might be helpful to mention some of the deficiencies and improper procedures encountered: no protocol was written until the study was well under way; animals were not permanently tagged to avoid mix-ups; changes were introduced in some laboratory methods during the study with inadequate documentation; there was sporadic monitoring and/or inadequate reporting of food consumption and animal weights; tumors were removed and the animals returned to the study; animals were recorded as dead, but subsequent records, after varying periods of time indicated the same animal was still alive (almost certain evidence of mix-ups); many animal tissues were autolyzed (decomposed) before any postmortem examinations were performed; and finally, of extreme importance, in the DKP study there was evidence, including pictures, that the diets were not homogeneous and that the animals could discriminate between feed and the included DKP. Almost any single one of these aberrations would suffice to negate a study designed to assess the safety of a food additive, and most certainly a combination of many such improper practices would, since the results are bound to be compromised.
>
> It is unthinkable that any reputable toxicologist, given a completely objective evaluation of data resulting from such a study, could conclude anything other than that the study was uninterpretable and worthless, and should be repeated.

In 1978, the journal *Medical World News* reported that the methanol content of aspartame is a thousand times greater than most foods under FDA control. In high concentrations, methanol, or wood alcohol, is a lethal poison that can cause blindness and damage the brain. Some of us humans can be uniquely sensitive to such materials, especially those who do not yet crawl or walk. While most of us can easily handle methanol and its more familiar cousin, ethanol—the alcohol we drink on social occasions—some of us can't handle it well at all.

At the end of September 1980, another FDA review board weighed in. Its three members voted unanimously against approving aspartame, noting that the FDA "has not been presented with proof of reasonable certainty that aspartame is safe for use as a food additive." Meanwhile, the grand jury investigation fizzled. So much time had elapsed that the authority to keep it going had expired. Expert legal advice—doubtlessly provided by former FDA officials who had started the investigation and now worked for the aspartame industry—had helped Searle run out the clock. Scientific evidence became irrelevant.

And then Donald Rumsfeld proved his worth. Searle's directors clearly had not hired him for his pharmaceutical expertise—he had none to speak of—but for his already legendary Washington connections. After the election of President Reagan in November 1980 these suddenly became much more powerful. Rumsfeld told a Searle sales meeting that he would get aspartame approved within the year. According to a 2006 article in the *Ecologist*, he vowed to "call in his markers" with the FDA.[39]

The day after President Reagan was inaugurated, January 22, 1981, Searle reapplied for FDA approval. Ignoring the recommendation of yet another review panel, the new FDA commissioner, Arthur Hull Hayes, approved aspartame for use in dry products on May 19, 1981. Within a year, that approval extended to liquids and vitamins.

On October 12, 1987, United Press International reported that more than ten American government officials who had been involved in the decision to approve aspartame were now working in the private sector with or for the aspartame industry. One of them was Commissioner Hayes, who had gone to work for Burton-Marsteller, the chief public relations firm representing Searle and Monsanto. (Monsanto purchased Searle in 1985. In this merger, Searle's aspartame business became a separate Monsanto subsidiary, the NutraSweet Company.)[40]

The U.S. military was not sanguine about aspartame's safety. Both the U.S. Air Force magazine *Flying Safety* and the U.S. Navy magazine *Navy Physiology* warned that aspartame can cause serious brain problems in pilots.[41]

Around 1995 the FDA stopped gathering adverse reaction reports. If you don't want to know, don't ask. By 1996, aspartame was approved for all uses.

What about all those studies finding aspartame safe? In 1996 Ralph G. Walton, a professor of clinical psychology at Northeastern Ohio University, surveyed them for the news show *60 Minutes*. Walton reviewed 165 separate studies published in medical journals over a twenty-year period. *All* of the studies that found aspartame safe happened to be sponsored by industry. Every single one that questioned its safety was produced by scientists without industry ties.

The *Ecologist* quotes the Bressler report directly:

> The question you have got to ask yourself is: why wasn't greater care taken? Why didn't Searle, with their scientists, closely evaluate this, knowing full well that the whole society, from the youngest to the elderly, from the sick to the unsick will have access to this product?

Aspartame is now one of the most commonly used ingredients in drinks, cakes, cookies and candies. There is no evidence at all that those who use it actually lose weight. There is some indication that it creates a sugar deficit, leading people to seek more sugar from other sources.

But leaving aside whether it has any benefits, is it safe? Do we have enough information to know? Remember that we test compounds in animals to find out how they might affect humans. Olney, who published original studies showing that rats exposed to aspartame developed brain abnormalities in the 1970s and 1980s, returned to the subject in 1996, asking whether patterns of brain cancer in adults in the early 1990s might reflect past use of aspartame.[42] He told me that even though his findings were written in the tentative tone of scientific inquiry, when this paper, questioning whether brain tumors could be tied with aspartame use, was accepted for publication by the *Journal of Neuropathology and Experimental Neurology*, attorneys for Monsanto asked the editors not to publish the work. In fact, brain cancer may have a latency as long as thirty years between the time of first exposure and the expression of the illness, so Olney's question was certainly premature. Given the rapid growth in recent years in the use of aspartame, if there is an impact on brain cancer or other cancers, it might not yet be evident.

Still, the National Cancer Institute and the American Cancer Society felt compelled to weigh in on the same question with a resounding no.

How did they reach this conclusion? They conducted a survey in 1995 and 1996 of drinking and eating patterns in half a million volunteers of the AARP, asking how many of them had come down with lymphoma or brain cancer five years later. The fact that no effect was found is hardly the last word on the subject.

Until recently, studies in lab animals were run for a period of two years. When working with rodents, scientists have generally ended the studies and the varmints' lives after 712 days, hoping in that time to get the animals to consume as much of whatever is being tested as a human would use in seventy years. But rats can live longer. It's entirely possible that by stopping studies at this point, we are missing an important part of the picture. What happens to animals or people in the last third of their lives? That's something that many of us want to know.

At the European Ramazzini Foundation in Bologna, Italy, a group of scientists led by Italy's leading toxicologist, Cesare Maltoni, came up with a different approach to testing animals to predict human impacts—an approach that is being adopted throughout the world today. For more than three decades, they have been letting rats and mice live out their natural lifetimes (generally three years) as they are exposed to various substances present in the industrial or general environment, to see whether these agents affected their chances of getting cancer. This lifespan protocol contrasts with that of other laboratories where rodents are killed at two years of age and examined for signs of tumors or other damage. Two years in a rat corresponds to about sixty years in humans, but more and more of us expect to live into our eighties.

What about aspartame? Maltoni died in 2001, but his work continued. Eighteen hundred Sprague-Dawley rats were allowed to eat aspartame from the age of eight weeks until their natural deaths about three years later. The study demonstrated for the first time that the artificial sweetener, when administered to rats in feed, caused a statistically significant, dose-related increase of lymphomas/leukemias and tumors of the renal pelvis and ureter in females and malignant tumors of peripheral nerves in males.[43] Moreover, these tumors occurred even at a daily dose well under that allowed in America or Europe, namely 50 mg/kg and 40 mg/kg respectively. This table from the European Ramazzini Foundation estimates the average amount of as-

Table 15-4 Average Daily Intake of Aspartame

Substance	Quantity/Day	Concentration of Aspartame Consumed
Diet soda (200 mg/can)	2 cans	400 mg
Yogurt (125 mg/yogurt)	2 yogurts	250 mg
Diet custard/pudding (75 mg/mousse)	1 serving	75 mg
Coffee with sweetener (40/mg packet)	4 cups	160 mg
Candy/chewing gum (2.5/candy)	10 candies	25 mg
Totals		**910 mg**

partame consumed from only a few of the 6,000 products in which it is present.

If a woman ate these foods and weighed 60 kilograms (approximately 132 pounds), she would consume an aspartame daily dose of 15.1 mg/kg of body weight; a child weighing 30 kg (approximately 66 pounds) with a similar daily intake would have an aspartame daily dose twice as high—30.3 mg/kg of body weight. This level is well over the dose that caused lymphomas/leukemias in the study. This study was the first to show that aspartame caused cancer in multiple organs.

In a letter to the journal that published Soffritti's work, the industry soundly rejects his study. It refers to an independent evaluation by the European Food Safety Authority Scientific Panel on Food Additives, Flavorings, Processing Aids, and Materials in Contact with Food. In fact, many members of this "independent" panel work directly for the same industry. They suggest that the findings of this three-year study are completely without merit because there were unusual patterns of cancer in these animals that have not generally occurred in other studies.

Let's look at what the Ramazzini Foundation did. Soffritti's team studied more than 1,800 animals for a period of three years. The European critics of Soffritti noted that the animals in both the control group and those that regularly drank aspartame suffered from inflammatory respiratory problems. They charge that the increased risk of cancer could have happened as a result of some underlying infection, but this completely misses one key point. The animals lived out their natural lifetimes. Animals, just like the rest of us, have to die of something. Pulmonary infections, like pneumonia, are called the old man's friend, because they can be a relatively painless way to go. But, even if these animals all developed lung problems as they aged, why did so many of those who drank aspartame develop cancers in so many different organs?

THERE IS NO QUESTION that medical and information breakthrough technologies make our fast-moving, fast-talking world easier to handle. Whether they may also make our lives more prone to cancer is a question that is simply not being asked. We presume that the things we rely on today to see through bodies, talk across the world in an instant, and keep our children from spinning out of control are safe; to do otherwise would require an entirely new way of looking at the world.

The absence of extensive information confirming that human health is endangered by any one of these technologies and medications lulls most of us into assuming that no such hazard exists. The lesson of this book is that we should all question this presumption. Highly profitable industries have no incentive to ask whether the products on which they depend may have adverse consequences. Nor is there any independent system in place to compel them to do so. As Searle's former CEO Donald Rumsfeld said in a very different context, absence of evidence is not evidence of absence. A lack of definitive evidence regarding human harm is not proof that no such harm occurs. Rather it shows the difficulties and roadblocks that surround efforts to develop information on the health effects of modern technologies and chemicals.

The presumption of innocence with which we accept new technologies today, like that with which the world greeted x-rays at the dawn of the twentieth century, makes no sense. Our naive faith and fascination with what's novel does a disservice to ourselves and our children and grandchildren.

If we are to learn about the long-term impact of essential components of new technologies and medicines, we need open systems of evaluation that currently don't exist. Drugs and technologies are created to fix problems. How can we know whether these rapid solutions may endanger our lives later on?

The short answer is, we can't.

If we insist that we cannot act to prevent future harm until we have proof of past harms, we are treating people like lab rats in uncontrolled experiments. If we say, let's let the experts decide, where do we get experts without baggage? The costs of experimental laboratory research are growing and debates about the value of various research methods are becoming ever more complex. These debates are sometimes fueled by those who have a knack for turning molehills of scientific minutiae into mountains of uncertainty. In a world where information on the health and safety of workers remains locked up in company files, wrapped in the protections of confidentiality, independent information and independent experts to make sense of it are an endangered species.

We have seen repeatedly how some people in industry, whether tobacco, asbestos, benzene or vinyl chloride, understood risks long before the rest of us were able to learn about them. We know of many instances where insurance companies tracked health hazards for years, as claims mounted and reports of various ailments accumulated, without letting workers know the dangers they faced. We also know that current laws discourage giving such information up. The federal Toxic Substances Control Act provides criminal penalties for anyone who has knowledge that someone endangers public health or the environment and does not report it. The result is that most companies no longer develop such knowledge or collect such information, so that they can't be charged with breaking this law.

With respect to long-standing workplace and environmental hazards, the idea of prosecuting those found guilty of past harms has just not worked. The crimes are of such long standing, the victims are so many, and retribution is so pointless that perhaps the best course is to emulate the South African Truth and Reconciliation Commissions (TRC) and offer amnesty in exchange for a better future. In 1994, to create a break with its deadly past, South Africa set up a stunning series of national public confessions. Nelson Mandela, the head of the African National Congress, and South African President F. W. de Klerk both understood that for the racist system of apartheid to die, it needed a proper burial. Without public acknowledgment of the brutality of the apartheid past, the country would never recover.

De Klerk's white supremacist National Party wanted blanket amnesty for the violence they had committed in the name of the law. The members of Mandela's African National Congress wanted revenge. If the country was not to be torn apart, it needed to create something that had never really been done before—a national commission for truth and reconciliation.

People came forward asking questions that they could not have uttered at any other time. What happened to my son or daughter or husband or wife or brother or sister or father or mother? Where did they go? Who killed them? How had they died? These were not easy questions to ask. They were even less easy to answer. The system was based on the premise that without answers, the country could never be brought together.

In her memoir *Every Secret Thing*, the South African writer Gillian Slovo writes that she had to struggle hard to accept this premise. Her mother, Ruth First, a white supporter of Mandela, was killed by a package bomb sent by the Nationalist Party. Her father, Joe Slovo, had fought Hitler in Germany and would fight with Mandela against the white supremacists in South Africa. The last effort of his life was to seek restorative justice for the man who had murdered his wife.[44]

Gillian Slovo explained her father's remarkable position. "My father, one of the architects of the final settlement, put it this way: the best revenge, he said, that I can think of for those men who murdered my wife is that they be made to live in peace in a system that they had fought so brutally against. The truth telling that this unleashed was painful, sobering, and so far has proved to have provided more healing than hurt. A nation that once was awash in the blood of its people is moving toward a more free and open society than it has ever known."[45]

This open approach went far beyond any of the efforts to mete out justice that arose after the end of World War II. National law, whatever it may have said or allowed, becomes irrelevant and is replaced by an almost biblical view of what is required to restore a nation. Not punishment and vengeance, but grace and forgiveness, become the grounds for renewal and restoration.

Those who witnessed the creation of South Africa's TRC call it a miracle. They note that what took place in postwar Germany and Japan, and in Central and Latin America after the fall of right-wing dictatorships, made it clear that direct and full prosecution, if carried out, would rip a country apart. Where a majority are guilty, punishment becomes unending. Where only a few are prosecuted, as was the case in Germany and Japan, this creates the delusion that the rest of the country bears no responsibility for the past. Because these nations had embraced violence against their own citizens as a matter of national policy and law, turning to the law to provide redress against this violence made no sense.

If persons in charge of major firms today learn that chemicals their workers are using will shorten their lives, and they fail to act on this knowledge, are these actions no less morally wrong than those of the South African leaders, Nazi supremacists or Japanese imperialists? Creating a harmful workplace and concealing that harm is surely a more subtle crime than forcing young girls to serve as "comfort women" or loading entire villages into boxcars for transport to death camps. But if we were to count the deaths caused, or if we could somehow reckon up the total human suffering, we would find ourselves in similar territory.

I have learned from others, whom I can't name at this point, that the files of many large multinational businesses could easily tell us about many more health risks associated with workplace exposures of the past. These companies are largely self-insured and pay for their workers' health care. They have complex information systems at hand to control the manufacture of chemicals, the ordering of materials, and the processing of health claims. Can you really imagine that such an organization does not know whether or not its workforce in Indonesia or Silicon Valley has greater risks of breast cancer and leukemia? Can you believe that Pratt & Whitney—one of the largest and most profitable makers of airplane engines in the world—does not know whether or not its workers have higher rates of brain cancer than the general population?

According to the company's website, Pratt & Whitney engines

power nearly half of the world's commercial fleet.[46] Every few seconds—more than 20,000 times a day—a Pratt & Whitney-powered airliner takes flight somewhere in the world. Their military engines power the air force's front line fighters today—the F-15 and F-16—and our F-119 and F-135 engines will power the front line fighters of the future—the F-22 Raptor and F-35 Joint Strike Fighter. Their rocket engines send payloads into orbit at 20,000 miles per hour. Is it believable that this same company can't tell us now whether the men and women who have made these engines over the past thirty years have more cancer than others?

I am not smart enough to know what kind of system will best identify and address the preventable causes of cancer in our environment. I just know that what we have been doing doesn't work. For every lawsuit that is won on behalf of persons harmed by cancerous activities, many more are never even filed. Lawsuits brought on behalf of those who believe their injuries were caused by their employers' bad actions succeed less and less often. In large part this is because recent court decisions have changed the rules of the game and the presumptions of evidence.

It may shock you to learn that of the 100,000 chemicals that are commonly used in commerce, most have not been studied as to their ability to affect our health. In 1983 and again in 1998, the National Academy of Sciences confirmed that we have no public record of the toxicity of three out of every four of the top 3,000 chemicals in use today. Despite declarations by industry of their intent to close this gap, in reality it would take dozens of years and billions of dollars to do so. It can take three weeks to approve a new chemical for use and thirty years to remove an old one.

In the past, experimental findings in rodents and small mammals were accepted as indications of human harm. Nowadays the field of experimental carcinogenesis itself has become wracked with debates about how to interpret findings. While drugs are still created based on animal experiments, the appraisal of commercial chemicals is littered with endless debates about mechanisms and appropriate models.

In a sense we have come full circle. In the 1930s, the world's leading cancer experts, using experimental information, detailed observations on highly exposed workers, and some public health statistics, identified many important causes of cancer in industry, nutrition and behavior. For the past seventy-five years, that evidence has been stretched, reviewed, revised, culled, pulled about and put back together again.

While Heraclitus said no one ever steps into the same river twice, he could not have had in mind the circular voyage the world of cancer research has taken. We have known for more than a century that it is inherently difficult to evaluate the extent to which a given exposure results in a given health problem. As we have seen repeatedly in this book, cancer-causing agents can produce many different types of cancer as well as an entire suite of other health problems through biological paths that can't be easily tracked.

Epistemology is the study of how we come to know what we believe we know. Plato pointed out that what we know is in a basic sense socially constructed at the intersection of our shared beliefs and presumed truths. Cancer research is no different from any other form of knowing. It relies on customs and practices. What can be considered known about cancer is profoundly economic and political and reflects the views and values of those who pay for the work, decide whether or not it should be carried out, and when and if it should ever become public.

The loggerhead at which science and law now abut may become a tipping point. Science works to establish the truth. Law aims to mete out justice. Because scientists know that certainty is never absolute, scientific knowledge is always hedged. There's always room for more. Law requires enough faith in precision to mete out justice. Carl Cranor and David Eastmond, two scholars on toxic injuries, succinctly describe the dilemma of how to meld scientific evidence to obtain justice regarding such harms:

> At this juncture, the point is not to propose a specific alternative, but to sketch the types of legal modifications that should be explored in order to induce corporations to engage in far more scientific research when it matters—not to win lawsuits but to protect society against the risks posed by their products. The proper role for scientists with regard to toxic substances should be to provide needed information about possible latent defects, not to cast deciding votes on liability because causation has been made a surrogate for morally responsible corporate behavior.[47]

As a first step to preserving the central aims of tort law, courts will need to recognize the wide variety of respectable, reliable patterns of evidence on which scientists themselves rely for drawing inferences about the toxicity of substances. The patterns of inferences presented above for carcinogens, arguably one of the most difficult of toxic sub stances with which scientists and courts must deal, serve as examples of some of the variety of inference patterns utilized in the scientific community. Courts, recognizing a wider variety of inferences, would then be able to better assess the sparse scientific evidence that is typically available.

. . . If scientific knowledge about the toxicity of a substance in humans could be accumulated instantaneously, there would not be the concern that science delayed or incomplete was justice denied. If scientists could instantaneously have the best human evidence of toxicity, they would not need to piece together animal, mechanistic, genetic, structure-activity, and other inferential evidence. If diseases could be identified at an early stage, left their signatures, or did not have long latency periods, there might be a lesser need for various kinds of non-human evidence. However, given the nature of the biological world and the recognition that science in its current stage of development does not have such capabilities, courts must recognize this and utilize scientifically reliable patterns of evidence that will permit plaintiffs to receive just treatment in tort cases. If this is not done or it is not adequate, more fundamental solutions to these issues will need to be found.[48]

In fact, the courts are moving in just the opposite direction. They are not piecing together information or tapping alternative methods of inferring facts and associations but allowing the absence of proof of human harm to be construed as evidence that there is no such harm.

A truth and reconciliation commission might provide the sort of revelations about toxic hazards that we all feel have to be at hand, but realize remain out of our reach. If one asks who should pay for this system, we may draw another lesson from World War II. During and after that war, an excess profits fee was placed on those industries that benefited from the conflict. There is no doubt that tobacco, alcohol, chemical and pharmaceutical manufacturing are industries that create risks and benefits, often to quite different groups of people. A fee can be levied on all those industries as a way to fund a truly independent and neutral forum where information can be safely exchanged on environmental health hazards.

Some will argue that creating a TRC-like institution to accept information on environmental health hazards would only allow people to get away with past activities that have injured or killed people. The tort system exists to redress wrongs and to exact financial penalties from those who have harmed others. Such people cannot be absolved by a new institution.

The harder we try to exact vengeance against those who have caused harm, the more incentive they have to conceal information, and the more harm will be done in the future. But new approaches to generate information on the risks of work and the environment can reduce the chances that current harms will result in future damages. If we create a place where industry can deposit information on health hazards of work and the environment, with the privacy of individuals appropriately shielded, the world will be better off for our having tried to do so.

The European community is trying to produce a better set of information on chemical hazards as part of the Re-registration, Evaluation, and Assessment of Chemical Hazards (REACH) program. It's too soon to know if this voluntary program will work, but some are concerned.

As Soffritti, the Italian scientist who continues Maltoni's pioneering approach in long-term animal carcinogenicity testing, recently told me, "The REACH program has passed the burden of proof for chemical safety to the industries that produce the agents. Unfortunately, history teaches us that industry reports do not necessarily contain all that is needed to evaluate the risk or safety of products. In looking at the ways that information can be manipulated, John Bailar notes that, 'There are many ways to distort the scientific truth without actually lying.'[49] Consequently, I would therefore be very cautious about using industry data as the principal basis for regulatory action. The structure of the REACH program means that independent data will be relegated to anecdotal status."

It is a huge dilemma. Another tack can be pursued as well. For those with known or suspected exposures in the workplace or environment, why not establish what are called "medical surveillance" programs to look for the ones who are ill? We know that there are some illnesses where early treatment can keep people from early deaths. A smart outfit that has put a dangerous product on the market should be interested in systems that would reach out to people at risk and help them. This approach may back us all down from a precipice, to a point where people who are going to get increasingly sick will have a chance to be helped through medical surveillance programs set up to find them before their illness is too far advanced.

A leading tort lawyer commented on this idea with guarded enthusiasm. "In my twenty years of work, not once have I had a client who was glad they had me as a lawyer because they had this really awful cancer they just were thrilled about."

She noted that perhaps an admission of knowledge on cancer-causing practices would be lifesaving as well as good for the soul. "The companies need to put the shoe on the other foot. If it were their family, they'd want a system that gave them an opportunity to look for ways to hold people accountable, and to help them stay well for as long as possible. Medical surveillance would remove the hazard and work with those who've been exposed with the goal of keeping people healthy."

Of course, even if we set up such a system and found ways to pay for it, we have to live with what cannot be undone. The systems currently in place to understand and control toxic substances do not work well enough. The penalties mandated by the Toxic Substances Control Act, requiring that anyone having knowledge that an activity threatens public health or safety has to report it, had just the opposite effect. Rather than creating information on public health threats, the act has discouraged companies from publicizing analyses of their workers' health.

Protected trade secrets are now defined so broadly that they sometimes encompass information on the health and safety of workers, including even workmen's compensation claims. The way out of this problem has got to be different from the way in. As Einstein noted in another matter, we can't solve the problems of the present by repeating the mistakes of the past.

But the past still offers a guide. In the eighteenth century, the philosopher Immanuel Kant argued that we have to act as if there were goodness, truth and justice, because by doing so, we compel these qualities to arise. We have learned that much evil in modern history is not natural but man-made, the result not of divine but of human will.

The move toward greener energy from tides and sun and wind will eventually turn the Middle East conflict into a relic of the days when the world depended on liquid fossil fuels that may become irrelevant in my grandchildren's lifetimes. Arnold Schwarzenegger played the Terminator, Conan the Barbarian and Conan the Destroyer, characters wired to end much of the world. In his latest reinvention, the wealthy, unsalaried governor of California appears as the great green giant, campaigning for a just and clean world, featured on the cover of *Newsweek* holding up the entire globe with one finger. For more than twenty years, since the death of her first husband, Sen. John Heinz, Teresa Heinz Kerry has marshaled the wealth of the Heinz Endowment to foster a green renaissance of the once smoky city of Pittsburgh. Formerly known as hell with the lid off, Pittsburgh today is a center of green building, chemistry, health care and operations in my own hospital system—all built to have a smaller footprint. Television programming of dizzying arrays, luxury and regular magazines and new web sites are proliferating to promote green living options that are not just for the true believers. Yes, such activities appear in the midst of commercials for Hummers and SUVs, but those are dropping faster than lead balloons. The environment is no longer a niche issue of radical chic, but a matter of broadly understood importance. Those of us who indict past failures have a duty to develop new solutions.

My parents, Brigadier General (retired) Harry B. Davis and Jean Langer Davis, shown at West Point, New York, three weeks before my father died of multiple myeloma in 1984.

Epilogue

Mother's Last

"What's the matter with her?"
"There's nothing the matter with her. She's dying."
—*ZORBA THE GREEK*

WHENEVER SOMEONE CALLS your cell phone and asks if it's you on the other end, you know something's wrong. My brother Stan's voice had more urgency than usual. "Hello, Dev? Is that you? I've got news. It's not good."

"What's going on? You sound just awful," I said.

My methodical, mostly unflappable brother was taking his time. This was not because he wanted to, but because he had to. "It's Mom. She has inoperable cancer. It's spread from her stomach to her liver."

Wouldn't it be nice if we could erase things in life we can't bear—just wipe them from the tape on which our life is recorded. Let's back that one up, make it go away. People make mistakes? Yes, they do. Doctors can be wrong. My hip swinging, disco dancing mother, who missed her granddaughter's college graduation so she could sail off to the Hermitage in St. Petersburg to see the recovered French Impressionists, could not be full of cancer. Must be somebody else.

Some people know a lot about heavy equipment, like my brothers Stan and Martin. I happen to know a lot about cancer. Those of us who spend our lives working on cancer don't get to go through the usual stages of denial and anger when the disease finally hits. We just have to deal.

Stomach cancer is one of the lousiest you can have. It usually isn't found until it has breached the soft, pink lining of our chief digestive organ. The symptoms creep up on you like a gentle fog that slowly turns insidious. At first it leaves a warm, full, tingly feeling throughout the center of the lower ribs that is not entirely unpleasant. Eventually the sprawling tumor of the stomach makes you feel like you've just eaten the worst meal of your life all the time.

I tell Stan I'll be on a plane to Pittsburgh the next day and hang up, still thinking my mother will somehow be fine. Then reason seeps into my brain and I remember what I'd like not to know. When Fred Rogers faced the same diagnosis, the famed soother of children on public television chose to go off quietly. He didn't last three months.

But hey, this is my mother—the woman who at age twenty-three drove herself, two children and a trailer containing all of our household goods through snow and hail storms from Pennsylvania to Texas, long before there were interstates. Ten years later, when Stan and I were eleven and twelve, the three of us nearly drowned in posthurricane seas in Miami. Swimming in storm-driven surf seemed like an exciting thing to do at the time. We got swept out to sea, were rescued by lifeguards who spotted us dangling from a barnacle-covered jetty, and got taken by ambulance to a local hospital. In our family, thrills, even those full of risk, are readily sought and never feared. Mom flew through three cardiac bypass operations. You think advanced cancer will get her down?

We sit in the modern purple-hued waiting area of the Hillman Cancer Center, the gorgeous new cancer pavilion at the University of Pittsburgh Medical Center, Shadyside. Soft seating fills the warmly lit room. Many are glad for the chance to sit and wait. They have little choice. Chemo must be worth a try, if only to try to purchase the two things that can never be bought—time and hope. There are always those who beat the odds, even when they are grim. My mom intends to be one of them.

The door opens and the person who really runs things shows up. Oncology nurse Connie Kinney begins to recite the listed side effects of the drugs Mom is to get. "Complications include dizziness, bleeding, and death," she says.

Mom cuts her off. "So do you have these medications here? Now?" She doesn't blink.

"Yeah," Connie says, "we could start you today. But maybe you want to take some time to get things together?"

My mother is feisty, like a boxer about to enter the ring against an opponent twice her size. "No thanks. Let's just get on with it."

My mother definitely did just that. The week before chemo started, she returned to her hometown of Donora, Pennsylvania, delighted and nonplussed to be accompanied by Andy Masich, from the Heinz History Center, and a film crew with Pittsburgh's public television station, WQED. She starred in a local public television documentary about growing up in Donora—a place that most folks had long forgotten. In our home town, eighteen people had dropped dead in one day from dirty air. No one has ever studied what happened to all those who lived under such conditions. Could her illness have anything to do with where she had lived as a girl? We couldn't really answer the question, because so many other things can contribute to cancer in someone who reaches her seventies. In that film, she and I are walking close together. This is not because our sentimental attachment to the past brought back warm memories. It's because Mom refused to be filmed walking with a cane. She held her head up and walked as straight as she could past the cameras. Fortunately, they only needed one take.

Eventually we got a chance for the closeness that can emerge with grave disease. It was clear something was different. Amazingly, after all those years of exercise and dieting, her wonderfully zaftig Jewish thighs and arms were nearly svelte. Her skin was still soft as a fresh peach, but the color was gray and mottled, even under a futile tan. She bruised easily.

Like many people with lots of drugs in their system, Mom was too impatient to read. Chemotherapy does that to the brain, makes even those with considerable memories short on attention and long on irritation. Our entire family began the hovering dance, hanging around, not knowing quite what to do but feeling that they should be there. None of us could bear to be away for long. Just in case anything might happen.

"Don't you have something important to do?" Mom asked me after we had come back from another round of chemo at the medical center. "Isn't there a big meeting you are supposed to be going to in Washington? Why are you here so much?"

"I think this is where I'm supposed to be," I replied. "Can't I get you something? Can't I do something for you?"

"Sure," she deadpanned. "You want to be useful? Go get me a glass of water." That was about as far as I could go. On some level, we both understood that our time had limits now. We just never said it out loud.

In the winter of 2003, we were regularly wheeling in and out of the new Hillman Cancer Pavilion. I was surprised to hear Marina Posvar, one of the building's experienced guides, repeat what seemed a radical idea—the goal of the basic researchers was to put the clinical providers out of business. This meant that somewhere somebody looking at the big picture of cancer research understood the critical need to keep cancer from happening altogether.

There's not a lot that a cancer patient gets to control in her life. My mother always loved a really good kosher hotdog. I knew that the stomach cancer that she had could come about after years of eating lots of the preservatives that are used to make hotdogs—nitrates in meat get transformed by stomach acids to nitrosamines, a well-established potent cause of cancer. At this point, one hotdog more or less after a lifetime of salami, smoked salmon, pastrami, corned beef, and those other smoked deli foods, coupled with all that coal smoke she had swallowed, probably wouldn't make much of a difference.

One day in April 2003, we went to watch my brothers at an industrial auction sale on the Southside. We had come directly from the cancer center, where she had gone through an uneventful session of chemotherapy and had not eaten. When we got to the auction site, she immediately spied a small cart over which a big red-lettered sign was raised, "HOT-DOGS." I tried without success to move her out of the line of sight. She went right to it like a mouse pouncing on a crumb of cheese.

"Boy, I've been wanting a hotdog for a long time," she said to the spry old man behind the cart. She looked at me, expecting me to say something. I just shrugged my shoulders. I knew it was hopeless.

"Lady, you should be careful about hotdogs," the vendor said. "You know hotdogs can cause cancer."

My mother smiled and shrugged. "That's okay. I already have cancer."

The Hillman Center was the place where she got the best care, but at that time it did not offer much advice about whether there were any things—like eating hotdogs or not—that she should or should not do to keep herself well as she dealt with this disease. I, of course, being the oldest daughter and sometimes an insufferable food fascist, had more advice to offer than anyone could possibly use. When Mom got a blinding headache after getting her nails done, I suggested it might have had something to do with the acetone, the formaldehyde, the toluene and the other dozens of chemicals that get used in manicures. But I couldn't really be sure.

At one point, I groused, "Why can't we tell people about the goods and bads of cancer? Why do people just have to learn by trial and error what makes them feel better or worse?"

One of the people I complained to was Dr. Shalom Kalnicki, a charming, outgoing Brazilian who led radiation oncology programs at the more than thirty facilities within the cancer institute. This is the specialty you hope you never need, but when you do, if you're really lucky, you get somebody like Shalom. An exuberant and erudite man, Kalnicki had interests in topics well outside of mainstream medicine. He knew firsthand that no matter how good care had become for cancer patients, a fundamentally new look had to be taken at keeping the disease at bay. Treatments were becoming more sophisticated, more expensive, and sometimes more successful, for those who could afford them.

"You know, Devra, I really think you should talk with Herberman," Dr. Kalnicki told me one day, when we were both waiting with my mother to finish her infusion.

"Shalom, I can't imagine that this makes a lot of sense," I replied. There was a lot going on at the time. When bad things are happening in life, a day can feel like a week, a week can feel like a month. We grow old fast.

"One thing I've learned in my years in Pittsburgh," Dr. Kalnicki pressed, "never, never underestimate Herberman. He's one of the smartest men in the world. When you're ready, you have to talk to him."

Listening to this conversation, my mother sensed an opportunity. An hour later, she made her move. "Now look," my mother sighed. "Shalom says you really have to talk with that guy Herberman at the cancer institute. Don't waste your time with me. Go see him. That's important."

That's how I came to sit across the table from Dr. Herberman that winter. I couldn't say no to my mother at that point. His office was filled with books and awards and photos with presidents, the pope and other assorted dignitaries. Herberman is a dapper, wiry, intense man, who looks considerably younger than his sixty-six years. A prize-winning researcher in his own right, Herberman built Pittsburgh's Cancer Institute, securing major funding from the National Cancer Institute and American Cancer Society along the way, by recruiting some of the world's top basic scientists and oncologists, and the major support of two Pittsburgh philanthropists, Henry and Elsie Hillman.

"Dr. Davis," he began with a low voice that belied the power that he held. "What would you do at a cancer institute to address the role of the environment for the disease?" I was caught off guard. I never imagined that someone in his position would ask such a question. After all, the cancer establishment is a highly profitable, multibillion-dollar multinational effort that finds and treats the disease. Frankly, I found the question disarming and naive.

"Dr. Herberman," I replied, "I really appreciate your taking the time to talk to me. But I don't want to waste your time."

He waved his hand. "Dr. Davis, I've read your book. I understand that these aren't easy issues. I know that some people would rather not hear what you have to say. But I want to know. What would you do if someone challenged you to come up with a plan to study and to do something about how the environment impacts cancer?"

I really couldn't quite believe I was being asked such a question. After all, Herberman ran one of the nation's busiest cancer treatment centers. The more cases that walked in the door, the more profits the enterprise would make. But, he was no ordinary medical business-man. A trained scientist, Herberman had begun to ask why so many of the physicians he had trained with were developing cancer themselves.

Two years later, Dr. Herberman confided that his curiosity about the environment had not been merely academic. I learned that he and his brother, Harvey, had both developed chronic lymphocytic leukemia. "I can't explain why my brother and I have a form of cancer that no one else in our family ever had. It just seems clear to me that after all we've learned, there has to be more that can be done to keep cancer from arising. That's your job to figure out. I just want it to happen."

The best-known causes of stomach cancer relate to what people eat. But other parts of my mother's life could also have triggered the disease. The rolling mills, zinc plants, coke ovens, home coal stoves and blast furnaces of Donora in the 1940s burned more coal each day than all the mills in Pittsburgh and contained asbestos insulation throughout. She grew up with coal smoke in her home, ate lots of heavily smoked salamis and hotdogs and corned beef, and had enough diagnostic radiation for a small village. So what caused her cancer? Probably all of the above. You can't go back and change what you ate thirty years ago. You can't pick your parents, or where they lived, or how they heated their home.

As my mother coped with the ebbing of her life, we continued to study the Torah (Old Testament) with a group of women who belonged to the library minyan at Squirrel Hill's Beth Shalom Chavera (friends). Three weeks before she died, we gathered in her apartment as part of a tradition of reviewing sacred texts, examining line by line a part of the Bible where Moses receives the Ten Commandments the second time. We laughed at an old joke: Moses comes down and says, "Look, I've got good news. He originally wanted to give me twenty commandments, but I negotiated it down to ten. The bad news is that adultery is still on the list."

The women who sat with us that day understood the nuances of Torah reading. Helen Feder asked one of those unanswerable questions: What was the most important commandment? One traditional answer seemed like a triumph to my mother: The Fifth Commandment is the most important, because if you honor your father and mother, you will follow in everything that they do.

Those of us who spend our lives writing, researching and talking don't handle silence very well. Mom knew I was writing a new book, this one about cancer. In better days, she had looked over parts of *When Smoke Ran Like Water,* offering detailed corrections. Now that was not an option. All the drugs and sickness made it hard for her to hold papers, let alone read them. Writing was out of the question. So I asked, "Do you want me to read to you from some of the next book?"

"I guess so," she replied, with just a hint of interest. I sat next to her at the laminated dining room table to which her wheelchair had been rolled. I knew I would not have much time. I also knew that though her life was near its end, she had no interest in talking about this. The closest I could come to doing so was to read her this story:

When my dad was diagnosed with what we were told was incurable cancer—multiple myeloma, with maybe months to live, they said—I wanted to tell him right away about how I came to know the world we enter when we leave this one. The oncologist explained that his bone marrow had shut down. Dad took good notes. On the small spiral notebook he carried in his shirt pocket, he neatly wrote in his precise machinist's print, "Usually people die within six months."

"Look, I have lived a long life," Dad said. He was 55.

In my family, the notion that there is anything after we die only surfaces as the punch line on a corny joke. The humorist Molly Katz notes that Jews do not die. They pass on, or are gone or lost. As in "We lost him." Of course, he'll never really be found again, but that's not the point.

Sometimes those who pass on meet anything—as in God forbid anything should happen.

Cancer is a metaphor, but it is also a brute. When Dad first faced his terminal diagnosis, anything did seem about to happen. Suddenly this man, who like all fathers loomed larger than life, faced a death sentence—at least that was how his doctor put it. Like many first-time cancer patients, my father gave new meaning to the word denial. He had started his working life hZucksteringucksteringucksteringucksteringucksteringucksteringucksteringucksteringucksteringucksteringucksteringucksteringucksteringucksteringuckstering newspapers at age five, become a drill sergeant by twenty two, and ended up running a prosperous auctioneering business in Pittsburgh by middle-age. For him, this diagnosis was just another barrier to cross.

My sister-in-law, Mimi, probably the best standup comic in the family, relays one gag. "A lecherous old guy sidles up to a beautiful young woman and asks 'Do you believe in the hereafter?'

She replies, 'Of course, I do.' The guy sneers, 'Then, honey, you know what I am here after.'"

With my dad, I could not bear all the things not spoken of at that point. But those years of jokes and denial provided some momentary distraction. Can't handle the angst? Just laugh it away. Can't handle gallows humor? Then you can hope or pretend that death is just not a possibility. According to legend, the Prophet Elijah might just show up at the end of each Sabbath, because he never really died. So every week we sing songs to welcome him just in case he decides to come back. To be Jewish means to believe that at least some folks achieve eternal life. So how about the rest of us?

Dad was more than lucky. His original diagnosis was spot on, but the prognosis was not. At the University of Pittsburgh Medical Center, he connected with brilliant doctors on the cutting-edge, including Dane Boggs, and got the sort of cancer care that you can find nowadays if you have the insurance or the money to pay for it. So I got to wait six years for the chance to tell him how and why I came to believe that we reach peace when we leave this earth.

One day after his latest round of high-dose chemotherapy, Dad and I sat in a small booth with turquoise-plastic covering in the eternally comforting Eat 'n Park Diner, just off Murray Avenue in Squirrel Hill. We split a piece of lemon meringue pie and sipped tea—one bag for two cups, a family tradition. An old black hearse that had been converted into a rock 'n' roll van covered with mandalas rolled by. At the site of this paradoxical vehicle, he began to chant softly, in a singsong voice that could never quite carry a tune but always sounded completely charming to me. "Did you ever think when a hearse went by, that you would be the next to die? The worms go in. The worms go out. The worms go in and out of your mouth."

As little kids, my brother Stan and I used to march around behind our father singing cleaned-up ditties from his army days. "Sound off! Sound off! Hit it again. Hit it again. Sound off. One-two-three-four, one-two, three-four."

Like many military guys, Dad did not really sing so much as set words to a drumbeat, the repeated measure of familiar expressions meant to keep everyone in line and on pace.

He repeated the verse, tapping the white Formica countertop, which gave the rhyme an almost aggressive beat. I chimed in, "The worms go in. The worms go out. The worms go in and out of your mouth."

We laughed. But we both knew this was one tune nobody ever wanted to march to.

I figured it was the time to tell him about how I came to believe that when our space and time in this world end, we enter eternity. On a sunny summer day a decade earlier, my life had nearly stopped. Hiking alone, I walked onto the porch of an old, rotted cabin when I should have known better. Black-hearted, brilliant-yellow daisies sprouting straight through wooden planks told anyone with half a brain to stay off. It may take years, but wildflowers can wend through planks that humans had best avoid. I had come way too close. Never even heard the buzzing, blurring, whirring hornets til it was too late. My left leg plunged straight through the weathered floor, smack into the secret nest of hundreds of stinging yellow jackets. The blinding pain hit deep to the bone.

I was able to hobble to the car, haul up my throbbing leg into the driver's side and drive in slow motion to the lakeside home of my good and smart friend Richard. He knew what I did not. In response to the venom of more than two dozen hornets, I was going into the sort of shock—throat swelling, heart racing—from which one sometimes does not return. Right after we reached the emergency room, I began to die.

At this point I stopped my reading to her and looked up. I was surprised to see that my mother was actually listening. She gave me one of those close lipped smiles that said it was okay and signaled with a nod of her head to continue.

Some Swiss scientists claim that the intensely beautiful sensation of those who nearly expire is nothing but neuronal stimulation heightened by a lack of oxygen that somehow triggers sensations of white light, clouds and tunnels. Maybe so. Back then I had never heard any of that. All I know is the day I almost died, I floated into the whitest, holiest, most comforting and shimmering radiance I have ever known. I came face to face with a beatific, white-robed, vaguely maternal, olive-skinned being. I gestured to my body on the table below and told her, "This is just lovely. Really wonderful. But, I'm not ready. I would like to get back."

I woke up on that table confused about what had happened. I wanted to think I had just gone through an especially vivid dream. For years afterwards, I never spoke about what had happened. Just kept it as one of those things that is too strange to mention and too unsettling ever to forget. More than a decade later, I read accounts of what are now called near-death experiences.

I knew that when the hornets nearly took my life I had felt the presence of something more beautiful than I could ever describe. I spoke of how I reached a point of peace far beyond words, as I watched my body below me.

Dad listened as I recounted my close call with death. He did not interrupt. That in and of itself was atypical. We drank our second cup of diluted tea. That day he even let me pay.

When I had finished reading, Mom looked up. Her lower lip protruded pensively. She shrugged her shoulders, put her chin to her left shoulder with her lips pursed downward. "Too many words." That was all she ever said about it.

On her final day, my mother gave the entire family one last gift. Her granddaughter Claire was to celebrate her bat mitzvah on Saturday, June 22, 2003. According to tradition, if Mom had died, the bat mitzvah would still have been held, but there would have been no party afterward. As the day of the bat mitzvah dawned, Mom's breathing was slow and shallow. Mom's skin was graying, even beneath the slight tan she was so pleased to have acquired. She had entered that deep coma where all organs slow down just before life ends.

Two weeks earlier, Claire, one of those children who are always well prepared, had chanted her complete haftorah at Mom's bedside. Throughout the day of the bat mitzvah, Mom hung on, unconscious. My daughter Lea and I sang soft Hebrew melodies in the afternoon at her bedside. We went to the party that night, where most of our family drank and danced harder than usual. Mom took her last breath just after the party ended, at 12:45 Sunday morning.

A month after my mother died, I was sitting with my dear friend Andrea Martin in her small terraced townhouse overlooking Marin Bay, just off the Golden Gate bridge, north of San Francisco. With our husbands, both named Richard, we watched what we all knew would be her last Fourth of July fireworks. My daughter, Lea, who has always had the capacity to bring peace to others, was there, as well as Andrea's suddenly grown-up daughter, Mather. My friend's high Slavic cheek bones had disappeared into a bloated face. Her eyes twinkled like a candle that was about to lose its flame. She had truly become a breast cancer survivor. She lay in bed dying from a completely new cancer of the brain that had invaded her skull and spread to her lungs.

Unlike my parents, Andrea could talk about what was going on at the end of her life. A natural comic, she made this terminal scene become another opportunity for what might be called lie-down comedy. Her home was filled with multicolored Tibetan prayer flags, Jewish and New Age talismans, hiking poles and other mementos from climbing expeditions. The walls were covered with poems and love letters from hundreds of friends and admirers, some from complete strangers whose lives she had inspired with her climbing expeditions for the clan of one-breasted women.

Prominently displayed were outrageous advertisements, so over the top that the San Francisco public transit authority removed them from buses. These movie-size posters featured knock-offs of *Cosmopolitan* magazine covers with gorgeous women with nearly perfect bodies and bare chests. "This is what breast cancer looks like," they declared. Where luscious, sometimes artificial breasts would have been, Andrea's double mastectomy scars had been superimposed. You knew you were dealing with someone powerful since she had managed to come up with public advertisements that got pulled in this most liberal American city.

Some children display striking qualities from their earliest days, evincing a wisdom and maturity that most of us only develop in adulthood. My daughter, Lea, has always been able to find a way to comfort those in physical or emotional disarray. As a kindergartner, she insisted on coming with me to the intensive care unit to sit with my father. Unfazed by all the tubes, she held her small hand on his, which was taped to a board and had intravenous tubes running from it. She sang quietly with my dad when he was running out of time. "Dovid, Melech Yisreal, Chai, Chai, V'kayom."

Now, as an adult who had just the month earlier chanted to my dying mother, she wanted to join me at Andrea's bedside. We drove together from Lea's urban outpost in Oakland, California, across the always inspiring Bay Bridge to Andrea's home. Andrea looked like she was sleeping. We sang the same prayerful melodies at her bedside that we had sung to my mother. We figured singing couldn't hurt. It certainly made us feel better. When we had finished one melody, Andrea opened one eye and gave us a look that said, "Enough with the songs, guys. I'm still here. So, talk to me."

I offered up one of our old standard jokes, "Hey, girlfriend, you know why they put nails in coffins?" I asked. "Sure," she replied. "To keep out the oncologists."

No oncologists were needed at this point. Andrea, her husband and daughter had sought the haven of hospice. This brought into her home a steady stream of massage therapists, soothing sounds and quiet moments. Andrea was fearless. We spoke about the ways in which our lives and those we love would remain connected even after death. I repeated the story of my mishap with the yellow jackets, when I almost died.

When I had finished, I said, "I can't really be sure that all this happened to me. I felt that I had come to a place more beautiful than anything I had ever known. I just can't prove it to you. But I believe that at the end we are surrounded by holiness."

Andrea looked to me. She frowned. She shook her head, "I'm so disappointed."

I felt just awful. Maybe I had just gone over the edge. "I'm so sorry. I don't want to let you down," I replied.

Andrea repeated herself, speaking slowly and clearly. "I'm really very disappointed in you." She let the words sink in. She was waiting for some response from me, but I couldn't figure out what. I began to get upset. "I am so disappointed," she said again.

She looked at me through her swollen face. Her eyes told me that a good joke was about to come. "Why can't you prove it? After all, you're the scientist!" She laughed.

That was our last laugh, and I will savor it for the rest of my life.

No one really knows what happens. I understand that. Now words are all that I have as I struggle to depict the memories of the ways I tried to assure my parents that at the end of life it really isn't all over. It's just all over here and now, in the space and time that we have known our entire lives on earth. And, perhaps, as my near-death experience suggested to me, we go to a place that nobody can ever know, where we achieve a peace beyond anything that can be experienced on this earth.

One of the many things I would argue with my mother about was life after death. She insisted there is nothing we can know at the end. "When you are dead, you are dead."

Of course, none of us ever comes back with perfect proof, so no one can be sure on this issue. Does dying in a state of peace with the world mean that we enter a realm where we will experience full peacefulness? My mother wanted to hang on to this world as long as possible.

In Judaism there is a tradition of honoring the memories of family members by visiting their graves just before the Jewish New Year begins. At this time we are also supposed to remember the millions who perished through the ages. Some cemeteries, like those of McKeesport's Orthodox Jewish community, have also created their own memorials to those who perished in the Holocaust. In the well-maintained Gemilas Chesed cemetery just up Ripple Road around a small bend off Pennsylvania Route 48, the Holocaust monument is just a bit larger than the typical tombstone. A large, mostly blank white marble slab on which a few names are inscribed sits in the midst of the cemetery, right behind the tombstones of my mother and father, giving their graves a central focus that they both would love.

For years, I had seen this mostly empty stone but never really looked at it. It loomed as something vaguely somnolent, unbidden, uninteresting but unavoidable. The Holocaust, a half century after it ended, remains an abstraction even to those of us who realize that it did not mark the end of inhumanity, but merely one of humanity's worst chapters. The very anonymity and enormity of genocide makes it easy for people to dismiss it as an aberration, as though such a calamity happened somewhere else to someone else and is of little relevance to anyone's life now. The emptiness of the large white stone space says it all. We do not even know the names of most of those who perished.

In the late summer of 2003, my brothers and sister and I lingered, lost in our own thoughts on a sunny Sunday morning at our parents' graves. As we finished our recitation of the Kaddish memorial prayer, a slow, ambling group of elderly people gathered near the mostly empty Holocaust memorial. They plodded with walkers and canes over still damp grass. At age eighty-eight, Joe Ungar could barely walk. He had been coming to this place for more than fifty years. Even though his legs were no longer working well, he had to be counted.

Rabbi Chinn, tall and commanding at age eighty-two, stood at the memorial and spoke about the need to remember not just those we knew and loved, but those without names. Some of those inscribed on the stone had one or two survivors, who had come to remember sisters, brothers, mothers and fathers dead for more than half a century. The rabbi recited every single name. Where it was listed, he mentioned the town and date of the deaths. "Samuel and Laib Rahlah Small, Moish, Hersh, Bella, Ari, Feige, Sima, Ruchel and Leah Weingarten. Kimmelman (Hrubeshauer Kalisz). Einhorn Friedman (Puschalan-Bratislav, Shmini Atzeres, 1941) Newman Family, Rozen Family, Glanternik Family, Kalisz Poland, Benovitz Family, Losha and their families, Liebish, Rivka and Milla, Eisenberg Family, Efraim and Blima Zell and children."

Wiping his brow, the rabbi continued, "Mirel Weinberger and children, Harry and Lina Weingarten." At this last name, a soft sigh and sob was heard. Tears welled in the squinting crinkled eyes of Maydala Mandelbaum, a woman well into her seventies who stood quietly, with legs planted at the monument like small trees that can bend but not break. She turned to me and to anyone else within her sight. "Sixty years ago, I lost everyone, when I was just a girl. My whole family was gone. No one survived. But I am here to remember them."

Some folks who had come to make the customary visit before the New Year to their deceased family members' gravesites began to leave. Golf games and workouts were waiting. Sensing that he might lose a critical opportunity, Rabbi Chin called out, "Do we have a minyan?" He referred to the need in traditional Judaism to gather ten men before it is possible to recite the traditional memorial prayer of Kaddish.

I chimed in, "Well, Rabbi, I have a man's job, so perhaps I can be counted."

"Thank-you for your offer. But we only have nine men, and we need another." He smiled and explained. "You are a woman of valor but cannot be a part of our minyan."

My brother Martin, who like me had begun to walk away, came back and stood as the tenth man. My parents would have been pleased at all the attention and delighted to have us hang back a bit longer to honor those who perished years ago. They always did love a crowd. Even in death, they reminded us about that which we might have preferred to forget.

I LIKE TO THINK that they also would love what this book has done. My mother would certainly take credit and be quite pleased that I ended up working precisely where she had sent me—at Pittsburgh's Cancer Institute. As Andrea and other friends in the breast cancer movement have taught me, cancer that you survive, like much breast cancer today, can be a defining moment. It also sometimes stinks. Those of us who confront a disease with such a lethal limit know that all life is precious and restricted in a way that few others can imagine. Do I wish that my mother had managed to escape? That my dad had been given more time? Of course I do. But I have come to understand that we are truly on the verge of major breakthroughs, not in just the science with which we understand how to prevent the illness, but in the ways that we can manage and prevent it.

The choice is ours to make. Do we wait for more certain proof about potential causes of cancer, both natural and man-made, and continue practices that have left one in three women and half of all men confronting this illness in their lifetimes? Do we act to avoid those risks that appear avoidable, while science continues its mission of amassing enough data to arrive at unquestionable conclusions? Do we acquiesce to the demand that we can only look at cancer hazards one at a time, even though life comes at us all at once as the ultimate mixture?

Understanding what we can do as a society to lessen the burden of cancer does not require a radical shift in the ways our brains work, but in the ways we make decisions as a society about how we work and organize our lives. We must continue to tally the toll of cancer, even as we seek to avert its growth. We also must stop demanding proof at the level of the individual and accept the limits of what we can know in science at this point.

The dramatic drop in smoking-related deaths in some groups around the world is no accident. This success is also exceptional, because it stems from controlling a well-defined hazard that was known to be a problem long before society finally acted. But those who forced us to wait for incontrovertible proof exacted a heavy price in premature deaths. Millions perished while the debate on tobacco lingered far longer than it should have, and millions of those in the developing world are headed toward certain death because of these delays.

Today, the debate on chemicals and radiation has been framed by the same terms and fanned by some of the same expert public relations strategies that kept us tied in knots on tobacco. We are repeatedly asked to prove that people have already been harmed before acting to stave off future damage. Rather than accepting the burden that the environment, as we have changed and are changing it, places on our health only after the evidence has become undeniable, as happened with tobacco and with some sickening buildings, we need to embrace cleaner, less polluting ways of organizing the world. In the nineteenth century, better housing, sanitation, and the end of child labor led to healthier and longer lives. This happened long before scientists understood the ways that germs festered in dark, dank environments and often led to disabling and lethal infectious diseases. Women, whether the Ladies Factory Inspectorate in England, the Women's Federations of America, or more modern community activists or philanthropists, championed many of these basic advances. Today women, and the men who support them, are working at the local, national and international levels, advocating parallel efforts to reduce the use of cosmetic pesticides in schools and homes, promote less polluting cleaning and grooming agents and make greener, cleaner and less toxic hospitals and communities. If we insist on having at hand absolute proof that harm has happened before we move to prevent or control damage, we are dooming future generations.

There is a story from Rabbi Tarfun in the Talmud that cannot be told too many times. It sums up how we should think about environmental health challenges today. A group of workers has been given a big, complicated job to do. They complain. "We do not have the right tools. The task is enormous. We will never be able to get it done."

The rabbi replies, "It is not for you to complete the task. But you must begin."

Notes

PREFACE

1. Margaret Atwood, *Negotiating with the Dead: A Writer on Writing* (Cambridge: Cambridge University Press, 2002), p. xxiv.

2. *Puki tlusty schudine, to chudy zdechnie.* Polish proverb.

3. John Bailar and H. L. Gornik, "Cancer Undefeated," *New England Journal of Medicine* 336, no. 22 (1997): 1569–1574.

4. Devra L. Davis, David Hoel, John Fox, and Alan D. Lopez, "International Trends in Cancer Mortality in France, West Germany, Italy, Japan, England and Wales, and the United States," *Lancet* 336 (1990): 474–481, expanded and reprinted in Devra L. Davis and David Hoel, eds., "Trends in Cancer Mortality in Industrial Countries," *Annals of the New York Academy of Sciences* 609 (1990).

5. *Report of the International Congress of Scientific and Social Campaign Against Cancer,* Brussels, 1936, vols. 1–3.

6. W. C. Heuper, *Occupational Tumours and Allied Diseases* (1942).

7. David Michaels, "Doubt Is Their Product," *Scientific American;* Robert Proctor, *Cancer Wars* (New York: Basic, 1995).

8. E. Kessler and P. W. Brandt-Rauf, "Occupational Cancers of the Nervous System," *Seminar in Occupational Medicine* 2 (1987): 311–314.

9. M. Mehlman, "Benzene: A Haematopoietic and Multi-Organ Carcinogen at Any Level Above Zero," *European Journal of Oncology* 9, no. 1 (2004): 15–36.

10. L. A. Peipins et al., "Radiographic Abnormalities and Exposure to Asbestos-Contaminated Vermiculite in the Community of Libby, Montana, USA," *Environ Health Perspect* 111, no. 14 (2003): 1753–1759; A. Schneider and D. McCumber, *An Air That Kills: How the Asbestos Poisoning of Libby, Montana, Uncovered a National Scandal* (New York: Putnam, 2004).

11. Rachel Carson, *Silent Spring* (Boston: Houghton Mifflin, 1962); Murray Bookchin, *Our Synthetic Environment* (New York: Knopf, 1962).

12. Larry Agran, *The Cancer Connection* (New York: St. Martin's, 1977); Samuel Epstein, *The Politics of Cancer* (New York: Sierra Club Books, 1980); James T. Patterson, *The Dread Disease: Cancer and Modern American Culture* (Cambridge: Harvard University Press, 1987); Janette Sherman, *A Delicate Balance* (New York: Taylor & Francis, 2000).

13. Proctor, *Cancer Wars.*

14. Robert N. Proctor, *The Nazi War on Cancer* (Princeton: Princeton University Press, 1999).

15. Sandra Steingraber, *Having Faith* (New York: Perseus Publishing, 2001) and *Living Downstream* (New York: Vintage Books, 1998).

16. Mitchell Gaynor, *Nurture Nature Nurture Health* (New York: Nurture Nature Press, 2005) and "The New War on Cancer: Against All Causes," *Explore* 1, no. 4 (2005).

17. D. Markowitz and D. Rosner, *Deceit and Denial: The Deadly Politics of Industrial Pollution* (New York: Milbank Memorial Fund, 2003).

18. *Combined Financial Statements as of and for the Year Ended August 31, 2005, American Cancer Society, Inc., and Affiliated Entities* (Atlanta: Ernst & Young, 2006), pp. 3–5. See http://www.cancer.org/downloads/AA/ACS%20Combined%20Financials%20FY%2005%20Final%20issued.pdf.

19. www.army.mil/cmh/reference/etocode.html.

20. Barry Castleman, "Asbestos Is Not Banned in North America," *European Journal of Oncology* 11, no. 2 (2006): 85–88.

21. M. Goldberg, E. Imbernon, P. Rolland et al., "The French National Mesothelioma Surveillance," *Occupational Environmental Medicine* 63 (2006): 390–395.

22. K. E. Shepherd, L. C. Oliver, and H. Kazemi, "Diffuse Malignant Pleural Mesothelioma in an Urban Hospital: Clinical Spectrum and Trend in Incidence over Time," *American Journal of Industrial Medicine* 16 (1989): 373–383. This retrospective analysis reviews the clinical experience of a major urban referral hospital with diffuse malignant pleural mesothelioma during the fourteen-year period from 1973 through 1986. In twenty-one cases (33 percent), there was no known history of asbestos exposure.

23. Agran, *Cancer Connection;* Epstein, *Politics of Cancer;* Patterson, *Dread Disease;* Proctor, *Cancer Wars;* Sherman, *Delicate Balance;* Bookchin, *Our Synthetic Environment;* Carson, *Silent Spring.*

The "First Total War," David A. Bell, Houghton Mifflin reviewed by Adam Gopnik, in the *New Yorker,* February 12, 2007. Gopnik notes that the French philosophes were profoundly opposed to war, "Condorcet . . . thought that war, like murder will one day number among those extraordinary atrocities which reflect and shame nature . . . : But in reaction to the failures of war to end . . . liberal idealism of the enlightenment gave vent to the notion of a 'war to end all wars' a decisive battle that would end all conflict."

"Wars to end all wars give way to wars that never end." Gopnik, p. 82. "The vision of war as redemptive continues to flourish." Bell quoted by Gopnik. "All wars are total to the people they kill." Adam Gopnik, "Slaughterhouse," *New Yorker* 82, no. 49 (2007): 85.

CHAPTER I

1. Environmental Working Group, Body Burden Studies, www.ewg.org/bodyburden/results.php?subject=bb1_sub1.

2. Amr S. Soliman, Melissa L. Bondy, Bernard Levin, Mohamed R. Hamza, Kadry Ismail, Sohair Ismail, et al. "Colorectal Cancer in Egypt Patients Under 40 Years of Age," *International Journal of Cancer* 71 (1997): 26–30.

3. David Biello, "Mixing It Up: Harmless levels of chemicals prove toxic together," *Scientific American,* May 10, 2006.

4. Lester Breslow, Larry Agran, and Devra Breslow, *A History of Cancer Control in the United States, 1946–1971,* Report to the National Cancer Institute on the Early History of the Cancer Control Program, Funded by the National Cancer Institute, Washington, D.C. Xerox, DHEW pub. no. (NIH) 79-1519, 1979; in possession of the author.

CHAPTER 2

1. I. Berenblum, "Cancer Research in Historical Perspective: An Autobiographical Essay," *Cancer Res.*, January 1977, 1–7.

2. W. Cramer, "The Importance of Statistical Investigations in the Campaign Against Cancer," *Report of the Second International Congress of Scientific and Social Campaign Against Cancer*, Brussels, 1936, p. 443.

3. Cramer, "Importance of Statistical Investigations," pp. 452–453.

4. Ibid., p. 444.

5. A. H. Roffo, "La etiologia fisica-quimica del cancer (sobre todo en relacion con las irradiaciones solares), *Ponencias Congreso International De Lucha Cientifica y Social Contra el Cancer*, Brussels, 1936, p. 76.

6. Roffo, "La etiologia fisica-quimica del cancer," p. 84.

7. U.S. Department of Health and Human Services, Public Health Service, National Toxicology Program, *Report on Carcinogens*, 10th ed.

8. Suessmann Mutner, "Moses ben Maimon," in *Encyclopaedia Judaica*, 2d ed.

9. George H. Nash, *The Life of Herbert Hoover*, vol. 1 (New York: Norton, 1983).

10. Donald Hunter, *The Diseases of Occupations*, 5th ed. (London: Hodder & Stoughton, 1973), p. 35.

11. Hunter, *Diseases of Occupations*.

12. V. B. Kamble, "X-rays: The Unknown Glimmer," www.vigyanprasar.gov.in/dream/mar2001/X-Rays.htm (accessed April 4, 2007).

13. See the National Library of Medicine Dream Anatomy website: www.nlm.nih.gov/exhibition/dreamanatomy/images/1200%20dpi/Z1.jpg.

14. Howard H. Seliger, "Wilhelm Conrad Rontgen and the Glimmer of Light," *Physics Today*, November 1995.

15. A. Schedel, "An Unprecedented Sensation—Public Reaction to the Discovery of x-rays," *Physics Education* 30 (1995): 324–347.

16. Schedel, "Unprecedented Sensation," pp. 324–347.

17. John L. Montgomery, "Diagnostic Imaging: Finding New Ways to See; Seeing New Ways to Cure," *Postgraduate Medicine*, January 1997, pp. 144–148, 155.

18. Nanny Froman, "Marie and Pierre Curie and the Discovery of Polonium and Radium," December 1, 1996, http://nobelprize.org/nobel_prizes/physics/articles/curie.

19. National Research Council, *Biographical Memoirs: National Academy of Sciences of the United States of America*, vol. 67 (Washington, D.C.: National Academy Press, 1995).

20. André F. Cournand, *From Roots to Late Budding* (New York: Gardner, 1985), pp. 165–182.

21. R. Forssmann-Falck, "Werner Forssmann: A Pioneer of Cardiology," *American Journal of Cardiology*, March 1, 1997, pp. 651–660.

22. Ibid.

23. National Safety Council, Chemical and Rubber Sections, *Final Report of the Committee, Chemical and Rubber Sections, National Safety Council, on Benzol, May 1926* (New York: National Bureau of Casualty and Surety Underwriters, 1926), p. 4.

24. Ibid., p. 121.

25. Ibid., p. 118.

26. P. Drinker, *API Toxicology Review: Benzene* (American Petroleum Institute, 1948).

27. B. Castleman, *Asbestos: Medical and Legal Aspects,* 5th ed. (New York: Aspen, 2005), p. 62.

28. W. B. Cannon, "Reflections on the Man and His Contributions," *International Journal of Stress Management,* April 1994, pp. 145–158.

29. Ibid.

30. D. Ferber, "Blocked Cancer Study Published," *Science,* October 2006, p. 579.

CHAPTER 3

1. Office of Strategic Services, *Hitler Source Book,* "Interview with Dr. Eduard Bloch," March 5, 1943. Available at www.nizkor.org/hweb/people/h/hitler-adolf/oss-papers/text/oss-sb-bloch-02.html.

2. Rudolph Binion, *Hitler Among the Germans* (New York: Elsevier, 1976); Sandy Macleod, "Mrs. Hitler and Her Doctor," *Australasian Psychiatry* 13, no. 4 (2005): 412.

3. Attachment theorists would have a field day connecting these traumas to a host of pathological traits. See Richard Bowlby, *Fifty Years of Attachment Theory: Recollections of Donald Winnicott and John Bowlby (Donald Winnicott Memorial Lecture)* (London: Karnac, 2004); Binion, *Hitler;* Ron Rosenbaum, *Explaining Hitler* (New York: Random House, 1999).

4. History Learning Site, "Adolf Hitler," www.historylearningsite.co.uk/adolf_hitler.htm.

5. Rudolph Binion, "Hitler's Concept of Lebensraum: The Psychological Basis," *History of Childhood Quarterly* 1, no. 2 (1973).

6. Ibid.

7. Robert N. Proctor, "The Nazi War on Tobacco: Ideology, Evidence, and Possible Cancer Consequences," *Bulletin of the History of Medicine* 71, no. 3 (1997): 463.

8. Office of Strategic Services, "Interview with Dr. Eduard Bloch."

9. Janet Browne, *Charles Darwin: Voyaging* (New York: Knopf, 1995).

10. Francis Galton, *Hereditary Genius: An Inquiry into Its Laws and Consequences* (London: Macmillan, 1869).

11. Edwin Black, *War Against the Weak: Eugenics and America's Campaign to Create a Master Race* (London: Turnaround, 2004).

12. Alexandra Minna Stern, *Eugenic Nation: Faults and Frontiers of Better Breeding in Modern America* (Berkeley: University of California Press, 2005).

13. Jeremiah A. Barondess, "Care of the Medical Ethos: Reflections on Social Darwinism, Racial Hygiene, and the Holocaust," *Annals of Internal Medicine* 129, no. 11 (1998): 891–898.

14. "Biography of Harry H. Laughlin," http://library.truman.edu/manuscripts/laughlin bio.htm.

15. "Europe as an Emigrant-Exporting Continent," Harry H. Laughlin testimony before the House Committee, including Immigration Restriction Act, www.eugenicsarchive.org/html/eugenics/index2.html?tag=1140.

16. Garland Allen, "Is a New Eugenics Afoot?" *Science* 294, no. 5540 (2001): 59–61.

17. Greta Jones, *Social Hygiene in Twentieth Century Britain* (London: Croon Helm, 1986), p. 180.

18. Bertrand Russell, *The Sanctity of Life and the Criminal Law*, as quoted by Professor Glanville Williams, the Rous Ball Professor of English law at Cambridge University, a fellow of the English Eugenics Society.

19. As early as 1903, the dedication to George Bernard Shaw's philosophy and comedy of *Man and Superman* satirized the charitable and hypocritical impulses of the eugenicists: "Being cowards, we defeat natural selection under cover of philanthropy: being sluggards, we neglect artificial selection under cover of delicacy and morality."

20. John Cornwell, *Hitler's Scientists* (New York: Penguin, 2003), p. 88.

21. Robert Proctor, *Racial Hygiene: Medicine Under the Nazis* (Cambridge: Harvard University Press, 1989) p. 97.

22. Thomas C. Leonard, "Protecting Family and Race: The Progressive Case for Regulating Women's Work," *American Journal of Economics and Sociology* 64, no. 3 (2005).

23. Allen, "Is a New Eugenics Afoot?"

24. André N. Sofair and Lauris C. Kaldjian, "Eugenic Sterilization and a Qualified Nazi Analogy: The United States and Germany, 1930–1945," *Annals of Internal Medicine* 132, no. 4 (2000): 312–319. By 1944, thirty states with sterilization laws had reported more than 40,000 eugenic sterilizations; of those sterilized, 20,600 were reported as insane and 20,453 as feebleminded (p. 61). In the pre-Nazi period, German eugenicists expressed admiration for U.S. leadership in instituting sterilization programs and communicated with their U.S. colleagues about strategies (p. 62). Despite waning scientific and public support and the history of the human rights abuses of Nazi Germany, state-sponsored sterilizations in the United States continued long after the war. Between 1943 and 1963, approximately 22,000 American citizens were sentenced to surgery to end their ability to reproduce in twenty-seven states (p. 60).

25. Sofair and Kaldjian, "Eugenic Sterilization."

26. Ibid.

27. Leonard, "Protecting Family and Race."

28. Ibid.

29. Paul Lombardo, "Facing Carrie Buck," *Hastings Center Report* 33 (2003): 14–17.

30. Stephen Jay Gould, "Carrie Buck's Daughter: A Popular, Quasi-Scientific Idea Can Be a Powerful Tool for Injustice—This View of Life," *Natural History,* July-August 2002.

31. Edwin Black, *War Against the Weak* (New York: Thunder's Mouth, 2003).

32. Neal Baldwin, *Henry Ford and the Jews* (New York: Public Affairs, 2001).

33. Black, *War Against the Weak.*

34. Gerhard L. Weinberg, ed., *Hitler's Second Book: The Unpublished Sequel to* Mein Kampf (New York: Enigma, 2003).

35. Cornwell, *Hitler's Scientists.*

36. Ernst Dormanns, "Die Vergleichende Geographisch-Pathologische Reichs-Carcinomastatistik, 1925–1933" (The Comparison of Geographic-Pathological Statistics of Germany on Cancer, 1925–1933), *Report of the Second International Congress of Scientific and Social Campaign Against Cancer* (Brussels, 1936), pp. 460–482.

37. Robert Payne, *The Life and Death of Adolf Hitler* (New York: Praeger, 1973), pp. 346–347.

38. Robert Lifton, *The Nazi Doctors: Medical Killing and the Psychology of Genocide* (New York: Basic, 1986), p. 217.

39. Ibid.

40. Jeremiah A. Barondess, "Reflections on Social Darwinism, Racial Hygiene, and the Holocaust," *Annals of Internal Medicine* 129, no. 11 (1998): 895, citing M. H. Kater, "Hitlerjugend und Schule im Dritten Reich," *Historische Zeitschrift* 228 (1979): 609–610.

41. Proctor, *Nazi War on Cancer*, p. 241.

42. Carmen Callil, *Bad Faith: A Forgotten History of Family, Fatherland, and Vichy France* (New York: Knopf, 2006).

43. Laurence Jourdan, "Gypsy Hunt in Switzerland: Long Pursuit of Racial Purity," *Le Monde Diplomatique*, October 1999, http://mondediplo.com/1999/10/11gypsy.

44. R. R. Reilly, "Eugenic Sterilization in the United States," in A. Milunsky and George J. Annas, eds., *Genetics and the Law* (New York: Plenum, 1985) p. 236, as cited in Jeremiah A. Barondess, "Reflections on Social Darwinism," pp. 891–898.

45. Gerhard Weinberg, personal conversation, March 19, 2007.

46. Ibid.

47. "Revealed: The Amazing Story Behind Hitler's Second Book," *Telegraph,* September 9, 2003.

48. Reilly, "Eugenic Sterilization."

49. Jonathan M. Samet, "Smoking Kills: Experimental Proof from the Lung Health Study," *Annals of Internal Medicine* 142, no. 4 (2005): 299–301.

50. Thomas E. Addison, "A Chronology of Tobacco in the Civilized World," *San Francisco Medicine,* July 1998, www.mindfully.org/Industry/Tobacco-ChronologyJul98.htm.

51. Angel H. Roffo, "Krebszeugendes Benzpyren, gewonnen aus Tabakteer," *Zeitung Krebsforschung* 49, no. 5 (1940): 88–97; Verhandlungen der deutschen pathologischen Gesellschaft (1923), p. 19, as cited in Rober Proctor, "Commentary: Schairer and Schoniger's Forgotten Tobacco Epidemiology and the Nazi Quest for Racial Purity," *International Journal of Epidemiology* 30 (2001): 31–34.

52. Robert Proctor, "Angel H. Roffo: The Forgotten Father of Tobacco Carcinogenesis," *Bulletin of the World Health Organization* 84, no. 6 (2006): 494–496.

53. Ibid.

54. S. Zimmermann, M. Egger, and U. Hossfeld, "Commentary: Pioneering Research into Smoking and Health in Nazi Germany: The Wissenschaftliches Institut zur Erforschung der Tabakgefahren in Jena," *International Journal of Epidemiology* 30, no. 1 (2001): 35–37.

55. Robert Proctor, *The Nazi War on Cancer* (Princeton: Princeton University Press, 1999).

56. F. H. Müller, "Tabakmissbrauch und Lungencarcinoma," *Eingegangen am 24*, December 1938, pp. 57–85.

57. See www.fhs.mcmaster.ca/pbls/writing/probono.htm.

58. Müller, "Tabakmissbrauch."

59. E. Schairer and E. Schoniger, "Lungenkrebs und Tabakverbrauch," *Zeitung Krebsforschung* 54 (1943): 261–269; *International Journal of Epidemiology* 30, no. 1 (2001): 31–34.

60. Ibid.

61. Proctor, *Nazi War on Cancer*, p. 10.

62. Ibid., pp. 196–197.

63. Richard Doll, "Commentary: Lung Cancer and Tobacco Consumption," *International Journal of Epidemiology* 30, no. 1 (2001): 30–31.

64. Proctor, *Nazi War on Cancer*, p. 217.

65. Warren Winkelstein, "Janet Elizabeth Lane-Claypon: A Forgotten Epidemiologic Pioneer," *Epidemiology* 17, no. 6 (2006): 705–706.

66. Hans A. Baer, Merrill Singer, and Ida Susser, *Medical Anthropology and the World System,* 2nd ed. (New York: Praeger Paperback, 2003).

67. Proctor, *Racial Hygiene*, p. 237.

68. Ibid.

69. Lifton, *Nazi Doctors.*

70. Richard Bessel, "Hatred After the War: Emotion and the Postwar History of East Germany," *History and Memory* 17, no. 2 (2005): 195–216.

71. See UMKC law school web site: www.law.umkc.edu/faculty/projects/ftrials/nuremberg/NurembergIndictments.html.

72. John Gimbel, *Science, Technology, and Reparations: Exploitation and Plunder in Post-war Germany* (Stanford: Stanford University Press, 1990)

73. See www.spartacus.schoolnet.co.uk/2WWv2.htm.

74. Michael Brian Petersen, "Engineering Consent: Peenemuende, National Socialism, and the V-2 Missile, 1924–1945," Ph.D. diss., https://drum.umd.edu/dspace/bitstream/1903/2861/1/umi-umd-2357.pdf.

75. Cornwell, *Hitler's Scientists*, p. 422.

76. For the signature of Erich Lepier with swastika, see *II Band: Erst Hälfte: Tafel* 2 *Abb.* 13, *Tafel 33 Abb.* 14, *Tafel 14 Abb.* 25, *Tafel* 15, *Abb.* 26, *Tafel 16 Abb.* 27, *Tafel 17 Abb.* 28, *Tafel* 18, *Abb.* 29, *Tafel 32 Abb.* 43, *Page* 351 *Abb.* 108, and *Tafel 65 Abb.* 4. For the signature of Karl Entresser with "SS" symbol, see *III Band: Tafel 9 Abb.* 14, and *Tafel 10 Abb.* 15. E. Pernkopf, *Topographische Anatomie des Menschen: Lehrbuch und Atlas der regionär-stratigraphischen Präparation* (Vienna: Urban & Schwarzenberg, 1952). For the signature of Franz Bratke with "SS" symbol, see *III Band: Tafel 9 Abb.* 14, and *Tafel 10 Abb.* 15. *Parasitologia* 42, no. 1–2 (2000): 53–58. For antimalarial drugs, see *Malaria and World War II: German Malaria Experiments, 1939–45*; as cited in Israel H. Seidelman, "Nazi Origins of an Anatomy Text: The Pernkopf Atlas," *Journal of the American Medical Association* 276, no. 20 (1997): 1633.

77. Seidelman, "Nazi Origins."

78. D. C. Angetter, "Anatomical Science at the University of Vienna 1938–45," *Lancet* 355 (2000): 1445–1457.

79. Seidelman, "Nazi Origins." Presse-Konferenz der Universität Wien zu den Recherchen über den Anatomieatlas "Topographische Anatomie des Menschen" von Eduward Pernkopf (1937, 1989) und das dazu eingeleitete Forschungsproject "Untersuchungungen zur Anatomischen Wissenschaft an der Universität Wien 1938–1945." Vienna, February 12, 1997. A. Ebenbauer and W. Schütz, "Origins of the Pernkopf *Atlas* (letter)," *JAMA* 277 (1997): 1123–1124. P. Malina, "Eduard Pernkopf's *Atlas of Anatomy* or: The fiction of pure science," *Wien. Klin. Wschr.* 110 (1998): 193–201. An English translation of an interim report of the Pernkopf Commission was published in *Wien. Klin. Wschr.* 109 (1997): 935–943. Senatsprojekt der Universität Wien, *Untersuchungen zur Anatomischen Wissenschaft in Wien: 1938–1945* (Vienna, 1998).

80. William E. Seidelman, "Nuremberg Lamentation: For the Forgotten Victims of Medical Science," *British Medical Journal* 313, no. 7070 (1997).

81. D. J. Williams, "The History of Eduard Pernkopf's *Topographische Anatomie des Menschen*," *Journal of Biomedical Communications* 2, no. 12 (1988).

82. Angetter, "Anatomical Science," p. 1456.

CHAPTER 4

1. Larry Agran, *The Cancer Connection* (New York: St. Martin's, 1977), p. 175.

2. Proctor, *The Nazi War on Cancer*, p. 1.

3. David Michaels, "When Science Isn't Enough: Wilhelm Hueper, Robert A. M. Case, and the Limits of Scientific Evidence in Preventing Occupational Bladder Cancer," *International Journal of Occupational and Environmental Health* 1 (1995): 278–288.

4. Autobiography, also cited in Michaels, 1995, pp. 282–283.

5. Robert Proctor, *Abhandlungen zur Geschichte der Medizin und der Naturwissenschaften* 81 (1997): 290–305.

6. Nriagu Jo, "Clair Patterson and Robert Kehoe's Paradigm of 'Show Me the Data' on Environmental Lead Poisoning," *Environ Res.,* August 1998, pp. 71–78.

7. William Kovarik, "The 1920s Environmental Conflict over Leaded Gasoline and Alternative Fuels" (paper presented to the American Society for Environmental History Annual Conference March 26–30, 2003, Providence, RI), www.radford.edu/%7Ewkovarik/papers/ethylconflict.html.

8. William Kovarik, "Ethyl-leaded Gasoline," *International Journal of Occupational and Environmental Health,* October-December 2005, pp. 384–397.

9. Roy Albert, Memoirs, p. 37, www.eh.uc.edu/ceg/pdf/albertsmemoirs.pdf.

10. Kehoe papers, University of Cincinnati, provided to author.

11. Jamie Lee Kitman, "The Secret History of Lead: Special Report," *The Nation,* March 20, 2000.

12. Joseph Borkin, *The Crime and Punishment of I. G. Farben* (New York: Free Press, 1978).

13. Joseph C. Robert, *Ethyl: A History of the Corporation and the People Who Made It* (Charlottesville: University Press of Virginia, 1984).

14. Ibid., p. 8.

15. Antony C. Sutton, *Wall Street and the Rise of Hitler* (Seal Beach, CA: 76 Press, 1976). http://reformed-theology.org/html/books/wall_street/chapter_04.htm.

16. United States Senate, Hearings before a Subcommittee of the Committee on Military Affairs, *Scientific and Technical Mobilization,* 78th Congress, 2nd sess., part 16 (Washington, D.C.: Government Printing Office, 1944), p. 939, cited in Sutton, *Wall Street and the Rise of Hitler.*

17. Edgar B. Nixon, ed., *Franklin D. Roosevelt and Foreign Affairs,* Volume III: September 1935–January 1937 (Cambridge: Belknap Press, 1969), p. 456, as cited in Sutton, *Wall Street and the Rise of Hitler.*

18. James Stewart Martin, *All Honorable Men* (Boston: Little Brown, 1950), cited in Sutton, introduction, p. 1.

19. Hanspeter Witschi, "Fritz Haber: 1868–1934," *Toxicological Sciences* 44 (2000): 1–2; and Witschi, "Some Notes on the History of Haber's Law," *Toxicological Sciences* 50 (1999): 164–168.

20. Michael Berenbaum, *A Promise to Remember: The Holocaust in the Words and Voices of Its Survivors* (Boston: Bullfinch, 2003), p. 40.

21. E. C. Dodds, L. Goldberg, W. Lawson, and R. Robinson, "Estrogenic Activity of Certain Synthetic Compounds," *Nature* 141, no. 3562 (1938): 247–248.

22. Office of Military Government for Germany, Field Information Agency Technical FIAT Review of German Science, 1939–1946; see also Proctor, *Nazi War on Cancer,* p. 273.

23. Cited by Herbert L. Needleman, "Clair Patterson and Robert Kehoe: Two Views of Lead Toxicity," *Environmental Research* 78 (1998): 79–85.

24. Medical records obtained from University of Cincinnati Medical Center Kehoe Papers Archive. In possession of author.

25. Needleman, op. cit.

26. Clair Patterson, http://oralhistories.library.caltech.edu/32.

27. Lorenzo Tomatis, "Experimental Chemical Carcinogenesis: Fundamental and Predictive Role in Protecting Human Health in the 1930s–1970s," *European Journal of Oncology* 11 (2006): 5–13.

28. Wilhelm Hueper, *Autobiography,* pp. 222–223.

29. Memorandum of understanding between Kehoe and Ethyl Corporation. In possession of author.

30. Ibid., p. 223.

31. Wilhelm Hueper, *Occupational Tumors and Allied Diseases* (Springfield, IL: Thomas, 1942).

32. L. W. Pickle, T. J. Mason, N. Howard, R. Hoover, J. F. Fraumeni Jr., Atlas of U.S. Cancer Mortality Among Whites: 1950–1980 (Washington, D.C.: United States Government Printing Office, 1990) p. 98.

33. W. C. Hueper, *Environmental Cancer* (Washington, D.C.: U.S. Government Printing Office, 1950).

34. Hueper, *Autobiography,* p. 203.

35. Agran, *Cancer Connection,* p. 180.

36. Hueper, *Autobiography,* p. 213.

CHAPTER 5

1. Breslow, *A History of Cancer Control,* p. 18.

2. Ibid.

3. W. H. Walshe, *Physical Diagnosis of Diseases of the Lungs* (London: Taylor & Walton, 1843).

4. Clark T. Sawin, "George N. Papanicolaou and the Pap Test," *Endocrinologist* 12, no. 4 (2002): 267–272.

5. General Federation of Women's Clubs, "About Us," www.gfwc.org/about_us.jsp?pageId=2090611881251062016982136.

6. The Women's Field Army of the American Society for the Control of Cancer, *There Shall Be Light!* (New York: New York City Cancer Committee, 1945).

7. "Cancer Foes Begin Nation-Wide Drive," *New York Times,* November 21, 1936, p. 19.

8. Baron H. Lerner, "Fighting the War on Breast Cancer Debates over Early Detection, 1945 to the Present," *Annals of Internal Medicine* 129, no. 1 (1998): 74–78.

9. Clarence Cook Little, *Civilization Against Cancer* (New York: Little & Ives, 1939).

10. Little, *Civilization Against Cancer,* 138.

11. Ibid., 117.

12. "Cancer Army."

13. *Time,* March 22, 1937, p. 56.

14. Little, *Civilization Against Cancer,* p. 124.

15. Ibid.

16. National Research Council, *Biographical Memoirs: National Academy of Sciences of the United States of America,* vol. 46 (Washington, D.C.: National Academy Press, 1995).

17. Eagle Pictures/American Society for the Control of Cancer, *Enemy X* (Washington, D.C.: U.S. Public Health Service, 1942).

18. Richard Rettig, *Cancer Crusade* (Princeton: Princeton University Press, 1977), p. 20.

19. G. N. Papanicolaou and Herbert Frederick Traut, *Diagnosis of Uterine Cancer by the Vagina Smear* (NewYork: Commonwealth Fund, 1943).

20. Collie Small, "AreYou Risking Cancer—Because of False Modesty?" *Reader's Digest*, February 1952, p. 11.

21. Leslie J. Reagan, "Engendering the Dread Disease:Women, Men, and Cancer," *American Journal of Public Health* 87, no. 11 (1997).

22. D. A. Boyes, M. K. Fidler, and D. R. Lock, "Significance of In Situ Carcinoma of the Uterine Cervix," *British Medical Journal* 1 (1962): 203–205.

23. BC Cancer Agency, "About Cervical Screening," www.bccancer.bc.ca/PPI/Screening/Cervical/About.htm.

24. http://muse.jhu.edu/journals/bulletin_of_the_history_of_medicine/v081/81.1toon.html#REF42.

25. H. S. Ahluwalia and Richard Doll, "Mortality from Cancer of the Cervix Uteri in British Columbia and Other Parts of Canada," *British Journal of Preventive and Social Medicine* 22, no. 3 (1968): 161–164.

26. Karen Canfell, Freddy Sitas, and Valerie Beral, "Cervical cancer in Australia and the United Kingdom: Comparison of Screening Policy and Uptake, and Cancer Incidence and Mortality," *Medical Journal of Australia* 185, no. 9 (2006): 482–486.

27. *Combined Financial Statements as of and for theYear Ended August 31, 2005,American Cancer Society, Inc., and Affiliated Entities* (Atlanta: Ernst &Young, 2006), pp. 3–5.

28. Rettig, *Cancer Crusade*, p. 21.

29. Ibid.

30. Ibid., p. 22.

31. Ibid.

32. Ibid.

33. Breslow, *History of Cancer Control*, p. 843.

34. See www.historylink.org/essays/output.cfm?file_id=7531.

35. Breslow, *History of Cancer Control*, p. 843.

36. Ibid.

37. S. Sato, G. Matsunaga, R. Konno, and A.Yajima, "Mass Screening for Cancer of the Uterine Cervix in Miyagi Prefecture, Japan: Effects and Problems," *Acta cytologica* 42, no. 2 (1998): 299–304.

38. Leif Gustafsson, Jan Pontén, Matthew Zack, and Hans-Olov Adami, "International Incidence Rates of Invasive Cervical Cancer After Introduction of Cytological Screening," *Cancer Causes and Control* 8 (1997): 755–763.

39. Mike Quinn, Penny Babb, Jennifer Jones, and Elizabeth Allen, "Effect of Screening on Incidence of and Mortality from Cancer of Cervix in England: Evaluation Based on Routinely Collected Statistics," *British Medical Journal,* April 1999, pp. 904–908.

40. Judith Siers-Poisson, "The Politics and PR of Cervical Cancer," *Counterpunch,* June 30/July 1, 2007 (http://www.counterpunch.org/siers06302007.html), confirms that Gardisil is another instance of overselling and understudying a women's health remedy.

41. Eric Schoch, "Unwanted Sex Appears Common in SomeTeen Relationships," *Medical News Today*, June 6, 2006, www.medicalnewstoday.com/medicalnews.php?newsid=44635.

42. See http://transcripts.cnn.com/TRANSCRIPTS/0606/09/lt.02.html.

43. E. J. Samelson, M. A. Speers, R. Ferguson, and C. Bennett, "Racial Differences in Cervical Cancer Mortality in Chicago," *American Journal of Public Health* 84, no. 6 (1994): 1007–1009.

44. Deborah S. Porterfield, Genevieve Dutton, and Ziya Gizlice, "Cervical Cancer in North Carolina: Incidence, Mortality and Risk Factors," *North Carolina Medical Journal* 64, no. 1 (2003): 11–17.

45. Rosalind J. Neuman et al., "Association Between DQB1 and Cervical Cancer in Patients with Human Papillomavirus and Family Controls," *Obstetrics & Gynecology* 95, no. 1 (2000): 134–140.

46. Katha Pollit, "Virginity or Death," *The Nation,* May 30, 2005, www.thenation.com/doc/20050530/pollitt.

CHAPTER 6

1. John E. Calfee, "The Ghost of Cigarette Advertising Past," *Regulation* 10, no. 2 (1986).

2. John Stauber and Sheldon Rampton, *Toxic Sludge Is Good for You!* (Monroe, ME: Common Courage, 1995), p. 2.

3. Jonathan Samet, "Smoking Kills," *Annals of Internal Medicine* 142 (2005): 291–301.

4. Jonathan M. Samet and Frank E. Speizer, "Sir Richard Doll, 1912–2005," *American Journal of Epidemiology,* May 1–3, 2006.

5. Lester Breslow, "Control of Cigarette Smoking from a Public Policy Perspective," *Annual Review of Public Health* 3 (1982): 129–151.

6. Cuyler Hammond, "Smoking and Lung Cancer: Pros and Cons," *CA: A Cancer Journal for Clinicians* 5 (1955): 88–94.

7. John Stauber and Sheldon Rampton, *Trust Us, We're Experts: How Industry Manipulates Science and Gambles with Your Future* (New York: Tarcher/Penguin, 2001); John Stauber and Sheldon Rampton, "How the American Tobacco Industry Employs PR Scum to Continue Its Murderous Assault on Human Lives," *Tucson Weekly*, November 22–29, 1995, www.tucsonweekly.com/tw/11-22-95/cover.htm.

8. Stanton A. Glantz et al., eds., *The Cigarette Papers* (Berkeley: University of California Press, 1996), p. 37.

9. Clarence Cook Little, "Some Phases of the Problem of Smoking and Lung Cancer," *New England Journal of Medicine*, June 15, 1961, pp. 1241–1245.

10. Robert N. Proctor, *Cancer Wars* (New York: Basic, 1995), pp. 101–132.

11. Breslow, *A History of Cancer Control,* p. 846.

12. Donald G. Cooley, "Smoke Without Fear," *True Magazine,* 1954, http://tobaccodocuments.org/landman/11310873-0908.html.

13. Council for Tobacco Research Continuation Funding for Drs. Seltzer and Mancuso: Request for Authorization, 1972, http://tobaccodocuments.org/bliley_pm/21277.html.

14. W. C. Hueper, "The Cigarette Theory of Lung Cancer," *Current Medical Digest,* October 1954, pp. 35–39.

15. *The Facts About Smoking* (Consumer Reports Books, 1954).

16. Robert Proctor, "Tobacco and Health," Expert Witness Report Filed on Behalf of Plaintiffs in: *The United States of America, Plaintiff, v. Philip Morris, Inc., et al., Defendants,* Civil Action No. 99-CV-02496 (GK) (Federal case), reprinted in *The Journal of Philosophy, Science & Law* 4 (March 2004).

17. Howard Kurtz, "American Council on Science and Health Brief in Formaldehyde Suit Financed by Chemical Manufacturer," *Washington Post,* June 3, 1984.

18. Ashbel C. Williams, "The Committee on Tobacco and Cancer of the American Cancer Society," *CA: A Cancer Journal for Clinicians* 17 (1967): 259–260.

19. Breslow, *History of Cancer Control,* p. 849.

20. Ibid., p. 854.

21. Thomas E. Addison, "A Chronology of Tobacco in the Civilized World," *San Francisco Medicine,* July 1998.

22. Addison, "Chronology of Tobacco."

23. *Smoking and Health: Summary and Report of the Royal College of Physicians of London on Smoking in Relation to Cancer of the Lung and Other Diseases* (London: Pitmann, 1962).

24. *I'll Choose the High Road* (American Cancer Society, 1962).

25. Ross Hammond and Andy Rowell, "Trust Us: We're the Tobacco Industry," *Campaign for Tobacco-Free Kids (USA), Action on Smoking and Health (UK),* May 2001.

26. Interview, *Newsday,* December 18, 1964, http://tobaccodocuments.org/ti/TIMN0110397-0404.html.

27. Proctor, *Cancer Wars,* p. 228.

28. Richard Doll, "Commentary: Lung Cancer and Tobacco Consumption," *International Journal of Epidemiology* 30 (2001): 30–31.

29. See http://ark.cdlib.org/ark:/13030/ft8489p25j.

30. Allen Brandt, *The Cigarette Century* (New York: Basic, 2007).

31. Howard Wolinsky and Tom Brune, "Smoking Gun: Playing Politics with Tobacco and the Public's Health," in *The Serpent on the Staff: The Unhealthy Politics of the American Medical Association* (New York: Putnam, 1994), pp. 144–174.

32. *New York Times,* December 23, 1971, p. 28.

CHAPTER 7

1. Stanton A. Glantz et al., eds., *The Cigarette Papers* (Berkeley: University of California Press, 1996), p. 109.

2. Ibid., p. 113.

3. Ibid.

4. "Tobacco Industry Hires Top Cancer Scientist to Head Its Research," *Wall Street Journal,* Midwest edition, June 16, 1954, as cited in Karen Miller, "Smoking Up a Storm: Public Relations and Advertising in the Construction of the Cigarette Problem, 1953–1954," *Journalism Monographs* 126 (1992): 22.

5. William E. Longo, Mark W. Rigler, and John Slade, "Crocidolite Asbestos Fibers in Smoke from Original Kent Cigarettes," *Cancer Research* 55 (1995): 2232–2235.

6. Lois Mattox Miller and James Monahan, "Wanted and Available: Filter Tips That Really Filter," *Reader's Digest,* August 1957, pp. 43–49.

7. Lois Mattox Miller and James Monahan, "The Facts Behind Filter-tip Cigarettes," *Reader's Digest,* July 1957, p. 6.

8. Ibid.

9. *I'll Choose the High Road* (American Cancer Society, 1962).

10. Richard Kluger, *Ashes to Ashes* (New York: Knopf, 1995), p. 423.

11. Ibid., p. 428.

12. Ibid.

13. Ibid., p.429.

14. Ibid., p. 452.

15. Tursi et al., "One-Fanged Rattler."

16. Kluger, *Ashes to Ashes,* p. 447.

17. Ibid., p. 448.

18. Peter Crawford, letter to P. Sheehy, December 29, 1986, www.library.ucsf.edu/tobacco/batco/html/12800/12823/otherpages/allpages.html.

19. Frank Tursi, Susan E. White, and Steve McQuilkin, "One-Fanged Rattler," *Winston-Salem Journal,* 1999, http://extras.journalnow.com/lostempire/tob20b.htm.

20. Mark S. Boguski, "The Mouse That Roared," *Nature* 420, no. 6915 (2002): 515.

21. Adapted from Devra Davis, *When Smoke Ran Like Water* (New York: Basic, 2002).

22. Austin Bradford Hill, "The Environment and Disease: Association or Causation?" *Proceedings of the Royal Society of Medicine* 58 (1965): 295–300.

23. Raymond Pearl, "Tobacco Smoking and Longevity," *Science* 87, no. 2253 (1938): 216–217; Jonathan M. Samet, "Smoking Kills: Experimental Proof from the Lung Health Study," *Annals of Internal Medicine* 142, no. 4 (2005): 299–301.

24. Allen Brandt, *The Cigarette Century* (New York: Basic, 2006).

25. Edward L. Bernays, *The Later Years: Public Relations Insights, 1956–1986* (Rhinebeck, NY: H&M, 1986), p. 11, as cited in John Stauber and Sheldon Rampton, *Toxic Sludge Is Good for You: Lies, Damned Lies, and the Public Relations Industry* (Monroe, ME: Common Courage, 1995), p. 32.

26. Edward Bernays, *Propaganda* (New York: Horace Liveright, 1928), p. 9; as cited in Stauber and Rampton, *Toxic Sludge,* p. 24.

CHAPTER 8

1. www.freerepublic.com/focus/f-vetscor/1479888/posts.

2. E. B. Krumbhaar, "The Role of the Blood and the Bone Marrow in Certain Forms of Gas Poisoning," *Journal of the American Medical Association* (1919): 39–41.

3. Alfred Gilman, "The Initial Clinical Trial of Nitrogen Mustard," *American Journal of Surgery,* May 1963.

4. Ibid.

5. John Curtis, "From the Field of Battle, an Early Strike on Cancer," http://yalemedicine.yale.edu/ym_su05/capsule.html.

6. Louis S. Goodman et al., "Nitrogen Mustard Therapy," *Journal of the American Medical Association* 132 (1946): 126–132.

7. Judith Robinson, *Noble Conspirator: Florence S. Mahoney and the Rise of the National Institutes of Health* (Washington, D.C.: Francis, 2001), p. xiv.

8. Richard Rettig, *Cancer Crusade: The Story of the National Cancer Act of 1971* (Authors Choice Press, 2005), p. 26.

9. Dr. Philip Randolph Lee, interview for History of Health Services Research Project, www.nlm.nih.gov/hmd/nichsr/lee.html.

10. M. Yamakido, S. Ishioka, K. Hiyama, and A. Maeda, "Former Poison Gas Workers and Cancer: Incidence and Inhibition of Tumor Formation by Treatment with Biological Response Modifier N-CWS," *Environmental Health Perspective* 3, no. 140 (1996): 485–488.

11. Ibid.

12. Haruko Taya Cook and Theodore F. Cook, *Japan at War* (New Press, 1993), p. 153.

13. Y. Yoshimi and S. Matsuno, *Dokugasuen Kankei shiryo II, Kaisetsu* (Jugonen senso gokuhi shiryoshu 1997), pp. 27–29.

14. Gilbert Beebe, "Lung Cancer in WWI Veterans: Possible Relation to Mustard Gas Injury," *Journal of the National Cancer Institute*, December 1960, p. 52.

15. Yamakido et al., "Former Poison Gas Workers," 485–488.

16. Karimi Zarchi, Ali Akbar, and Kourosh Holakouie Naieni, "Long-term Pulmonary Complications in Combatants Exposed to Mustard Gas: A Historical Cohort Study," *International Journal of Epidemiology* 33, no. 3 (2004): 579–581.

17. "Middle East Cancer Incidence Through 2001," *Journal of the National Cancer Institute*, July 19, 2006, p. 957.

CHAPTER 9

1. Elizabeth Economy, *The Rivers Run Black* (Ithaca, NY: Cornell University Press, 2005).

2. Arden Pope et al., "Lung Cancer, Cardiopulmonary Mortality, and Long-Term Exposure to Fine Particulate Air Pollution," *Journal of the American Medical Association* 287, no. 9 (2002): 1132–1141.

3. Occupational Safety and Health Administration (OSHA), U.S. Department of Labor, "Substance safety data sheet for ethylene oxide (non-mandatory)—1910.1047 App A," www. osha.gov/pls/oshaweb/owadisp.show_document?p_table=STANDARDS&p_id=1007.

4. See www.atsdr.cdc.gov/toxprofiles/phs187.html#bookmark07.

5. International Agency for Research on Cancer, "Summaries and Evaluations: Ethylene Oxide," 1994, www.inchem.org/documents/iarc/vol60/m60-02.html.

6. Agency for Toxic Substances and Disease Registry, Centers for Disease Control, "Tox-FAQs for Ethylene Oxide," www.atsdr.cdc.gov/tfacts137.html.

7. Kyle Steenland, Elizabeth Whelan, James Deddens, Leslie Stayner, and Elizabeth Ward, "Ethylene Oxide and Breast Cancer Incidence in a Cohort Study of 7576 Women (United States)," *Cancer Causes and Control* 14, no. 6 (2003): 531–539.

8. K. Gunnarsdottir, T. Aspelund, T. Karlsson, and V. Rafnsson, "Occupational Risk Factors for Breast Cancer Among Nurses," *International Journal of Occupational and Environmental Health* 3 (1997): 254–258; J. Hansen and J. H. Olsen, "Cancer Morbidity Among Danish Female Pharmacy Technicians," *Scandinavian Journal of Work, Environment & Health* 20 (1994): 22–26; L. I. Levin, E. A. Holly, and J. P. Seward, "Bladder Cancer in a 39-year-old Female Pharmacist," *Journal of the National Cancer Institute* 85 (1993): 1089–1090; S. A. Petralia, M. Dosemeci, E. E. Adams, and S. H. Zahm, "Cancer Mortality Among Women in Health Care Occupations in 24 U.S. States, 1984–1993," *American Journal of Industrial Medicine* 36 (1999): 159–165.

CHAPTER 10

1. *Fifth Annual Report of the Registrar-General* (1842), appendix, pp. 206–207; and *Tenth Annual Report of the Registrar-General* (1847), p. xvii; as cited by UCLA School of Public Health, "William Farr: Campaigning Statistician," www.ph.ucla.edu/epi/snow/farr/farr_mist.html.

2. Devra Lee Davis and Brian H. Magee, "Cancer and Industrial Chemical Production," *Science* 206, no. 4425 (1979): 1356.

3. Devra Lee Davis, "Cancer in the Workplace: The Case for Prevention," *Environment* 23, no. 6 (1981): 30–31.

4. Devra Davis and Abraham Lilienfeld, "Increasing Trends in Some Cancers in Older Americans: Fact or Artifact?" *Toxicology and Industrial Health* 2, no. 1 (1986): 127–144.

CHAPTER 11

1. James S. Michaelson, Elkan Halpern, and Daniel B. Kopans, "Breast Cancer: Computer Simulation Method for Estimating Optimal Intervals for Screening," *Radiology* 212 (1999): 551–560.

2. John W. Gofman, *Radiation and Human Health* (San Francisco: Sierra Club Books, 1981); Rosalie Bertell, *No Immediate Danger: Prognosis for a Radioactive Earth* (Summertown, TN: Book Publishing Company, 2000).

3. Cited in Barron Lerner, *The Breast Cancer Wars: Hope, Fear, and the Pursuit of a Cure in Twentieth-Century America* (New York: Oxford University Press, 2001).

4. Philip Strax, *Early Detection: Breast Cancer Is Curable* (New York: Harper & Row 1974), p. xiii.

5. See http://cms.komen.org/komen/AboutBreastCancer/EarlyDetectionScreening/EDS3-3-3?ssSourceNodeId=292&ssSourceSiteId=Komen and http://www.natlbcc.org/bin/index.asp?strid=496&depid=9&btnid=1.

6. Jane E. Brody, "Health; Personal Health," *New York Times,* August 2, 1990.

7. See http://thomas.loc.gov/cgi-bin/bdquery/z?d102:HR06182:@@@L&summ2=m&.

8. Devra Lee Davis and Susan M. Love, "Mammographic Screening," *Journal of the American Medical Association* 271, no. 2 (1994): 152–153.

9. Environmental Research Foundation, "The Truth About Breast Cancer, Part 2," *Rachel's Environment & Health News*, November 12, 1997, www.rachel.org/bulletin/bulletin.cfm?Issue_ID=548.

10. Devra Lee Davis and H. Leon Bradlow, "Can Environmental Estrogens Cause Breast Cancer?" *Scientific American* 273, no. 4 (1995): 167–172.

11. See www.medicalnewstoday.com/medicalnews.php?newsid=70889.

12. Cornelia J. Baines, "Are There Downsides to Mammography Screening?" *Breast Journal* 11, Suppl. 1 (March-April 2005): S7–10.

13. Ibid.

14. See www.natlbcc.org/bin/index.asp?strid=560&depid=9&btnid=1.

15. See http://cms.komen.org/Komen/AboutBreastCancer/EarlyDetectionBSEFAQs.

16. Lucille Adams-Campbell, K. Makambi, J. Palmer, and L. Rosenberg, "The Gail Model as a Diagnostic Indicator in African-American Women: Truth or Consequence," *American Society of Clinical Oncology,* 2004.

17. R. G. Ziegler et al., "Migration Patterns and Breast Cancer Risk in Asian-American Women," *Journal of the National Institute of Cancer* 85, no. 22 (1993): 1819–1827.

18. Alex Berenson and Andrew Pollack, "Doctors Reap Millions for Anemia Drugs," *New York Times,* May 9, 2007.

19. Guy B. Faguet, *The War on Cancer: An Anatomy of Failure, a Blueprint for the Future* (New York: Springer 2006).

20. Iain Chalmers, *Archie Cochrane (1909–1988)* (Oxford: James Lind Initiative 2006), www.jameslindlibrary.org/trial_records/20th_Century/1940s/cochrane/cochrane_biog.html.

21. Ruthann A. Rudel, Kathleen R. Attfield, Jessica N. Schifano, and Julia Green Brody, "Chemicals Causing Mammary Gland Tumors in Animals Signal New Directions for Epidemiology, Chemicals Testing, and Risk Assessment for Breast Cancer Prevention," *Cancer* 109, no. S12 (2007): 2635–2666.

CHAPTER 12

1. Randolph Jonakait, "Text, Texts, or Ad Hoc Determinations: Interpretation of the Federal Rules of Evidence," *Indiana Law Journal* 71, no. 3 (1996), www.law.indiana.edu/ilj/volumes/v71/no3/jonakait.html.

2. David Michaels, "Doubt Is Their Product," *Scientific American* 292, no. 6 (2005): 96–101.

3. Andrew Schneider and David McCumber, *An Air That Kills: How the Asbestos Poisoning of Libby, Montana, Uncovered a National Scandal* (New York: Putnam, 2004).

4. Barry Castleman, *Asbestos: Medical and Legal Aspects,* 5th ed. (New York: Aspen, 2005), chap. 1.

5. Isaac Berenblum, "Cancer Research in Historical Perspective: An Autobiographical Essay," *Cancer Research* 37, no. 1 (1977): 1–7.

6. David S. Egilman and Candice M. Hom, "Corruption of the Medical Literature: A Second Visit," *American Journal of Industrial Medicine* 34 (1998): 401–404.

7. Castleman, *Asbestos.*

8. Ibid., pp. 49–59.

9. Ibid.

10. Barry Castleman, personal communication, October 2006.

11. Evarts A. Graham, "Remarks on the Aetiology of Bronchogenic Carcinoma," *Lancet* 263, no. 6826 (1954): 1305–1308, as cited in Allan M. Brandt, *The Cigarette Century* (New York: Basic, 2006), p. 129.

12. Castleman, *Asbestos*, pp. 82–83.

13. Ibid., p. 82.

14. Ibid., p. 84.

15. Ibid.

16. Sarnia profile brochure, www.city.sarnia.on.ca/pdf/SARNIA_PROMO_JUN06.pdf.

17. Laurie Kazan-Allen, "Lies + Subterfuge = Canada's Asbestos Policy," *International Ban Asbestos Secretariat*, www.btinternet.com/~ibas/Frames/f_lka_lies_sub_can_asb_pol.htm.

18. Martin Mittelstaedt, "Dying for a Living," March 13, 2004, www.mesothel.com/pages/dying.htm; James T. Brophy, "The Public Health Disaster Canada Chooses to Ignore," February 20, 2007, www.mirg.org/mesothelioma-news/2007/02/20/the-public-health-disaster-canada-chooses-to-ignore.

19. See www.city.sarnia.on.ca.

20. Mittelstaedt, "Dying for a Living."

21. Margaret M. Keith and James T. Brophy, "Participatory Mapping of Occupational Hazards and Disease Among Asbestos-Exposed Workers from a Foundry and Insulation Complex in Canada," *International Journal of Occupational and Environmental Health* 10, no. 2 (2004): 144–153.

22. M. Landsberg, "The Tragic Legacy of Sarnia's White Death," *Toronto Star*, November 14, 1999.

23. James Brophy, personal correspondence, February 15, 2007.

24. Ibid.

25. Bruce Livsey, "The Asbestos Files," *Eye Weekly*, October 28 1999, http://www.eye.net/eye/issue/issue_10.28.99/news/asbestos.html.

26. Environment and Development Desk, Department of Information and International Relations, Central Tibetan Administration, "Resource Extraction: State of the Environment," *Tibet 2003: State of the Environment*, July 2003, www.tew.org/tibet2003/t2003.resource.ext.html.

27. Ontario Ministry of the Attorney General, *Report of the Royal Commission on Matters of Health and Safety Arising from the Use of Asbestos in Ontario*, 1984, 3:817.

28. Castleman, *Asbestos;* Colin L. Soskolne and David V. Bates, "Canada's Double Standard on Asbestos," *Edmonton Journal*, April 26, 2006.

29. Barry Castleman, "Asbestos Is Not Banned in North America," *European Journal of Oncology* 11, no. 2 (2006): 85–88.

30. P. G. Barbieri, S. Lombardi, A. Candela, C. Pezzotti, and I. Binda, "Incidence of Malignant Mesothelioma (1980–1999) and Asbestos Exposure in 190 Cases Diagnosed Among Residents of the Province of Brescia," *La Medicina del lavoro* 92, no. 4 (2001): 249–262. Only seven cases of asbestosis were diagnosed in the MM cases, whereas thirty-one cases of pleural abnormalities were observed, but only seventeen of these were observed in workers occupationally exposed to asbestos. G. Gorini, S. Silvestri, E. Merler, E. Chellini, V. Cacciarini, and A. S. Seniori Costantini, "Tuscany Mesothelioma Registry (1988–2000): Evaluation of Asbestos Exposure," *La Medicina del lavoro* 93, no. 6 (1993): 507–518. The article describes the incidence of pleural mesothelioma cases in Tuscany to analyze their possible past asbestos exposures. In 24 percent of the interviewed cases (15 percent of males; 74 percent of females) no known asbestos exposure was identified. A. C. Pesatori and C. Mensi, "Peculiar Features of Mesothelioma Occurrence as Related to Exposure Patterns and Circumstances in the Lombard Region, Italy," *La Medicina del lavoro* 96, no. 4 (2005): 354–359. They concluded that the high proportion of cases with unknown exposure underlines the need to explore new tools and sources to ascertain asbestos exposure. V. Gennaro, D. Ugolini, P. Viarengo, L. Benfatto, M. Bianchelli, A. Lazzarotto, F. Montanaro, and R. Puntoni, "Incidence of Pleural Mesothelioma in Liguria Region, Italy (1996–2002)," *European Journal of Cancer* 41, no. 17 (2005): 2709–2714. In this study, the incidence of pleural malignant mesothelioma in the Liguria region of Italy in the presence of asbestos exposure was investigated. Asbestos exposure was unlikely or unknown for 57.5 percent of females and 15 percent of males. V. Ascoli, C. C. Scalzo, F. Facciolo, M. Martelli, L. Manente, P. Comba, C. Bruno, and F. Nardi, "Malignant Mesothelioma in Rome, Italy, 1980–1995: A Retrospective Study of 79 patients," *Tumori* 82, no. 6 (1996): 526–532. The study confirmed that mesothelioma risk is present in several job titles of the construction industry, and it is no longer confined to workers employed in the manufacture or application of asbestos products. The occurrence of malignant mesothelioma in patients with unexpected occupational and nonoccupational exposures indicates the need for further investigation on previously underestimated exposures. Exposure to asbestos was assessed for 45.5 percent of patients; another 45.5 percent had unknown exposure.

31. International Society for Environmental Epidemiology, *Ethics and Philosophy,* www.iseepi.org/about/ethics.html.

32. American College of Epidemiology, *Ethics Guidelines,* http://acepidemiology2.org/policystmts/EthicsGuide.asp.

33. Barry Castleman and Grace E. Ziem, "Business Ethics and Threshold Limit Values: Response to Letter to the Editor," *American Journal of Industrial Medicine* 28, no. 2 (1995): 299–300.

34. Lennart Hardell et al., "Secret Ties to Industry and Conflicting Interests in Cancer Research," *American Journal of Industrial Medicine* 50, no. 3 (2007): 227–233.

35. Hardell et al., "Secret Ties"; Devra Davis, *When Smoke Ran Like Water* (New York: Basic, 2002), pp. 136–138.

36. Michaels, *Doubt Is Their Product.*

37. Brief of Amicus Curiae Concerned Scientists in Opposition to Defendants-Appellants' Application for Leave to Appeal, Chapin v. A & L Parts, Inc., no. 257917 (Mich. App. Jan. 30, 2007).

38. Heather H. Nelson and Karl T. Kelsey, "The Molecular Epidemiology of Asbestos and Tobacco in Lung Cancer," *Nature* 21, no. 48 (2002): 7284–7288.

39. D. Reid and C. Buck, "Cancer in Coking Plant Workers," *British Journal of Industrial Medicine* 13, no. 4 (1956): 265–269.

40. "Making the Invisible Visible: Thomas Sugrue Recounts the Story of the Urban Crisis," *University of Pennsylvania School of Arts and Sciences Newsletter,* Fall 1998.

41. Carol K. Redmond, "Study of Mortality Among Retirees," p. 99.

42. James E. Bowman and Robert F. Murray, *Genetic Variation and Disorders in People of African Origin* (Baltimore, MD: Johns Hopkins University Press, 1990).

43. Carol K. Redmond, "Cancer Mortality Among Coke Oven Workers," *Environmental Health Perspectives,* October 1983, pp. 67–73; Redmond et al., "Long Term Mortality of Steelworkers," *Journal of Occupational Medicine* 14 (1972): 621–629.

44. Carol K. Redmond and Patricia P. Breslin, "Comparison of Methods for Assessing Occupational Hazards," *Journal of Occupational Medicine* 17, no. 5 (1975): 313–317.

45. John William Shepherd, "A Select Bibliography of the History of Coal Mining in the State of Pennsylvania," Catholic University of America, 1999–2006, http://libraries.cua.edu/achrcua/coalbib.html.

46. "*Daubert:* The Most Influential Supreme Court Ruling You've Never Heard Of," Project on Scientific Knowledge and Public Policy, Tellus Institute, 2003.

47. United States Court of Appeals, *Ethyl Corp v. EPA,* April 14, 1995, www.ll.georgetown.edu/federal/judicial/dc/opinions/94opinions/94-1505a.html.

48. John Carey, "Medical Guesswork," *BusinessWeek Online,* May 29, 2006, www.businessweek.com/magazine/content/06_22/b3986001.htm.

CHAPTER 13

1. See www.ejnet.org/rachel/rhwn292.htm.

2. *Six Case Studies of Compensation for Toxic Substances Pollution: Alabama, California, Michigan, Missouri, New Jersey and Texas: A Report prepared under the supervision of the Congressional Research Service of the Library of Congress for the Committee on Environment and Public Works U.S. Senate at the request of Senators John C. Culver and Robert T. Stafford,* June 1980, serial no. 96-13, 96th Congress, 2nd sess. (Washington, D.C.: U.S. Government Printing Office, 1980).

3. "Toxic Town," interview with Russell Bliss, CNN News, June 26, 1997, www.cnn. com/US/9706/26/times.beach/transcript/index1.htm.

4. Padma Tadi-Uppala, *Biomarkers of Genotoxicity Induced by DDT and Risk for Breast Cancer in Madison County Alabama: Annual Summary* (Huntsville, AL: Oakwood College, 2001).

5. National Academy of Sciences, Board on Environmental Studies and Toxicology, *Environmental Epidemiology: Public Health Hazardous Wastes* (Washington, D.C.: National Academy Press, 1991), 76.

6. Ibid.

7. Ibid., p. 21.

8. National Research Council Committee on Environmental Epidemiology, *Environmental Epidemiology,* vol. 2, *Use of the Gray Literature and Other Data in Environmental Epidemiology* (Washington, D.C.: National Academy Press, 1997), p. 168.

9. M. Perales, "Smeltertown: A Biography of a Mexican American Community, 1880–1973" (Ph.D. diss., Stanford University, 2003).

10. See www.denix.osd.mil/denix/Public/Library/EMS/Documents/finalreport4.html.

11. See www.historyoftechnology.org/eTC/v47no1/allen.html.

12. http://www.atsdr.cdc.gov/HAC/PHA/marinesp/msp_p2.html.

CHAPTER 14

1. Jack Anderson and Les Whitten, February 11, 1974.

2. Barry Castleman, "Regulations Affecting Use of Carcinogens," in N. Irving Sax, ed., *Cancer-Causing Chemicals* (New York: Van Nostrand Reinhold, 1981), p. 88.

3. Ibid.

4. M. Lamy and P. Maroteaux, "Acro-osteolyse dominante," *Archives françaises de pédiatrie* 18 (1961): 693–702; and J. A. Ross, "An Unusual Occupational Bone Change," in A. M. Jelliffe and B. Strickland, eds., *Symposium Ossium* (London: Livingstone, 1970).

5. See www.pbs.org/tradesecrets/program/vinyl.html.

6. D. K. Harris and W. G. F. Adams, "Acro-Osteolysis Occurring in Men Engaged in the Polymerization of Vinyl Chloride," *British Medical Journal* 3 (1967): 712–714.

7. R. H. Wilson, W. E. McCormick, C. F. Tatum, and J. L. Creech, "Occupational Acro-osteolysis: Report of 31 Cases," *Journal of the American Medical Association* 201 (1967): 577–581.

8. Pietro L. Viola, A. Bigott, and A. Caputo, "Oncogenic Response of Rat Skin, Lungs, and Bones to Vinyl Chloride," *Cancer Research* 31 (1971): 516–522.

9. John Creech (deposition) as quoted in Gerald Markowitz and David Rosner, *Deceit and Denial* (Berkeley: University of California Press, 2002), p. 173.

10. Centers for Disease Control and Prevention, *Morbidity and Mortality Weekly Report* 46 (1997): 97–101.

11. Larry Agran, *The Cancer Connection* (New York: St. Martin's, 1977), p. 167.

12. Ibid., pp. 47–49.

13. Ibid., p. 50.

14. Nicholas A. Ashford, "The Use of Technical Information in Environmental, Health, and Safety Regulation: A Brief Guide to the Issues," *Science, Technology, & Human Values* 9 (1984): 130–133.

15. Jon Weiner, "Cancer, Chemicals and History," *The Nation*, February 7, 2005.

16. Gerald Markowitz and David Rosner, *Deceit and Denial* (Berkeley: University of California, 2002), p. 193.

17. Lila Guterman, "Peer Reviewers Are Subpoenaed in Cancer Lawsuit against Chemical Companies," *Chronicle of Higher Education,* November 19, 2004.

18. Richard Doll, "Effects of Exposure to Vinyl Chloride: An Assessment of the Evidence," *Scandinavian Journal of Work and Environmental Health* 14 (1988): 61–78.

19. Ibid.

20. Richard S. Doll, Deposition of William Richard Shaboe Doll, Ross V. Conoco, Inc., Case No. 90-4837. LA 14th Judicial District Court, London UK, January 27, 2000, cited in Sass et al., notes, p. 811.

21. Jennifer B. Sass, Barry Castleman, and David Wallinga, "Vinyl Chloride: A Case Study of Data Suppression and Misrepresentation," *Environmental Health Perspectives* 113, no. 7 (2005): 809–812.

22. George Roush, letter to Richard Doll, April 29, 1986, from Doll archives, Oxford University; copy provided by Martin J. Walker.

23. Alice Hamilton, "Naptha and Benzol Poisoning," in Kober and Hanson, eds., *Occupational Intoxications* (1916), pp. 136–144.

24. Ibid., p. 142.

25. Alice Hamilton, "Industrial Poisons Encountered in the Manufacture of Explosives," *Journal of the American Medical Association* 68, no. 20 (1917).

26. P. Drinker, *API Toxicology Review: Benzene* (American Petroleum Institute, 1948).

27. Dina Capiello, "Oil industry Funding Study to Contradict Cancer Claims: Research Will Analyze Effects of Benzene on Workers in China," *Houston Chonicle*, April 29, 2005.

CHAPTER 15

1. Vicki Wolf, "Brian Yeoman: Compassionate Futurist," www.cleanhouston.org/heros/yeoman.htm.

2. Ibid.

3. Clifford Pugh, "New Building Stands Out Amidst the Texas Medical Center's Sterile Architecture," *Houston Chronicle,* 2005, www.houstonarchitecture.info/haif/lofiversion/index.php/t1106.html.

4. "Green Light for Australian Ban on Old-Style Bulb," *Guardian Unlimited,* February 21, 2007, www.guardian.co.uk/australia/story/0,,2017669,00.html.

5. Joshua E. Muscat et al., "Handheld Cellular Telephone Use and Risk of Brain Cancer," *Journal of the American Medical Association* 284, no. 23 (2000): 3001–3007.

6. Joachim Schüz et al., "Cellular Telephone Use and Cancer Risk: Update of a Nationwide Danish Cohort," *Journal of the National Cancer Institute* 98, no. 23 (2006): 1707–1713.

7. L. Hardell et al., "Case-Control Study on the Use of Cellular and Cordless Phones and the Risk for Malignant Brain Tumours," *International Journal of Radiation Biology* 78, no. 10 (2002): 931–936.

8. S. Lonn, A. Ahlbom, P. Hall, and M. Feychting, "Mobile Phone Use and the Risk of Acoustic Neuroma," *Epidemiology* 15, no. 6 (2004): 653–659.

9. Roberta B. Ness, James S. Koopman, and Mark S. Roberts, "Causal System Modeling in Chronic Disease Epidemiology," *Annals of Epidemiology*, February 26, 2007 (EPub).

10. Joachim Schüz et al., "Cellular Phones, Cordless Phones, and the Risks of Glioma and Meningioma (Interphone Study Group, Germany)," *American Journal of Epidemiology* 163, no. 6 (2006): 512–520.

11. J. Schuz, E. Bohler, G. Berg, B. Schlehofer, I. Hettinger, K. Schlaefer et al., "Cellular Phones, Cordless Phones, and the Risks of Glioma and Meningioma (Interphone Study Group, Germany)," *American Journal of Epidemiology* 163, no. 6 (2006): 512–520; Anna Lahkola et al., "Mobile Phone Use and Risk of Glioma in 5 North European Countries," *International Journal of Cancer* 120, no. 8 (2007): 1769–1775.

12. Ed Edelson, "Men with Breast Cancer at High Risk of Second Tumor," *HealthDay News*, January 25, 2007, www.hon.ch/News/HSN/601257.html.

13. Christopher J. Portier and Mary S. Wolfe, eds., "Assessment of Health Effects from Exposure to Power-Line Frequency Electric and Magnetic Fields (NIEHS Working Group Report)," June 1998. http://www.niehs.nih.gov/emfrapid/html/WGReport/Working Group.html.

14. A. Ahlbom et al., "A Pooled Analysis of Magnetic Fields and Childhood Leukaemia," *British Journal of Cancer* 83, no. 5 (2000): 692–698.

15. http://www.who.int/mediacentre/factsheets/fs263/en/index.html.

16. Lorenzo Tomatis, personal communication.

17. Janette Sherman, interview by author, December 11, 2006.

18. "Cardinal Health Launches New PET Marketing Tools," November 26, 2006, http://nps.cardinal.com/nps/PETFoundations/CurrentNews.asp.

19. E. J. Hall, "Lessons We Have Learned from Our Children: Cancer Risks from Diagnostic Radiology," *Pediatric Radiology* 32, no. 10 (2002): 700–706.

20. E. Stephen Amis, Priscilla F. Butler, Kimberly E. Applegate et al., "American College of Radiology White Paper on Radiation Dose in Medicine," *Journal of the College of Radiology* 4 (2007): 272–284.

21. Zhores Medvedev, *The Legacy of Chernobyl* (New York: Norton, 1992).

22. David Brenner, Carl D. Elliston, Eric J. Hall, and Walter E. Berdon, "Estimated Risks of Radiation Induced Fatal Cancer from Pediatric CT," *American Journal of Roentgenology* 176 (2001): 289–296; See also Rosalie Bertell, Lynn Howard Ehrle, and Inge Schmitz-Feuerhake, "Pediatric CT Research Elevates Public Health Concerns: Low-Dose Radiation Issues Are Highly Politicized," *International Journal of Health Services* 37, no. 3 (2007): 419–439.

23. Roni Caryn Rabin, "With Rise in Radiation Exposure, Experts Urge Caution on Tests," *New York Times*, June 19, 2007.

24. Society for Pediatric Radiology and National Cancer Institute, "Radiation & Pediatric Computed Tomography: A Guide for Health Care Providers," Summer 2002, http://www.cancer.gov/cancertopics/causes/radiation-risks-pediatric-CT.

25. http://www.radiologyinfo.org/en/safety/index.cfm?pg=sfty_xray.

26. Atomic Bomb Survivors Relief Department, Social Affairs Bureau, City of Hiroshima, "Summary of Relief Measures for Atomic Bomb Survivors (2003)."

27. Caroline Richmond, "Alice Mary Stewart," *British Medical Journal* 325 (2002): 106.

28. Alice M. Stewart, J. W. Webb, B.D. Giles, and D. Hewitt, "Preliminary Communication: Malignant Disease in Childhood and Diagnostic Irradiation In-Utero," *Lancet* 2 (1956): 447.

29. http://www.hpa.org.uk/radiation/publications/documents_of_nrpb/abstracts/absd4-4.htm.

30. Gayle Green, *The Woman Who Knew Too Much* (Ann Arbor: University of Michigan Press, 1999), 91.

31. Janette Sherman, "Bullets, Bombs, and Nuclear Power Plants," *San Francisco Bay View,* April 11, 2007.

32. Eugenio Picano, "Informed Consent and Communication of Risk from Radiological and Nuclear Medicine Examinations: How to Escape from a Communication Inferno," *British Medical Journal* 329, no. 7470 (2004): 849–851.

33. Picano, "Informed Consent," pp. 849–851.

34. V. L. Castrol, "Evaluation of Prenatal Aldrin Intoxication in Rats," *Archives of Toxicology* 66, no. 2 (1992): 149–152.

35. Deborah C. Rice, "Neurotoxicity of Lead, Methylmercury, and PCBs in Relation to the Great Lakes," *Environmental Health Perspectives* 103, Suppl. 9 (1995): 71–87.

36. S. Bonassi, A. Znaor, M. Ceppi, C. Lando, W. Chang, N. Holland, et al., "An increased micronucleus frequency in peripheal blood lymphocytes predicts the risk of cancer in humans," *Carcinogenesis* 28, no. 3 (2007): 625–631; and H. Norppa, S. Bonassi, I.-L. Hansteen, L. Hagmar, U. Stromberg, P. Rossner, et al., "Chromosomal aberrations and SCEs as biomarkers of cancer risk," *Mutation Research* 600 (2006): 37–45.

37. S. Walitza, B. Werner, M. Romanos, A. Warnke, M. Gerlach, and H. Stopper, "Does Methylphenidate Cause a Cytogenetic Effect in Children with Attention Deficit Hyperactivity Disorder?" *Environmental Health Perspectives* 115, no. 6 (2007): 936–940.

38. "FDA Handling of Research on NutraSweet Is Defended," *New York Times,* July 18, 1987, p. 50.

39. Pat Thomas, "Aspartame—The Shocking Story of the World's Bestselling Sweetner," *The Ecologist,* September 2005, p. 35–51.

40. See www.nutrasweet.com/company.asp.

41. U.S. Air Force, "Aspartame Alert," *Flying Safety* 48, no. 5 (1992): 20–21.

42. John W. Olney et al., "Glutamate-Induced Brain Damage of Infant Primates," *Journal of Neuropathology and Experimental Neurology* 31, no. 3 (1972): 464–488; John W. Olney et al., "Brain Damage in Mice From Voluntary Ingestion of Glutamate and Aspartate," *Neurobehavioral Toxicology and Teratology* 2, no. 2 (1980): 125–129; John W. Olney, "Excitotoxic Food Additives: Functional Teratological Aspects," *Progress in Brain Research* 73 (1988): 283–294; and John W. Olney et al., "Increasing Brain Tumor Rates: Is There a Link to Aspartame?" *Journal of Neuropathology and Experimental Neurology* 55, no. 11 (1996): 1115–1123.

43. M. Soffritti et al., "First Experimental Demonstration of the Multipotential Carcinogenic Effects of Aspartame Administered in the Feed to Sprague-Dawley Rats," *Environmental Health Perspectives* 114, no. 3 (2006): 379–385; M. Soffritti et al., "Aspartame Induces Lymphomas and Leukaemias in Rats," *European Journal of Oncology* 10 (2005): 107–116.

44. Gillian Slovo, "Making History: South Africa's Truth and Reconciliation Commission," May 12, 2002, www.opendemocracy.net/democracy-africa_democracy/article_818.jsp.

45. Ibid.

46. "Pratt & Whitney: A United Technologies Company," www.pw.utc.com/vgn-ext-templating/v/index.jsp?vgnextoid=fb654e15c86fb010VgnVCM1000000881000aRCRD.

47. Carl F. Cranor and David A. Eastmond, "Scientific Ignorance and Reliable Patterns of Evidence in Toxic Tort Causation: Is There a Need for Liability Reform?" *Law and Contemporary Problems,* Autumn 2001, p. 5.

48. Ibid.

49. John Bailar, "How to Distort the Scientific Record Without Actually Lying," *European Journal of Oncology* 11, no. 4 (2007): 217–224.

Acknowledgments

DURING THE TIME this work was being carried out, my husband and I lost our parents and also became grandparents. Any errors this book contains are solely my responsibility. That there are not more of them is the result of generous efforts of a number of patient colleagues, including Devra and Lester Breslow, Brian McKenna, Annie Sasco, Barry Castleman, Rafael Tarnopolsky, John Topping, Leon and Hattie Bradlow, Katherine Henderson, Janette Sherman, Olivann Hobbie, Moe Mellion, Peggy Bare, and Michael Nussbaum. As he did with my previous book, William Frucht, executive editor at Basic Books, provided invaluable ballast, vision, and good will. My husband, Richard, remains the anchor of my life.

In Pittsburgh, I am privileged to work with and for people who have an unusual passion and tolerance for argument, affording me uncommon freedom to explore issues that many would rather leave undisturbed. Ron Herberman, a compact man and tough negotiator, has a big heart and grand vision. Maryann Donovan is that exceptional scientist with the capacity to drill deeply into big ideas yet have the soul and compassion of an artist. Ellen Dorsey, Jeff Lewis and Teresa Heinz Kerry have consistently asked me impossible questions, refused to accept irresolute answers, and provided generous financial and intellectual support for this effort. Jeanne Rizzo, Michael Lerner, Gary Cohen, Charlotte Brody, Pete Myers, Richard Clapp, Gina Solomon and other leaders of the Collaborative on Health and Environment (www.cheforhealth.org); Christopher Gavigan and Elizabeth Steward of Healthy Child, Healthy World; Michael Jacobson of the Center for Science in the Public Interest; and many colleagues of the Environmental Working Group have provided advice, inspiration, review and help throughout the process. Yvonne Cook and Christina Wild of the Highmark Foundation supplied resources and vision for the Highmark

Healthy Places, Healthy People program of the Center for Environmental Oncology. Nicolas Beldecos of the DSF Charitable Trust also supported scientific research at the Center described in this work.

Eula Bingham, David Servan-Schreiber, Leon and Hattie Bradlow, Carol and Myron Mehlman, Rachel and Shalom Kalnicki, Jewel Crawford, Christopher De Rosa, James Huff, Ronald Melnick, Kristine Thayer, Greg Dinse, David Umbach, Jerold Mande, Noel Raskin and Deborah Axelrod, Sheldon Samuels, Phil Landrigan, Ellen Silbergeld, Dan Wartenberg, Marion and Mike Taube, Elihu Richter, Jack Spengler, David Steinman, Colin Soskoline, and Michael Ducey not only supplied some surprising key documents and critical ideas but steered me to primary materials, many of which had never seen the light of day. Martin Walker, James Brophy, Guy Dauncey, Susan Luck, Liz Armstrong, Anne Wordsworth, Larry Plumlee, Lovell Jones, Ken Geiser, Joel Tickner and Margaret Keith offered out-of-print material as well as government, corporate and personal records relevant to several chapters, and directed me to doors I did not know existed. Others, who have asked not to be named, with the National Cancer Institute, the Centers for Disease Control, the Occupational Safety and Health Administration, the National Institute of Occupational Safety and Health, the Environmental Protection Agency, the U.S. Geological Survey, the Department of Energy, the Pennsylvania Department of Environmental Protection, the Department of Health, and agencies that cannot be mentioned, were equally helpful.

Bob Weinberg introduced me to one of his remarkable relative, Gerhard Weinberg, a distinguished diplomatic historian who provided invaluable insights on European history during the 1930s and 1940s. Lorenzo Tomatis, the erudite former head of the International Agency for Research on Cancer (IARC) of the World Health Organization and a founder of International Society of Doctors for the Environment, regularly encouraged my efforts and directed me to troves of documents in several languages, as did Annie Sasco, the fearless former chief of cancer prevention at IARC, now at the University of Bordeaux with INSERM, Intitut National de la Santé et la Research Médicale (the French National Institute for Health and Medical Research).

Bernard Goldstein, one of the few people to have achieved prominence on both the academic and regulatory sides of environmental health, shared early reports and recollections from his work in hematology. My good friend and neighbor Morris Mellion, past president of the American Academy of Family Practice Medicine, gave me candid, critical readings of many iterations of this book and shared his unique perspective on medical practice, customs and folklore.

Attorneys and legal scholars Amanda Hawes, Carl Cranor, Lisa Heinzerling, Jeff Tuckfelt, Marilyn Park, Michael Nussbaum, Russellyn Carruth, and budding legal analysts Mary Katherine Nagle, Lea Morgenstern and Leon Kababbie, parsed through my glosses on constitutional law, tort, and punitive and distributive justice and patiently explained arcane doctrines.

Anyone who has reached the age of some distinction as I have recognizes the debt owed to teachers and mentors who are no longer alive but whose spirit infuses this book. My Taylor Allderdice High School Advanced Placement History teacher, Nelly Norkus, taught me that making sense of the past advances the future. At the University of Pittsburgh, Robert Colodny first introduced me to the marvels of the history of science and showed me what it meant to fight for an idea. Richard Rubenstein proved that the divine can be found in even the darkest cracks of humanity, and Richard Tobias tutored me in keeping things as simple as possible but not more so, and Burkhart Holzner and Leonard Berkowitz explained the patterns underlying social behavior.

At the University of Chicago, Joseph M. Kitagawa and Edward Shils were demanding mentors. Marvin Schneiderman, one of the world's top biostatisticians at the National Cancer Institute, taught me to find humor in adversity, not to be afraid to tell the truth, not to be felled by cowardly attacks, and that you could really save money by changing light bulbs. Abe Lilienfeld, my post-doctoral mentor at Johns Hopkins University, showed me the thrill that eventually comes with finding different ways to reach the same conclusion and proved that a good scientist can be a true Mensch. A founder of the Environmental Mutagen Society, Marvin Legator shared his trusting nature, naïve optimism and killer instinct for the definitive experiment—attractive traits that sometimes make life easier and always make it worthwhile. The former Dean of University of British Columbia School of Medicine, David Bates, wrote poignant poetry and precise pathology reports with equal vigor, and encouraged my efforts to infuse scientific writing with soul.

Carnegie Mellon University's Barbara Lazarus and the speed-reading, indefatigable attorney Daniel S. Berger urged me to be clear and firm and to find new ways around old problems. Bella Abzug and Andrea Martin demanded and gave clarity and candor to all our joint endeavors. Cesare Maltoni and Irving Selikoff, founders of the Collegium Ramazzini, set the standard for open-sourcing of scientific information, as did physician researchers Olav Axelson, Saul Heller, Richard Remington and Dave Rall, whose gentle demeanors belied nerves of steel. My good friend and neighbor Mike Fitzgerald always encouraged me to do the right thing and to do things right, and did the same in his own tragically shortened life.

The preparation of this work was helped by a number of students and friends. Matthew Zurenski provided research and editing skills and a dedication to the effort far beyond his tender years, as did Leanne Ganter. Georgetown University professor Tim Beach and Pitt Honors College dean Alex Stewart identified a number of other exceptionally capable students with whom I have worked, including Sarah Chlebowski, Stephanie Leung, Tim Moreland, Michelle Aurelio, Raina Sharma, Michele Meyer, and Courtney Wilson. Several students who worked on earlier parts of the research have since gone on to greater glory and much more profitable endeavors, including Mary Kathryn Nagle, Rose Mellion, Monica Han, Elizabeth Reitano, Hilary Stainthorpe, John Topping Jr., Elizabeth Topping, Ora Sheinson, Michelle Gottlieb, Jean Kuo and Anna Ciesielska, each of whom made important contributions at various stages.

At the University of Pittsburgh Cancer Institute and the University of Pittsburgh Graduate School of Public Health and Medical Center, Marcia Barr, Frank Bontempo, Suzanne Lentzsch, Sharon McDermott, Stephen B. Thomas, Talal El-Hefnawy, John Kirkwood, Jean Latimer, Steve Grant, Nathan Behary, Robert Sobol, Rick Wood, Roger Oxendale, Ellen Mazo, Sam Jacobs, Ken Foon, Allison Robinson, Frank Lieberman, Ken McCarty, Vijaya Gandhi, Dan Volz, Linda Robertson, Michael Shaw, Mark Macri, Brandon McKenzie, Anthony Frisoli, Melissa Wickline, Adam Mohr, Ron Balassanian, C. Diane Colbert, Sara Werner, Alan Melton, Evelyn Talbott, Jane Cauley, Joel Weissfeld,

Joseph Schwerha, Donald Burke, Arthur Levine, Maggie Chapman, Bruce Pitt, Monica Han, Marcia Schwab, Alexandra Sotack, Terrae Davis, Samantha Malone, Jen Powers, JW Wallace, Charlie Nash, Roberta Ness, Don Burke, Sara Werner, Mike Lotze, Suzanne Lentzsch, Frank Bontempo, Judy Balk, Joshua Rubin, Jonathan Weinkle, Stan Marks, Ken Foon, Gayle Tissue, Jennifer Raetz, Sandra Danoff, Jules Heisler, John Innocenti, Eric Bechmann, Harvey Borovitz, Dorothy Mann, Alyce Katsur, Nancy Ferri, Donald Koch and Doug Romoff provided strategic support and critical problem-solving for which I am deeply grateful. At the eleventh hour, Brad Cisar spent more time than either of us would like to admit, fine tuning the graphics that grace this book. My debt to Elaine Ellenberger for keeping my life under control cannot be overstated.

Deborah Axelrod, Judy Balk, Talal El-Hefnawy, H. Leon Bradlow, Mitchell Gaynor, Dan Volz, Patricia Eagon, Frank Houghton, Nitin Telang, Michael Zeligs, Lovell Jones, Daniel S. Sepkovicz, Jean Latimer, Stephen Grant, Hope Nemiroff, Sheldon Feldman, and Michael Osborne have worked with me in developing research and testing theories on hormones and cancer.

For providing intellectual and spiritual support for my research, good counsel, lots of great food and wine and steady comfort over the years, I am indebted to my children, Lea and Aaron, and to Aaron's prolific wife who has also served as an unrelenting copy editor, Donielle Morgenstern; my sister, Sara Davis Buss and her husband Jay Buss; my brothers and their wives, Martin and Ann Davis and Stanford and Marian Davis; my sister-in-law and brother-in-law, Beth Morgenstern and Andrew Polsky; my cousins Mark, Michael, Sally, Jeff and Sondra Tuckfelt; my nieces and nephews, Molly and Justin Braver and Amanda and Leonard Davis; Dan and Carol Berger; Gordon and Peggy Bare; Susan Blumenthal; Michelle Bell; Joseph Cannon; Luis Cifuentes; Joseph DeCola; Fred Dobbs; Rukio Doi; Joycelynn Elders; Tony Fletcher; Harold Freeman; Mitchell and Cathy Gaynor; Gang Ke; David Gee; Wade Greene; Elizabeth Sullivan; Nancy and Stewart Smith; Andy Haines; John and Olivann Hobbie; Gary Hook; Jeff Cohan; Lisa Premo; Yiping Hu; Peter Infante; Karen Folger Jacobs; Mark Jacobs; Beverly Jackson Jones; Tracy Woodruff; Donna Karan; Rachel Goldstein; Phil Landrigan; Ronnie Levin; Barbara Seaman; Hope

Nemiroff; Karen Miller; Gloria Steinem; Christiane Northrup; Joel Schwartz; Lizbeth Lopez; Doree Lynn; Ed Markey; Adelle Morgenstern; Bill Godshall; Greg Hartley; David Walls-Kaufman; Martin Evans; Marcia Male; Avis Miller; Karen Miller; Ronald and Cathy Muller; Kirsten Niblaeus; Hope Nemiroff; Michael Nussbaum; Cheryl Osimo; Kathleen Piche; Laurie Thal; Nancy Taylor; Julia Brody; Jayne Ottman; John Pan; Penelope Pereira; Frank Press; Ruthann Rudel; David Saperstein; Jill and Richard Sheinbaum; Janette Sherman; Diane Shrier; Ellen Silbergeld; Lisa Simpson; Tiger Steuber; Stacey Olivito; Warren Stone; Daniel Swartz; David Suzuki; Elizabeth Sword; Jeffrey Tuckfelt; Mark and Sondra Tuckfelt; Jay Schulkin; David Walls-Kaufman; Sophia Wakefield; Shawnna Willey; Jeff Wohlberg; and Steve Wolin. At Carnegie Mellon University, a number of colleagues encouraged work that laid the foundation for this book when I worked at the Heinz School for Public Policy and Management, including Indira Nair, Cliff Davidson, Mark Kamlet, Jared Cohon, Lester Lave and Cathy Ribarchak. My agent, Al Zuckerman, of Writers House, kept me focused on the story line within the grand picture. Christine Marra of Marrathon Production Services and Jane Raese, Chrisona Schmidt, Donna Riggs, and Jeff Georgeson ensured accuracy and completeness through rapid and thorough copy editing, design and typesetting, and processes of nearly infinite complexity that make this book a reality.

I have learned much from my colleagues at a number of institutions, some of which have provided direct support for my research, including the American Association for the Advancement of Science, the American Holistic Medical Association, the American Medical Women's Association, Baylor College of Medicine, the American Public Health Association, the New York Academy of Sciences, Breast Cancer Fund, the Coalition of Organizations on the Environment and Jewish Life, Jackson Haverim, Beth Shalom Library Minyan, Adas Israel Synagogue's Havera, the Canadian Cancer Society, Cancer Care, Ontario, Carnegie Mellon University's H. John Heinz III School of Public Policy and Management, the Children's Health and Environment Coalition, the Climate Institute, the Coalition of Organizations on the Environment and Jewish Life, Collegium Ramazzini, the Department of Energy, the Conservative Women's League, the Environment Ministry of Sao Paulo, the U.S. Environmental Protection Agency, the European Environment Agency, Hadassah, Harvard University, Health Care Without Harm, the Intergovernmental Panel on Climate Change, Johns Hopkins University School of Public Health, the Jennifer Altman Fund, Keep Yellowstone Nuclear Free, the Susan G. Komen Foundation, the London School of Hygiene and Tropical Medicine, Health Canada, Na'amat, the National Institutes of Environmental Health Sciences, the National Cancer Institute, National Institutes of Health, the National Religious Partnership on the Environment, the National Renewable Energy Laboratory, the Oberlin College Environmental Studies Program, the Pan American Health Organization, the Pittsburgh Jewish Community Center, Prevent Cancer Now, Ottawa, the Rockefeller Brothers Fund, Rockefeller Family Financial Services, the Silent Spring Institute, the Stern College of Yeshiva University, the United Jewish Federation of Pittsburgh Environmental Committee, the United Nations Development Program, the United Nations Environment Program, the Women's Community Cancer Project, the World Bank, the World Health Organization, the World Resources Institute, the Louisiana Environmental and Network.

As will be apparent to anyone who looks at the website of key sources used in this report, the heroes of the research for this effort are the librarians, including those of Teton County Library, including Carol Conners and others, Doris Haag, at the University of Cincinnati Medical Center, Kehoe Papers Archives, the archives and film collection of the National Library of Medicine, the Medical and Historical Collections of the University of Pittsburgh Medical and College Libraries, and the generous spirited archivists and reference experts of the Carnegie Library of Pittsburgh, one of the nation's oldest public libraries and one of the few that remains opens on Saturday and Sunday.

I doubt that it would have been possible to complete this work in the days before the search engines of www.google.com and www.ask.com, or without the good graces of the Library of Congress Photographic Services Department, access to the archives of the American Association for the Advancement of Science, http://jstor.org. Those who set up and maintain EPA's web sites, and who staffed and operated its now closed libraries, deserve medals, as do scores of its dedicated, underrecognized employees, and those of the National Institute of Occupational Safety and Health, the Centers for Disease Control and Prevention, and the Occupational Safety and Health Administration.

Permission for the use of photographs and excerpts has been granted as follows: *Scientific American* for the excerpt from Groff Conklin's 1949 article, "Cancer and Environment, Larry Agran for the excerpt from his book *Cancer Connection.* The 1958 *Life* magazine cover (Chapter 1) was provided by Emil Bizub; the Pernkopf's atlas (Chapter 3) image was provided by David Williams of Purdue University; the Herb Block cartoon on cancer prevention, 1977 (Chapter 8), by the Herb Block Foundation; the photo of the asbestosis-afflicted beautician (Chapter 12) by Bill Ravanesi, from his photo-journalism tour de force, *Breath Taken: The Landscape & Biography of Asbestos* (Center for Visual Arts in the Public Interest, 1991) was provided by Barbara Landreth of the National Institute for Occupational Safety and Health Library in Morgantown, West Virginia; and the Smeltertown cemetary photo (Chapter 13) by Rick Provencio and Heather McMurray of the Sunland Park Environmental Group. The University of Cincinnati Medical Center's Kehoe Papers Archive generously provided copies of thousands of pages of documents, some of which are reproduced on the web site for this book.

Financial support for this work has been provided by the Heinz Endowments, the University of Pittsburgh Cancer Institute, the University of Pittsburgh Medical Center Endowment, the National Cancer Institute through a comprehensive cancer center grant to the University of Pittsburgh Cancer Institute, the Highmark Foundation, the Pittsburgh Foundation, the tobacco industry through its settlement with the state of Pennsylvania, the DSF Charitable Trust, the Sheila and Milton Fine Foundation, the Winslow Foundation, Amgen, and the Devra Lee Davis Charitable Foundation, a newly established foundation that will distribute any profits from this work for research on environmental oncology.

To the Holy One, who has allowed me to reach this season and maintain hope despite the world's terrifying struggles, I give whatever thanks can be rendered by mere words alone. The Talmud tells us: when there is no one there, you be the one.

In my life there is one who is always there.

For Richard

My rock and roll
My fire and ice
My day and night
My moon and sun
My dark and bright
My up and down
My all around
My black and white
My every where
My every thing
My private place
My blessed space
My my my my

Index